Nonlinear Evolution and Chaotic Phenomena

NATO ASI Series

Advanced Science Institutes Series

A series presenting the results of activities sponsored by the NATO Science Committee, which aims at the dissemination of advanced scientific and technological knowledge, with a view to strengthening links between scientific communities.

The series is published by an international board of publishers in conjunction with the NATO Scientific Affairs Division

A	**Life Sciences**	Plenum Publishing Corporation
B	**Physics**	New York and London
C	**Mathematical and Physical Sciences**	Kluwer Academic Publishers Dordrecht, Boston, and London
D	**Behavioral and Social Sciences**	
E	**Applied Sciences**	
F	**Computer and Systems Sciences**	Springer-Verlag
G	**Ecological Sciences**	Berlin, Heidelberg, New York, London,
H	**Cell Biology**	Paris, and Tokyo

Recent Volumes in this Series

Volume 170—Physics and Applications of Quantum Wells and Superlattices
edited by E. E. Mendez and K. von Klitzing

Volume 171—Atomic and Molecular Processes with Short Intense Laser Pulses
edited by André D. Bandrauk

Volume 172—Chemical Physics of Intercalation
edited by A. P. Legrand and S. Flandrois

Volume 173—Particle Physics: *Cargèse 1987*
edited by Maurice Lévy, Jean-Louis Basdevant, Maurice Jacob, David Speiser, Jacques Weyers, and Raymond Gastmans

Volume 174—Physicochemical Hydrodynamics: Interfacial Phenomena
edited by Manuel G. Velarde

Volume 175—Superstrings
edited by Peter G. O. Freund and K. T. Mahanthappa

Volume 176—Nonlinear Evolution and Chaotic Phenomena
edited by Giovanni Gallavotti and Paul F. Zweifel

Series B: Physics

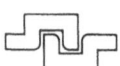

Nonlinear Evolution and Chaotic Phenomena

Edited by

Giovanni Gallavotti

Centro Linceo Interdisciplinare "B. Segre"
Accademia dei Lincei
Rome, Italy

and

Paul F. Zweifel

Virginia Polytechnic Institute and State University
Blacksburg, Virginia

Plenum Press
New York and London
Published in cooperation with NATO Scientific Affairs Division

Proceedings of a NATO Advanced Study Institute on
Nonlinear Evolution,
held June 8-19, 1987,
in Noto, Sicily, Italy

Library of Congress Cataloging in Publication Data

NATO Advanced Study Institute on Nonlinear Evolution (1987: Noto, Sicily)
Nonlinear evolution and chaotic phenomena / edited by Giovanni Gallavotti and Paul F. Zweifel.
p. cm.—(NATO ASI series. Series B, Physics; v. 176)
"Proceedings of a NATO Advanced Study Institute on Nonlinear Evolution held June 8-19, 1987, in Noto, Sicily, Italy"—T.p. verso.
"Published in cooperation with NATO Scientific Affairs Division.
Includes bibliographical references and index.

ISBN-13:978-1-4612-8294-5 e-ISBN-13:978-1-4613-1017-4
DOI: 10.1007/978-1-4613-1017-4

1. Chaotic behavior in systems—Congresses. 2. Particles (Nuclear physics)—Congresses. 3. Statistical mechanics—Congresses. I. Gallavotti, Giovanni. II. Zweifel, P. F. (Paul Frederick), 1929- III. North Atlantic Treaty Organization. Scientific Affairs Division. IV. Title. V. Series.
Q172.5.C45N38 1987 88-9415
530.1—dc19 CIP

Softcover reprint of the hardcover 1st edition 1988
A Division of Plenum Publishing Corporation
233 Spring Street, New York, N.Y. 10013

PREFACE

This volume represents the proceedings of a NATO Advanced Study Institute held at Noto, Sicily June 8-19, 1987. The director was Giovanni Gallavotti, Roma, with co-directors Marcello Anile, Catania and P. F. Zweifel, Virginia Tech. Other members of the scientific organizing committee included Mitchell Feigenbaum, Rockefeller University and David Ruelle, IHES.

The attendance at the school consisted of 23 invited speakers and approximately 80 "students", the term student being in quotation marks because many of them were of post-doctoral or even professorial status, although there were also a goodly number of actual graduate students in attendance also. Because of the disparate background of these "students", it was felt advisable to include at the conference special tutorials each afternoon, in which the contents of the morning's lectures were reviewed and clarified as necessary. These tutorials, organized by Gallavotti, involved various of the speakers, organizers, and other senior members of the school, and contributed in no little way to the overall success of the school. The organizers of the school would like to take this opportunity to thank all of those who assisted in these sessions, and to assure them that the results were definitely worth the effort.

Also contributing to the success of the school were a number of contributed papers, presented during the course of the afternoon tutorials. Three of those papers are included in these proceedings; they are the papers of DiFrancesco; Gallimbeni, Miari and Sertorio (presented by Sertorio); and Vittot. Again, the organizers would like to express their thanks to all those who gave contributed papers, and especially to those who supplied manuscripts. (A listing of the contributed papers for which manuscripts were not received appears at the end of the proceedings.)

Of the 23 invited speakers, three (McCormick, Libchaber, and Franceschini) never supplied manuscripts. After waiting for six months, the editors have decided not to delay publication any further. Summaries of the missing papers (prepared by the author of this preface are included in the proceedings.

The paper by Koch and Wittwer was presented by Wittwer; that of Krug and Spohn by Spohn and that of Keating and Mondragon by Michael Berry, also of Bristol. In point of fact, Berry described his own work--Keating and Mondragon acted as recorders. In a similar way, the talks of Lanford were written up by Sam Robinson, University of North Carolina-Wilmington; those of Feigenbaum by Edmond Rusjan, Virginia Tech; and those of Swinney by P. F. Zweifel.

We have attempted to arrange the order of the paper in these proceedings in some sort of logical fashion, i.e. by subject matter rather than chronologically according to the oral presentations. The results are probably chaotic, understandably enough, given the topic of the school, but since we have learned from a number of the speakers that there is actually order in chaos, it hopefully will not matter.

The organizers would like to take the opportunity to express their appreciation to all of the individuals who assisted in the organization of the school, especially Orazio Muscato, Daniela Auditore and Gloria Henneke. In addition, Ms. Henneke made heroic contributions to the preparation of these proceedings. Without her efforts, there would have been no proceedings.

P. Z. Zweifel

Center for Transport Theory and Mathematical Physics
Virginia Polytechnic Institute and State University
Blacksburg, Virignia 24061

CONTENTS

ADDENDA

CONFERENCE SUMMARY

Giovanni Gallavotti

Centro Linceo Interdisciplinare 'B. Segre'
Accademia dei Lincei
Via della Lungara 10
00100 Roma, Italia

This is a rather faithful transcript of the concluding remarks: no real editing has been attempted, the purpose being to try to convey the flavour the comments made had at the end of the school.

The uniform circular motion of Greek science dominated the scene until Poincaré. Newton produced firm *a priori* basis for predicting its frequencies in the most fundamental mechanical systems like the planets, rigid bodies, oscillators, waves, etc.

Perturbation theory was greatly developed to understand the quasi-periodic structure of more complex systems; however its growth led to the discoveries, by Poincaré, of the existence of motions fundamentally different from the quasi-periodic ones.

The work of Poincaré was properly understood only by few scientists and became widely appreciated only only after the work of Kolmogorov, Arnold, Moser, of Fermi, Pasta, Ulam and of Lorenz, Ruelle, Takens. We have witnessed in the last twenty years the diffusing awareness of the existence of substantially complicated motions and growing efforts towards their classification.

The conceptual necessity of the existence and relevance of non quasi-periodic motions, put forward by Ruelle led, directly or indirectly, to many discoveries. Most of them have been at least mentioned in this school.

The period doubling phenomenon and the various scaling structures associated with it have been studied in detail beginning with Feigenbaum who has presented their illuminating connection with the Statistical Mechanics of one dimensional lattice systems.

Among the many related topics, we had the opportunity of hearing Sullivan's characterization of the stable manifold of the Feigenbaum fixed point and Epstein's theory of the existence of the fixed point. Collet presented us a discussion on the effects of noise on the bifurcations heralding the chaotic transition under the period-doubling scenario, discussing also the related problems of the effects of noise on the motion on a strange attractor. We saw that, although a chaotic system can also be defined as a system which does not become more chaotic if noise is added to it, it is nevertheless possible to isolate and describe some quantitative effects of noise on such a system.

The frequency-locking phenomena have been analyzed by Cvitanović, who proposed a thermodynamic interpretation of the fractal structure of the set of frequency-locking intervals in circle maps while Lanford proposed detailed and stimulating conjectures in the same direction.

The concrete realization of such phenomena in models and experiments associated with the theory of fluid motions, both in numerical models and in actual laboratory experiments, has been treated in the lectures by Arecchi, Franceschini, Libchaber, McCormick, and Swinney.

Of course the period-doubling transition from quasi periodic to chaotic behaviour is not the only phenomenon; we have in fact seen in the above lectures the most varied bifurcation structures leading to chaos from periodic or quasi periodic motions.

One has the impression that we are building a wide phenomenology in which all the phenomena which can be listed as *a priori* possible are seen under suitable circumstances.

However it is quite clear that we still miss some organizing principle which could allow us to foresee qualitatively and quantitatively the various phenomena, the order of their appearance and their magnitude. The basic hope of Ruelle that there ought to be, for the description of the long time behaviour of dissipative systems, something analogous to Statistical Mechanics for conservative systems has not yet been realized and we are still waiting for a synthesis; but I would say that there is no evidence against it.

On the other hand the conservative systems are also not well understood and from Benettin, Galgani and Giorgilli we have seen that Classical Statistical Mechanics is not in very good shape in its predictive power of the statistical properties of the most simple systems (like lattices of non linear oscillators) which indicates that the understanding of the chaotic structures in dissipative systems may be accompanied by the understanding of the above questions in classical conservative mechanics.

In particular the very serious problem of the size of the time scales involved in the foundations of Statistical Mechanics is still open; we have seen here some attempts to attack it on a quantitative basis. But we also saw how still unsolved are some old problems of Classical Mechanics like the integrability problem: Siggia introduced us to a modern viewpont on integrability of analytic systems and Wisdom has shown us how the new computers and the methods developed to classify dynamical systems seem to have opened the way to reliable and reproducible theories of heavenly bodies moving in conic sections about the Sun. We shall not forget the new "Orrery" of Wisdom and the breathtaking sensation at the jump to the past at the time of the cathastrophic depletion of the asteroid belts.

In this respect it appears that it is time to try again to apply the recent theoretical results on Mechanics to Celestial Mechanics, going beyond the too hasty statement that perturbation theory cannot be of use in Celestial Mechanics because it leads to useless bounds. Probably it will be soon possible to obtain rigorous control of the simplest quasi periodic motions in the three body problem and research on this question seems to be necessary both for theoretical and applicative purposes.

The phenomena observed and described in Classical Mechanics are becoming so complicated that Classical Mechanics is acquiring the level of complexity of Quantum Mechanics. The course of Berry provided us with a clear mathematical formulation of quantum chaos in terms of the distribution of the spacing of the energy eigenvalues: a possible classification is beginning to emerge in terms of models for the spectra (like the real random matrix model or the complex random matrix model). It also provided us with the surprising connection between the spectra of random matrices and the zeroes of the Riemann zeta function, showing the value of pursuing the analysis further and the mathematical potential of a theory which has still to develop simple soluble examples and has to be imagined by the force of the mind.

The theory of chaos in quantum systems has been examined from a more practical point of view by Casati and Bayfield who show the existence of regimes, so far believed to be governed by quantum mechanics, which are equally well described classically and quantum mechanically as well as regimes where the quantum mechanics predictions differ sharply frm the corresponding classical ones and yet are at the same time anomalous, i.e. do not follow the conventional wisdom.

Essentially it has emerged, from Casati's seminars, that Quantum Mechanics suffers from the persistence of prejudices, very similar to those which plagued Classical Mechanics and delayed the understanding of the structure of perturbation theory.

The various techniques used in the above theories have wider applications, as explained to us in short talks by Spohn, Wittwer, Dürr: they gave us an idea of even wider connections between the topics of the courses and seemingly unrelated questions, bringing us a quick insight on some of the major problems of Quantum Field Theory and Statistical Mechanics.

Finally the course of Derrida on cellular automata has provided a beautiful example on how a physicist is able, following some simple ideas, to reach non-trivial conclusions on very hard problems. Most of us have probably been impressed with the philosophy of his course, always aiming at finding concrete and detailed soluble models to illustrate general ideas behind the intuition for the conjectures on harder problems; the old tautology that deep Physics can also emerge from simple Mathematics received here an interesting confirmation.

We come out with some more understanding of the basic ideas through which more complicated problems should be attacked, the first being the spin glass theory and the meaning of the corresponding Parisi's theory for the Sherington-Kirkpatrick model.

I wish to thank all the participants, lecturers, organizers and sponsors for their efforts, and I hope to share again a similar experience with you all.

Noto 18 June 1987

COMPLICATED OBJECTS ON REGULAR TREES

Mitchell Feigenbaum

Rockefeller University
Department of Physics
New York, New York 10021

Our goal in these lectures will be to study what kind of description is possible for complicated objects that arise from or can be associated with dynamical systems. In particular we will try to understand the structure of Cantor sets and strange attractors and the focal point will be cruder--thermodynamic understanding.

Let us motivate our discussion by considering some examples.

Example 1 (critical period doubling)

Consider a function g which satisfies the functional equation

$$g(x) = \alpha\, g(g(\frac{x}{\alpha})) \tag{1}$$

with $\alpha \approx -2.5$ and $g(0) = -1$ (Fig.1). We call the point $x = 0$ the critical point for g, because the slope of g changes sign at that point. By the trajectory coming out of the critical point we mean the sequence of points

$$x_r = g^{r+1}(0). \tag{2}$$

Since only the index r matters, we label the points on the trajectory by this index. Point 1 lies to the right of point 0 and all the other points lie between them. An intelligent approximate way of describing the trajectory is by covering it by intervals. The crudest approximation would be to cover it by only one interval $[x_0,x_1]$. A better approximation is the covering with two intervals $[x_0,x_2]$ and $[x_3,x_1]$. We can continue refining the covering indefinitely and on the n-th level the trajectory is covered by 2^{n-1} disjoint intervals.

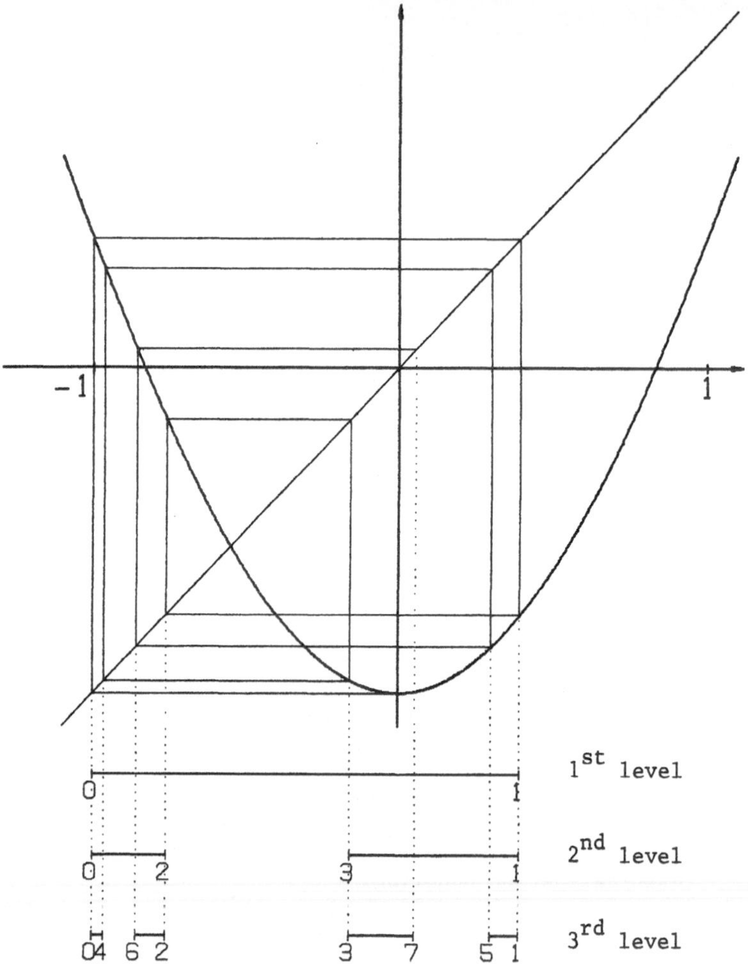

Fig. 1 Critical period doubling (only indices r are written in place of x_r)

The most interesting feature is the ordering of the points on the trajectory in the x direction. If we know the ordering at the nth level, then we can construct the ordering at the next level by a simple rule: take the numbers at the nth level, multiply them by 2 and you get the left half of the $(n+1)^{st}$ level. To get the right half, multiply the nth level by 2, flip it over so that the last number becomes the first and add 1. The reason that this simple rule holds is that g is a rather special function, namely it satisfies (1), which says that after two time steps (composing g twice) the effect of the map is just a magnification. Not only the order, but also coordinates of the points at the $(n+1)^{st}$ level can be easily generated from the points at the nth level: For odd points

$$x_{2r+1} = g^{2r+2}(0) = (g^2)^{r+1}(0) = \alpha^{-1} g^{r+1}(0) = \alpha^{-1} x_r , \tag{3}$$

while for even points

$$\alpha g(x_{2r}) = \alpha (g^2)^{r+1}(0) = x_r . \tag{4}$$

So we get the two rules

$$x_{2r}^{(n+1)} = (\alpha g)^{-1} (x_r^{(n)}) \tag{5}$$

and

$$x_{2r+1}^{(n+1)} = \alpha^{-1} (x_r^{(n)}) , \tag{6}$$

that enable us to construct any level from the previous level. These rules are called the "presentation function" and are shown in Fig.2.

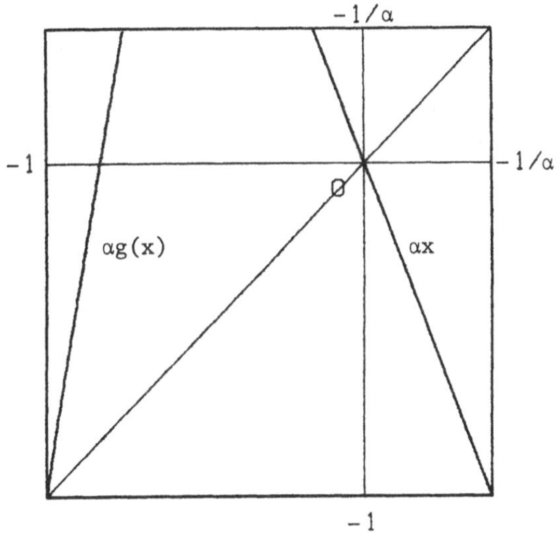

Fig.2 Presentation function for the fixed point of the period doubling process.

Notice that since the functions αx and $(\alpha g)(x)$ are shown on Fig.2, rather than their inverses, we can recover the trajectory by starting at the critical point, taking its inverse image, taking the inverse images of the inverse images, and so on.

To summarize, in this example we have a set of points that is being successively described by a series of intervals. Later we will see that there is a piece of thermodynamic information that one can deduce about these complicated sets.

The idea of constructing sets by starting at a critical point and taking successive inverse images can be generalized to other unimodular maps. The simplest example is the following.

Example 2 (the tent map)
The presentation function in this case is a set of two lines, one with slope 2 and the other with slope -2 (Fig.3), so it has the right flipping property. In consequence, the ordering of the points is the same as in Example 1. However, new intervals are not contained in the previous ones.

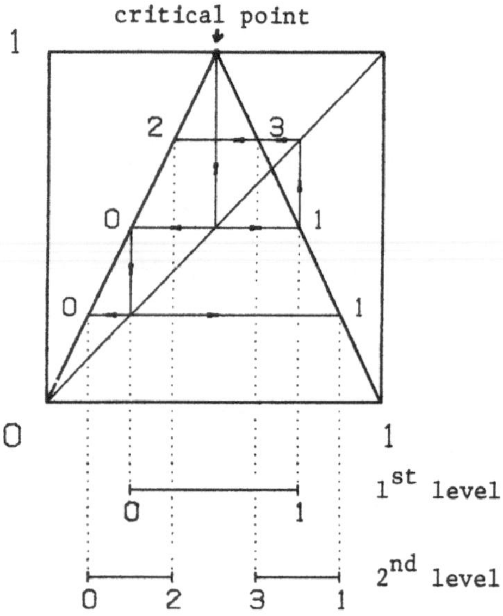

Fig.3 The tent map

Example 3. (Farey tree)
Starting at the point 1/2, successively construct the levels by applying presentation functions $F_0 = x/1+x$ and $F_1 = 1/1+x$ (Fig.4). This is of course the same as taking inverse images of the point 1/2 under the maps

F_0^{-1} and F_1^{-1}, which corresponds to the viewpoint taken in the previous example.

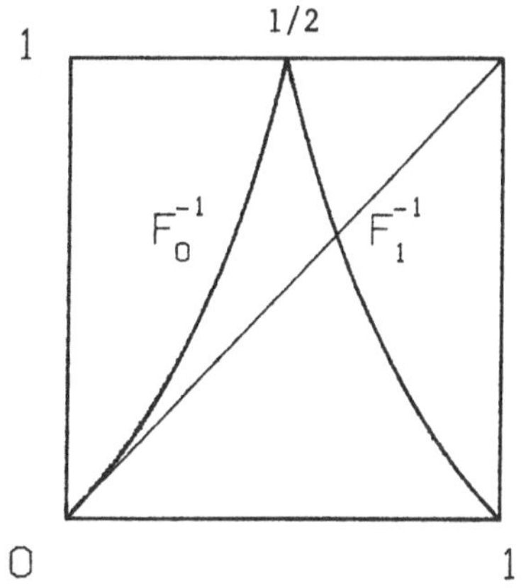

Fig.4 A dynamical way of generating the Farey tree

The first few numbers generated by this dynamics are

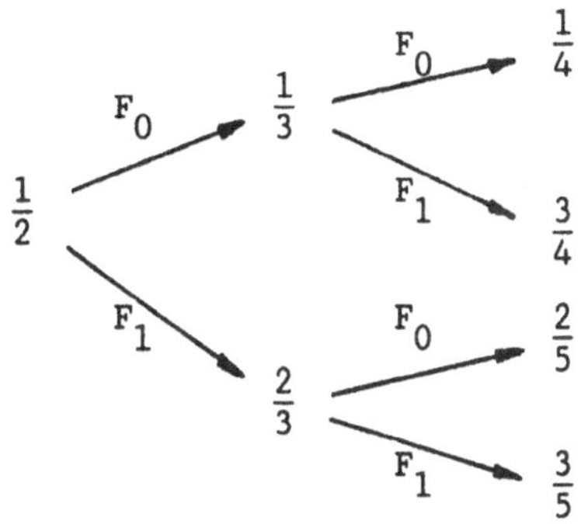

The result is the Farey tree, which is an ordering of the rational numbers between 0 and 1 on a regular binary tree, defined as follows. Start with fractions 0/1 and 1/1 (which are not in the Farey tree) and define the root of the tree to be their Farey sum

$$\frac{p}{q} \oplus \frac{r}{s} = \frac{p+r}{q+s}\,, \tag{7}$$

which in this case equals 1/2. Proceed from n^{th} level to $(n+1)^{st}$ level by taking the Farey sum of each number on the nth level by its nearest smaller number in the tree and nearest larger number in the tree (using 0/1 and 1/1, if necessary), as illustrated by the first few levels:

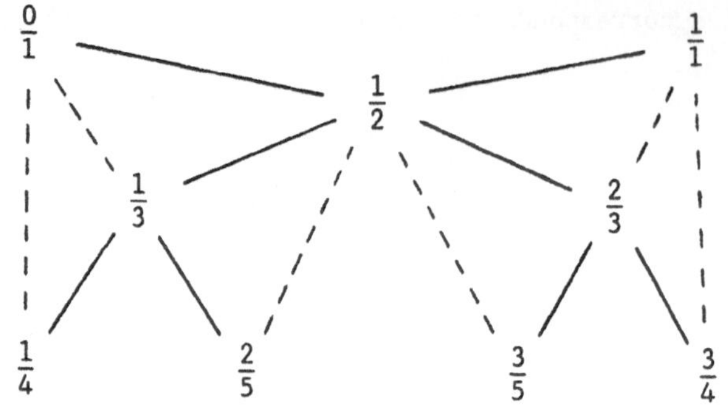

<u>Example 4</u> (Julia set)

Consider the mapping from the complex plane into itself, given by

$$z \rightarrow z^2 + \frac{1}{4} \quad . \tag{8}$$

If we iterate this mapping, each point in the complex plane will either approach the fixed point 1/2 or move out towards infinity. The Julia set is defined as a boundary of the region that moves towards infinity (Fig.5); so it is a repeller for the mapping. Since (8) implies $-z \rightarrow z^2 + 1/4$, the Julia set is symmetric about the origin. Because 1/4 is a real number it is also symmetric about the real axis. The inverses of the mapping (8), namely $\pm\sqrt{z-1/4}$ have attraction properties and they map the Julia set into itself. Replace $-\sqrt{z-1/4}$ by $\sqrt{-\bar{z}-1/4}$ so that both F_+ and F_- , defined by

$$z \begin{array}{l} \xrightarrow{F_+} \sqrt{z-1/4} \\ \xrightarrow[F_-]{} \sqrt{-\bar{z}-1/4} \end{array} \quad , \tag{9}$$

map the upper right quadrant into itself.

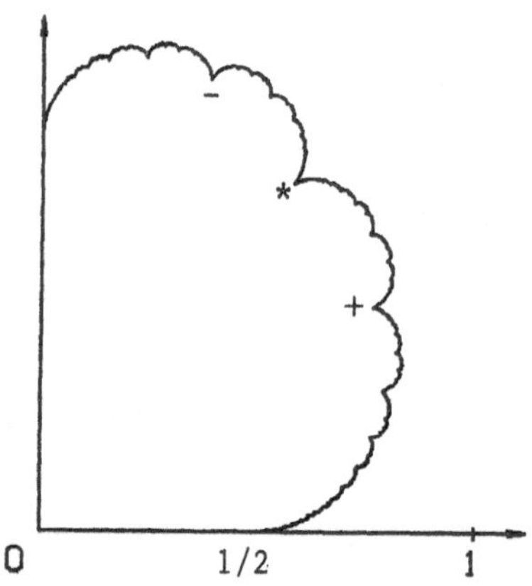

Fig.5 Julia set (symmetric around real and imaginary axes)

Starting at the point * and applying F_+ and F_- , we get two points, which we label by + and -. Proceeding this way, we get 2^n points in the nth level, labeled by the string of n pluses and minuses. If we understand the label as a binary digit, we get the same ordering as in previous examples. The reason is that the mapping F_+ is orientation preserving while the mapping F_- is orientation reversing.

The common feature of the examples above is that objects arising from a dynamics or abstractly could all be organized on a complete binary tree (binary can be generalized to n-ary). Let us study this closely on the example of the Farey tree:

<u>Example 3</u> (revisited)

Label the links by 0 and 1 and index the points by picking the zeros and ones from the bottom to the top.

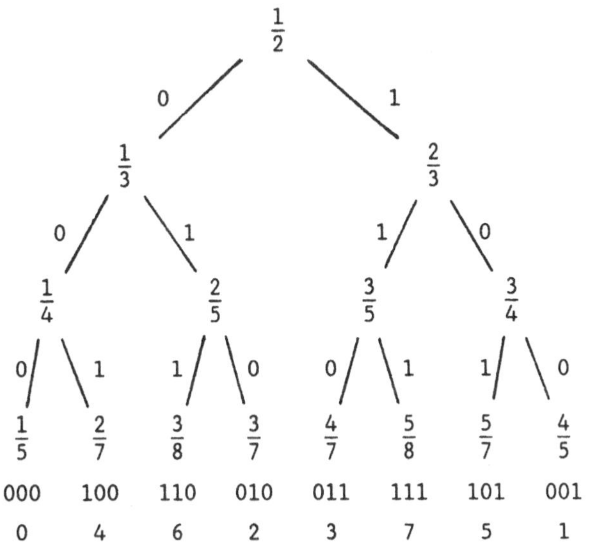

Notice that points with higher index have large denominators and that we get the same sequence 0 4 6 2 3 7 5 1 as in the period doubling example. Consider this sequence as an orbit of period 8. By doing this I can view it as coming from a dynamics (from a unimodal map). If the ordering is "smart", it will come from a reasonable map.

One reason for interest in the Farey tree is mode locking, a phenomenon in which the ratio of frequencies of two coupled oscillators (called rotation number) likes to be rational. Namely if the rotation number is initially rational, it will remain constant when we change the frequency of one oscillator (the other oscillator will adjust its frequency). The smaller the denominator of the rotation number, the bigger will be its range of constancy (Fig.6).

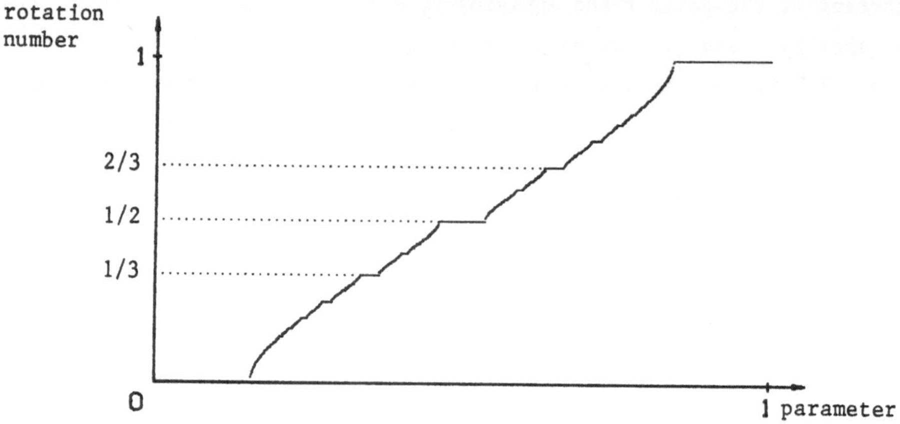

Fig.6 Devil's staircase arising in mode locking

So there is strong mode locking at 1/2 and at 1/3. The strongest mode locking in between these two appears at their Farey sum, which is 2/5 in this case. Hence strong mode locking corresponds to small index in the Farey tree.

The natural way of indexing is by continued fractions:

$$[c_1, c_2, \ldots, c_k] = \cfrac{1}{c_1 + \cfrac{1}{c_2 + \cdots + \cfrac{1}{c_k}}} \tag{10}$$

In this language, the presentation functions F_0 and F_1 for the Farey tree have the following form

$$[c_1, \ldots, c_k] \begin{array}{l} \xrightarrow{F_0} [c_1+1, c_2, \ldots, c_k] \\ \xrightarrow{F_1} [1, c_1, c_2, \ldots, c_k] \end{array} \tag{11}$$

They are transformations to the next level and they increase the sum of the c's by 1, so

$$\sum_1^k c_i = n+2 \tag{12}$$

for every number on the nth level. Starting from 1 and using the functions F_0 and F_1 , we can build up all the continued fractions in the following way:

$$[c_1] \quad = F_0^{c_1-1}[1]$$

$$[c_1,c_2] \quad = F_0^{c_1-1}F_1F_0^{c_2-1}[1] \tag{13}$$

$$\vdots$$

$$[c_1,\ldots,c_k] = F_0^{c_1-1}F_1F_0^{c_2-1}F_1\ldots F_1F_0^{c_k-1}\ [1]\ .$$

Identifying $[1] = x_0^{(0)}$, we immediately get a formula for the index:

$$r = \text{index}\ ([c_1,\ldots,c_k]) = 2^{c_1-1}(1+2^{c_2}(1+\ldots+(1+2^{c_k-1})\ldots)). \tag{14}$$

There is a dynamical motion that produces this ordering, so we may think of index r as time. Each level is a periodic orbit of a unimodal map.

Let us now turn to investigate the scaling properties. The idea is, that F_0 and F_1 are generally contractive, so any composition of thereof is also contractive. Hence x_n is the image of the starting point x^* under a contractive mapping:

$$x^{(n)}(\epsilon_n,\ldots,\epsilon_0) = F_{\epsilon_0} \circ F_{\epsilon_1} \circ\ldots\circ F_{\epsilon_n}(x^*). \tag{15}$$

Here $x^{(n)}$ is indexed by a sequence of binary digits. At the nth layer we have 2^{n+1} points organized in 2^n pairs of closest points (intervals). Estimate the length of the interval

$$\Delta^{(n)}(\epsilon_{n-1},\ldots,\epsilon_0) = F_{\epsilon_0}\ldots F_{\epsilon_{n-1}}(F_1(x^*)) - F_{\epsilon_0}\ldots F_{\epsilon_{n-1}}(F_0(x^*))$$

$$\sim F'_{\epsilon_0}(F_{\epsilon_1}F_{\epsilon_2}F_{\epsilon_3}\ldots)F'_{\epsilon_1}(F_{\epsilon_2}F_{\epsilon_3}\ldots)\ldots\ . \tag{16}$$

Having in mind the idea of applying the statistical mechanics formalism and obtaining some thermodynamic information, we raise $\Delta^{(n)}$ to the power β (inverse temperature). Since the number of intervals grows exponentially, we expect their lengths to decay exponentially. To compensate for this we multiply by z^n. We are interested in the sum of such terms over all the intervals. This can be approximated as follows:

$$\sum_{\{\epsilon\}} z^n |\Delta^{(n)}|^\beta \sim \ldots \underbrace{\sum_{\epsilon_r} z\, |F'_{\epsilon_r}(F_{\epsilon_{r+1}}\ldots)|^\beta \underbrace{\sum_{\epsilon_{r-1}} z|F'_{\epsilon_{r-1}}(F_{\epsilon_r} F_{\epsilon_{r+1}}\ldots)|^\beta \ldots}_{\mu_r(F_{\epsilon_r} F_{\epsilon_{r+1}}\ldots)}}_{\mu_{r+1}(\underbrace{F_{\epsilon_{r+1}}\ldots}_{x})} \quad . \tag{17}$$

Assuming weak dependence on the far digits, we get the equation

$$\mu_{r+1}(x) = \sum_{\epsilon} z\, |F'_{\epsilon}(x)|^\beta \mu_r(F_{\epsilon}(x)). \tag{18}$$

For z equal the reciprocal of the largest eigenvalue $\lambda_>$,

$$z^n \sum_{\{\epsilon\}} |\Delta^{(n)}|^\beta \sim (z\lambda_>)^n \sim 1, \text{ independent of } n, \tag{19}$$

so we get an eigenvalue equation

$$\lambda(\beta)\mu(x) = \sum_{\epsilon} |F'_{\epsilon}(x)|^\beta \mu(F_{\epsilon}(x)) \ , \tag{20}$$

which is a key result. For the Farey tree example this equation gives

$$\lambda(\beta)\mu(x) = \frac{1}{(1+x)^{2\beta}} \ [\mu(\frac{x}{1+x}) + \mu(\frac{1}{1+x})] \ , \tag{21}$$

which can be rewritten as

$$\lambda(\beta)\ \phi(x-1) = \phi(x) + \frac{1}{x^{2\beta}} \phi(\frac{1}{x}) \ . \tag{22}$$

All the thermodynamics is in the leading eigenvalue. However, it is also clear that this cannot give the complete information.

Thinking about dynamically constructed sets leads us to the concept of close return times, which we next define. Consider a starting point x(t). Suppose that, following the dynamics, the first time that we return close to x(t) is after $T_1 = 1$ second. We define the sequence of return times T_n by waiting each time for the first new closer approach, i.e.

$$|\underset{\sim}{x}(t) - \underset{\sim}{x}(t+T_n)| \text{ decreases with } n \ . \tag{23}$$

We make a profound assumption that these return times grow geometrically, say

$$T_n \sim a^n , \quad \text{for } a > 1. \tag{24}$$

This is motivated by a variety of problems. For critical period doubling the return times are 2^n and for golden mean rotation they are the nth Fibonacci numbers, which grow like $[1/2(1+\surd 5)]^n$. More generally, any object that can be organized on a k-ary tree, has $a = k$. Next we decompose the orbit into approximate orbits by defining the nth data set to be

$$\{x_t : t = 0, \ldots, T_{n+1} - 1\} . \tag{25}$$

Notice that the nth data set is almost periodic with period T_{n+1} and it becomes more periodic with increasing n. Provided t is small enough so that $t + T_n$ is still within the data set, then $x(t+T_n)$ is the closest point to $x(t)$ of all the points in the orbit. So there is a natural smallest distance between pairs of points. For the nth data set we have a set of distances

$$\Delta_t^{(n)} = \underset{\sim}{x}(t) - \underset{\sim}{x}(t+T_n) , \tag{26}$$

which we can view as an approximate covering of the orbit. If we parameterize the index t in the data set by

$$t = \epsilon_1 T_1 + \ldots + \epsilon_n T_n , \tag{27}$$

then the fraction through the "orbit" of period T_{n+1} is

$$\begin{aligned} \frac{t}{T_{n+1}} &= \epsilon_n \frac{T_n}{T_{n+1}} + \epsilon_{n-1} \frac{T_{n-1}}{T_{n+1}} + \ldots \sim \\ &\sim \frac{\epsilon_n}{a} + \frac{\epsilon_{n-1}}{a^2} + \ldots \\ &= .\ \epsilon_n \epsilon_{n-1} \cdots \quad \text{in base "a"} . \end{aligned} \tag{28}$$

The epsilons here are integers. If $T_n \sim 2^{n-1}$, then $\epsilon_i = 0,1$ (base 2). For golden mean rotation, where $a \approx 1.62$, epsilons must in addition obey some selection rules which guarantee that there is a unique representation (28) for each t up until some large Fibonacci number T_{n+1}. Observe that the epsilons in (28) appear in "reversed" order. The most important is the controlling digit, the coefficient of the largest possible return time. In one dimension we can define the trajectory scaling function σ by

$$\Delta_t^{(n)} \equiv \sigma_n\left(\frac{t}{T_{n+1}}\right) \Delta_t^{(n-1)} , \tag{29}$$

or, rewriting using (28)

$$\Delta^{(n)}(\epsilon_n,\dots,\epsilon_1) = \sigma_n(\epsilon_n,\epsilon_{n-1},\dots,\epsilon_1)\, \Delta^{(n-1)}(\epsilon_{n-1},\dots,\epsilon_1) . \tag{30}$$

Assuming a smooth dynamics, f, the distance one step later looks like the derivative times the distance at the previous time:

$$\Delta_{t+1}^{(n)} = x(t+1) - x(t+1+T_n) =$$
$$= f(x_t) - f(x_{t+Tn}) \sim f'(x_t)\, \Delta_t^{(n)} . \tag{31}$$

Since these derivatives cancel when we compute the scaling function, we get

$$\sigma\left(\frac{t}{T_{n+1}} + \frac{1}{T_{n+1}}\right) \approx \sigma\left(\frac{t}{T_{n+1}}\right) . \tag{32}$$

This means that the scaling function doesn't change very much as t varies over small intervals, or in other words σ depends only on the leading few epsilons. By a similar argument it follows that σ is invariant under coordinate transformations. In general the scaling function depends on the level, but we assume that

$$\sigma_n(\tau) \xrightarrow[n \to \infty]{} \sigma(\tau) , \quad \tau \in [0,1) . \tag{33}$$

If we know the scaling function, we can infer from a short measurement of the system its long time behaviour, by successively computing level after level. So this is a method of solving the equations of motion where the initial data is a short measurement of the system.

New intervals are related to old ones by a multiplicative transformation. So we would like to think of σ as a linear transformation that gives new intervals from old intervals. The problem is that there are more new intervals than old intervals. The way to fix this is by noting that σ depends only on the leading r+1 epsilons in the r-th "good" approximation.

Consider intervals to the power β and sum them up. First we sum over the epsilons we don't care about:

$$\psi^{(n)}(\epsilon_n,\dots,\epsilon_{n-r+1}) = \sum_{\epsilon_1,\dots,\epsilon_{n-r}} |\Delta^{(n)}(\epsilon_n,\dots,\epsilon_1)|^\beta . \tag{34}$$

Using the definition of the scaling function, we have

$$\psi^{(n)}(\epsilon_n,\ldots,\epsilon_{n-r+1}) = \sum_{\epsilon_{n-r}} \sigma^{\beta}(\epsilon_n,\ldots,\epsilon_{n-r})\psi^{(n-1)}(\epsilon_{n-1},\ldots,\epsilon_{n-r}), \tag{35}$$

or

$$\psi^{(n)}(\epsilon_r,\ldots,\epsilon_1) = \sum_{\epsilon_0} \sigma^{\beta}(\epsilon_r,\ldots,\epsilon_1,\epsilon_0)\psi^{(n-1)}(\epsilon_{r-1},\ldots,\epsilon_0)$$

$$= \sum_{\epsilon'_{r-1},\ldots,\epsilon'_0} T_{(\epsilon_r,\ldots,\epsilon_1),(\epsilon'_{r-1},\ldots,\epsilon'_0)}\, \psi^{(n-1)}(\epsilon'_{r-1},\ldots,\epsilon'_0) . \tag{36}$$

Equation (36) defines the linear transformation T to be

$$T_{(\epsilon_r,\ldots,\epsilon_1)\,(\epsilon'_{r-1},\ldots,\epsilon'_0)} = \sigma^{\beta}(\epsilon_r,\ldots,\epsilon_1\epsilon'_0)\delta_{\epsilon_{r-1},\epsilon'_{r-1}}\cdots\delta_{\epsilon_1,\epsilon'_1} . \tag{37}$$

Since $\psi^{(n)} = T\psi^{(n-1)}$, we have asymptotically

$$\psi^{(n)} = T^n\,\psi^{(o)} \sim \lambda_>^n\,\psi_> \tag{38}$$

where $\lambda_>$ is the largest eigenvalue

$$T\,\psi_> = \lambda_>\psi_> . \tag{39}$$

So we only need to sum over the rest of the epsilons

$$\sum_{\epsilon_1,\ldots,\epsilon_n} |\Delta^{(n)}(\epsilon_n,\ldots,\epsilon_1)|^{\beta} = \sum_{\epsilon_r,\ldots,\epsilon_1} \psi^{(n)}(\epsilon_r,\ldots,\epsilon_1) \sim$$

$$\sim \lambda_>^n(\beta)\Big(\sum_{\epsilon_r,\ldots,\epsilon_1} \psi_>(\epsilon_r,\ldots,\epsilon_1)\Big) \tag{40}$$

If we define the free energy $F(\beta)$ by

$$\lambda_>(\beta) = a^{-F(\beta)} \quad \text{i.e.} \quad F = \frac{\ln\lambda_>}{\ln a^{-1}} , \tag{41}$$

then

$$\sum_{\{\epsilon\}} |\Delta^{(n)}(\{\epsilon\})|^{\beta} \sim a^{-nF(\beta)} \sim T_n^{-F(\beta)} . \tag{42}$$

The idea is that the pieces will geometrically decrease in size with different exponents which are some average values of scaling

$$\Delta^{(n)}(\epsilon_n,\ldots,\epsilon_1) \sim \sigma(\epsilon_n,\ldots,\epsilon_1)\sigma(\epsilon_{n-1},\ldots,\epsilon_1)\ldots$$

$$\sim \sigma^n_{\text{eff}}(\epsilon_n,\ldots,\epsilon_1) \tag{43}$$

$$\sim a^{-nh(\epsilon_n,\ldots,\epsilon_1)} \equiv a^{-H(\epsilon_n,\ldots,\epsilon_1)} .$$

Equation (43) defines the Hamiltonian H. So in this step we constructed a statistical mechanical system with the number of degrees of freedom equal to the level of construction of the set. In this interpretation the sum of the intervals to the β power is the canonical partition sum

$$\sum_{\{\epsilon\}} |\Delta^{(n)}(\{\epsilon\})|^{\beta} \sim \sum_{\{\epsilon\}} a^{-\beta H(\{\epsilon\})} \sim a^{-nF(\beta)} . \tag{44}$$

Example 5 (the simplest possible)
Consider a two level Cantor set, where at each step an interval is broken into two, one being σ_0 times the starting interval and the other σ_1 times the starting interval (Fig.7).

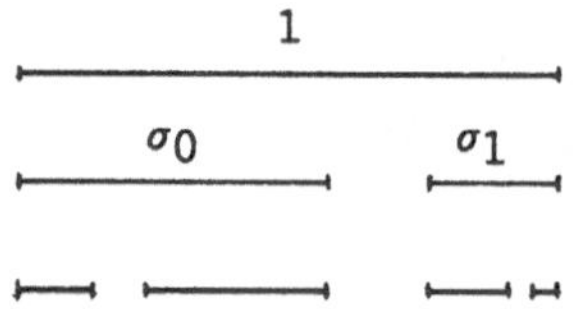

Fig.6 Two level Cantor set

In this case the scalings only depend on the leading epsilon

$$\sigma(\epsilon_n,\ldots,\epsilon_1) = \sigma(\epsilon_n) = \begin{cases} \sigma_0 & \epsilon_n = 0 \\ \sigma_1 & \epsilon_n = 1 \end{cases} . \tag{45}$$

An interval is the product of these scalings

$$\Delta^{(n)}(\epsilon_n,\ldots,\epsilon_1) = \sigma(\epsilon_n)\sigma(\epsilon_{n-1})\ldots\sigma(\epsilon_1) =$$

$$= \sigma_0^n \left(\frac{\sigma_1}{\sigma_0}\right)^{\sum_1^n \epsilon_i} \sim 2^{-H(\{\epsilon\})} . \tag{46}$$

The Hamiltonian

$$H = n\left\{ \frac{\ln\sigma_0}{\ln\frac{1}{2}} + \left[\frac{1}{n}\sum_1^n \epsilon_1\right] \frac{\ln\frac{\sigma_1}{\sigma_0}}{\ln\frac{1}{2}} \right\}$$

corresponds to n noninteracting spins in an external magnetic field. This example corresponds to period doubling in the lowest approximation.

Example 6 (next approximation to period doubling)
Consider a scaling that depends on two epsilons

$$\sigma(\epsilon_n,\epsilon_{n-1},\ldots,\epsilon_1) = \sigma(\epsilon_n,\epsilon_{n-1}) = \begin{cases} \sigma(0,0) = \sigma_0 & \sigma(0,1) = \sigma_1 \\ \sigma(1,0) = \sigma_2 & \sigma(1,1) = \sigma_3 \end{cases}$$

$$= \sigma_0 \left(\frac{\sigma_1}{\sigma_0}\right)^{\epsilon_{n-1}} \left(\frac{\sigma_2}{\sigma_0}\right)^{\epsilon_n} \left(\frac{\sigma_0\sigma_3}{\sigma_1\sigma_2}\right)^{\epsilon_{n-1}\epsilon_n} . \tag{47}$$

An interval is a product of such terms

$$\Delta^{(n)}(\epsilon_n,\ldots,\epsilon_1) = \sigma(\epsilon_n,\epsilon_{n-1})\sigma(\epsilon_{n-1},\epsilon_{n-2})\ldots\sigma(\epsilon_2,\epsilon_1) =$$

$$= \sigma_0^{\,n-1} \left(\frac{\sigma_1}{\sigma_0}\right)^{\sum_1^{n-1}\epsilon_i} \left(\frac{\sigma_2}{\sigma_0}\right)^{\sum_2^{n}\epsilon_i} \left(\frac{\sigma_0\sigma_3}{\sigma_1\sigma_2}\right)^{\sum_1^{n-1}\epsilon_i\epsilon_{i+1}} \tag{48}$$

and its logarithm is the Hamiltonian

$$H_{n+1} = n \left\{ \frac{\ln\sigma_0}{\ln\frac{1}{2}} + \left[\frac{1}{n}\sum_1^n \epsilon_i\right] \frac{\ln\frac{\sigma_1}{\sigma_0}}{\ln\frac{1}{2}} + \left[\frac{1}{n}\sum_2^{n+1}\epsilon_i\right] \frac{\ln\frac{\sigma_2}{\sigma_0}}{\ln\frac{1}{2}} + \right.$$

$$\left. + \left[\frac{1}{n}\sum_1^n \epsilon_i\epsilon_{i+1}\right] \ln\left[\frac{\sigma_0\sigma_3}{\sigma_1\sigma_2}\right] \Big/ \ln\frac{1}{2} \right\} , \tag{49}$$

which corresponds to the one dimensional nearest neighbour Ising model in an external magnetic field.

Higher approximations to period doubling involve second nearest neighbour interactions, third nearest neighbour interactions, ternary interactions and so on but with coefficients that exponentially become unimportant.

Notice that asymptotically

$$\frac{1}{n}\sum_i \epsilon_i \sim \frac{1}{n}\sum_i \epsilon_{i+1} , \tag{50}$$

so the two terms in the Hamiltonian corresponding to the external magnetic field combine into

$$\left(\frac{1}{n}\sum_i \epsilon_i\right) \ln \left[\frac{\sigma_1 \sigma_2}{\sigma_0^2} \right] \Big/ \ln \frac{1}{2} \,, \tag{51}$$

and logarithmically only the product $\sigma_1 \sigma_2$ enters independently.

The matrix T is very sparse, so it is useful to think of it in terms of Markov diagrams as follows:

1. For each configuration $\epsilon_r, \ldots, \epsilon_1$ draw a node for ψ:

$$\psi(\epsilon_r, \ldots, \epsilon_1) \quad \rightarrow \quad \boxed{\epsilon_r, \ldots, \epsilon_1} \;. \tag{52}$$

2. Connect the nodes with non-zero σ values:

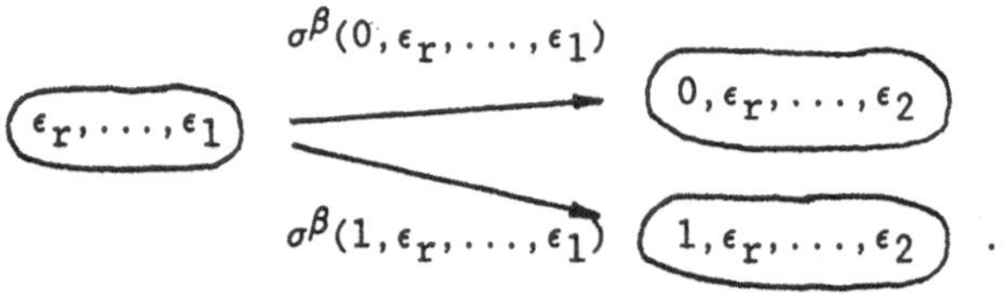

(53)

The advantange of the Markov diagram is that there are not n^2 links but decidedly fewer due to the sparsity of the matrix.

To obtain $|\Delta^{(n)}(\epsilon_n, \ldots, \epsilon_1)|^\beta$ we find the corresponding path of length n on the directed graph and multiply the sigmas to the power β encountered on the links. To get the partition sum we need to sum over all such paths. It is better to multiply each σ^β by z and sum over all paths

$$\sum_{n=0}^{\infty} z^n \sum_{\{\epsilon\}} |\Delta^{(n)}(\{\epsilon\}|^\beta \sim \sum_n z^n \sum_{\{\epsilon\}} T^n \psi(o) \sim$$

$$\sim \frac{\Sigma \text{ cofactors}}{\det (1\text{-}zT)} \,. \tag{54}$$

Notice that this sum is singular at $z = \lambda^{-1} = a^{F(\beta)}$. So in order to find the thermodynamics all we need to do is find the smallest singularity of (54).

<u>Example 6</u> (revisited)

First we draw the Markov graph

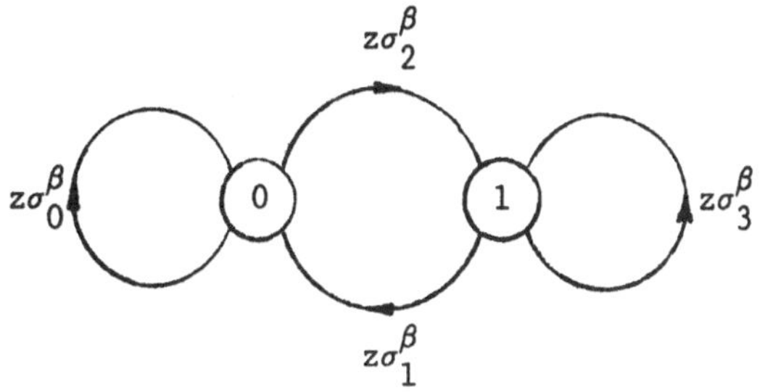

$$. \qquad (55)$$

We have to sum up all the paths in any way we want. One way is to first eliminate node 1. Since we can go around the σ_3 loop any number of times we want to, we get the sum of a geometric series and the reduced graph is

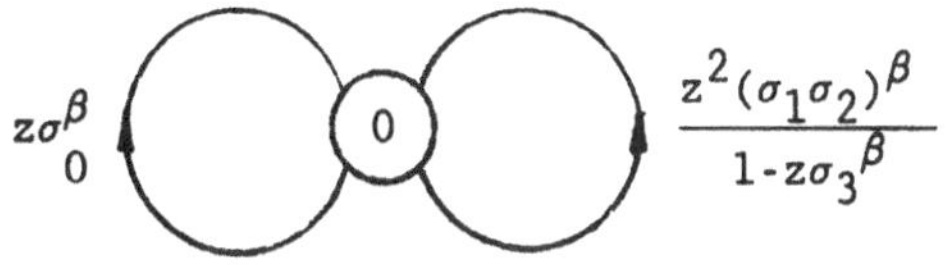

$$. \qquad (56)$$

When we have two loops in parallel, we can simply add them, so (56) reduces to

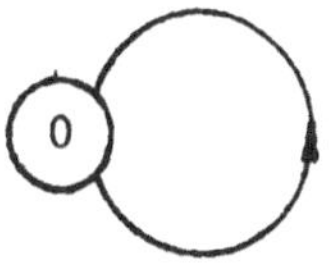

$$z\sigma_0{}^\beta + \frac{z^2(\sigma_1\sigma_2)^\beta}{1-z\sigma_3{}^\beta} =$$

$$= \frac{1-z\sigma_3{}^\beta}{(1-z\sigma_0{}^\beta)(1-z\sigma_3{}^\beta)-z^2(\sigma_1\sigma_2)^\beta} \quad . \qquad (57)$$

The condition that the characteristic determinant (the denominator in (57)) vanishes gives

$$0 = 1 - z(\sigma_0{}^\beta+\sigma_3{}^\beta) - z^2(\sigma_1{}^\beta\sigma_2{}^\beta - \sigma_0{}^\beta\sigma_3{}^\beta) \ . \qquad (58)$$

The case $\beta = 0$ corresponds to having a 1 on each link in the graph, which produces the topological determinant of the graph. Condition (58) gives

$$z = 2^{-1} = 2^{F(0)} \quad . \qquad (59)$$

The result $F(0) = -1$ holds quite generally, since equation (42) implies

$$T_n = \sum_n 1 \sim T_n{}^{-F(0)} . \qquad (60)$$

Since

$$z(0) = a^{F(0)} = a^{-1} , \quad (61)$$

the topological determinant of the graph must have a^{-1} as a zero.

Example 7 (golden mean rotation)

The recurrence times are the Fibonacci numbers satisfying

$$F_{n+1} = F_n + F_{n-1} , \quad F_0 = 0, F_1 = 1 . \quad (62)$$

Explicitly,

$$F_n = \frac{1}{\sqrt{5}} [\rho^{-n} - (-\rho)^n] , \quad (63)$$

where ρ is the golden mean ratio, i.e.

$$\rho^{-1} = \frac{\sqrt{5} + 1}{2} , \quad -\rho = \frac{1-\sqrt{5}}{2} . \quad (64)$$

Asymptotically,

$$F_n \sim \rho^{-n} . \quad (65)$$

Consider the golden mean rotation on the circle, i.e.

$$\theta_{t+1} = \theta_t + \rho \mod 1 . \quad (66)$$

Assuming $\theta_0 = 0$, we have

$$\theta_t = \rho^t \mod 1 . \quad (67)$$

Hence, using (63)

$$\theta_{F_n} = \rho F_n \mod 1 = (\rho F_n - F_{n-1}) \mod 1 =$$

$$= - (-\rho)^n \mod 1 \to 0 . \quad (68)$$

There are selection rules in the representation

$$t = \epsilon_n F_n + \ldots + \epsilon_1 F_1 , \quad (69)$$

since $\epsilon_{i-1} = \epsilon_i = 1$ is degenerate with $\epsilon_{i+1} = 1$, $\epsilon_i = \epsilon_{i-1} = 0$. So, $\epsilon_i = 0,1$, but two consecutive epsilons cannot equal 1. Graphically this means that the loop $\sigma(1,1)$ is forbidden

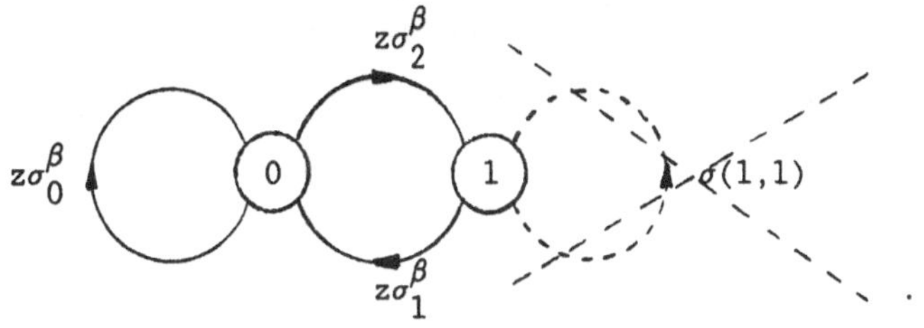

The characteristic determinant is singular when

$$0 = 1 - z\sigma_0^\beta - z^2(\sigma_1\sigma_2)^\beta \; . \tag{70}$$

For $\beta = 0$, the smallest zero of equation (70) is $z = \frac{\sqrt{5}-1}{2} = \rho = a^{-1}$ as expected. For subcritical and critical rotations, the orbit densely covers the circle, so that the set has dimension 1. Hausdorff dimension is essentially defined through the consistency condition, that the thermodynamic sum (42) has no exponential dependence on n: it is that β for which F = 0 and hence z = 1. Thus (70) must be satisfied by z = 1 at $\beta = 1$, yielding

$$\sigma_1\sigma_2 = 1 - \sigma_0 \; , \tag{71}$$

which holds exactly for subcritical golden mean rotation. Condition (70) now simplifies to a one parameter formula

$$0 = 1 - z\sigma_0^\beta - z^2(1-\sigma_0)^\beta \; . \tag{72}$$

It is worth emphasizing that equation (61) places serious constraints on the topology of those graphs that can serve for a process of return time ratio a. For dynamical processes whose return time structure is compatible with low order graphs, this restriction on the characteristic polynomial so marks the thermodynamics as to allow the reconstruction of some dynamical information from empiric thermodynamic information .

Let us review what we have presented so far. For regular (viz. binary) objects, the presentation function machinery immediately yields the thermodynamics through the functional eigenvalue equation (20). Further notion, of return times and scalings, led to finite order approimations (37-39) naturally represented by certain Markov diagrams. (These approximations are exact for scaling functions depending on precisely a finite number of epsilons.) An advantage to this machinery is its deeper intuitive connection between thermodynamics and strong dynamical information (i.e. the structure of the return times.) As we saw in the last example of golden mean rotation, this machinery easily (by merely erasing certain links on the Markov graph) embraces the possibility of objects constructed on incomplete trees. In fact, taking a cue from the Markov ideas, it is easy to work out how the presentation function machinery can be adapted to incomplete trees. All one need do is to return to the iterated sum derivation in (17) and note that the iteration of (18) will produce arbitrary sequences of F_ϵ's. This is easily remedied by keeping some leading F_ϵ's explicitly and restricting the sum to epsilons allowed by the grammar defined by the selection rules. Should the grammar be expressed through g+1 epsilons then g leading F_ϵ's must be explicitly kept. Finally, $\mu(x)$ becomes a vector through the definition

$$\mu(F_{\epsilon_1}F_{\epsilon_2}\ldots F_{\epsilon_g}(x)) = \mu_{\epsilon_1,\ldots,\epsilon_g}(x) \ . \tag{73}$$

This generalization of (20) is now an exact eigenvalue equation, possessing the finite Markov graphs as approximations. The reader should be able to explicitly render this.

In conclusion, we have explored a variety of formalisms, mostly concerned with regular objects, but easily adapted to incompletely generated ones. While we emphasized thermodynamic notions, the ideas are deeply related to the microscopic structure of dynamically generated strange sets. It is so far only for these dynamical "multifractals" that our understandings are well enough developed to allow us to consider a non-trivial set of interesting objects possessing an infinite spectrum of different scalings, rather than artificially constructed ones with properties not naturally generalizing to those objects of still greater complexity that we ultimately aim to understand.

RENORMALIZATION GROUP METHODS FOR CIRCLE MAPPINGS

Oscar E. Lanford III

IHES
91440 Bures-sur-Yvette
France

Lecture notes from a series of lectures given in the NATO Advanced Study Institute in Nonlinear Evolution and Chaotic Phenomena, Noto, Italy, June 8-19, 1987.

I. *INTRODUCTION*

The topic of this series of lectures is the theory of circle mappings with emphasis on recently developed ideas on the application of renormalization group methods to the analysis of parameter dependence of circle mappings. Circle mappings arise naturally in the study of flows which admit invariant two-dimensional tori. For example consider a surface of section or Poincaré map for a flow with a two-dimensional invariant torus. The first return map is a continuous and invertible mapping of the circle (the intersection of the surface of section with the torus) into itself.

Two-dimensional attracting tori arise in the study of the dynamics of two relaxation oscillators (a relaxation oscillator is a dynamical system with an attracting invariant circle in the phase space). For two such uncoupled oscillators the phase space is a product and has an attracting invariant two-dimensional torus. If one "turns on" a coupling between the oscillators the torus has some stabilty and one could study the circle mapping on the surface of section as the coupling varies.

These tori also appear in the study of the Hopf bifurcation for maps, an elementary bifurcation process in which one passes from a flow having an attracting periodic orbit to a situation where the orbit is no longer attracting but in the immediate vicinity of the orbit there is an attracting two-dimensional torus. This is a standard step on the route from periodic to chaotic behavior and one would like to understand the motion on the torus as the parameter is varied past the bifurcation point.

Insight into both applications mentioned above can be gained by the study of how the behavior of circle mappings changes on varying a parameter. We note that the applications above have a dissipative rather than Hamiltonian flavor. The ideas in these lectures, although having some connection to the theory of Hamiltonian systems, are intended to be applied in dissipative systems with attracting invariant tori rather than tori with more subtle stability properties as in Kolmolgorov-Arnold-Moser theory.

II. *CLASSICAL THEORY OF CIRCLE MAPPINGS*

As a first step in studying mappings of the circle into itself, we lift the mapping to the real line. We view the circle S^1 as $\mathbb{R}/\mathbb{Z}$ and define a <u>circle mapping</u> as a continuous strictly increasing function $f:\mathbb{R}\to\mathbb{R}$ satisfying the periodicity condition $f(x+1)=f(x)+1$. We remark that the periodicity condition is equivalent to the requirement that $f(x)-x$ be periodic with period 1. An important invariant associated with a circle mapping f is the <u>rotation number</u> $\rho(f)$ defined by $\rho(f)=\lim_{n\to\infty}(f^n(x)-x)/n$ where f^n denotes the nth iterate of f.

Theorem: (Poincaré): If f is a circle mapping and $x\in R$, then $\lim_{n\to\infty}(f^n(x)-x)/n$ exists and is independent of x.

For the proof of the theorem we require the following lemma whose proof relies mainly on the fact that $f(x)-x$ is a continuous periodic function on $\mathbb{R}$.

Lemma: If f is a circle mapping, the $\max_x(f(x)-x)\le\min_x(f(x)-x)+1$.

Proof: By the remark preceding the lemma, we can choose x_0 and x_1 such that $x_0\le x_1<x_0+1$ and $f(x_0)-x_0=\min_x(f(x)-x)$, $f(x_1)-x_1=\max_x(f(x)-x)$.

Since f is increasing we have $f(x_1)\le(f(x_0+1)=f(x_0)+1$ and hence $f(x_1)-x_1\le$ $f(x_1)-x_1\le f(x_0)-x_0+1$.

Proof of the Theorem: For n=1,2,...we define $M_n=\max_x(f^n(x)-x)$.

The sequence $\{M_n\}$ is subadditive, i.e. $M_{n_1+n_2}\le M_{n_1}+M_{n_2}$ and hence, by a standard elementary theorem, the sequence $\{M_n/n\}$ is convergent. We now apply the above lemma to f^n (any iterate of a circle mapping is also a circle mapping) and obtain that $\min_x(f^n(x)-x)/n$ converges as $n\to\infty$ to the same limit as M_n/n. Hence, $\lim_{n\to\infty}(f^n(x)-x)/n$ exists and is clearly independent of x.

We now list several facts concerning rotation numbers. We note first that by the definition of $\rho(f)$ if f_1 and f_2 are circle mappings with $f_1(x) \le f_2(x)$ for all $x \in \mathbb{R}$, then $\rho(f_1) \le \rho(f_2)$. Secondly we remark that the lemma provides us with the uniform estimate $|\rho(f) - (f^n(x) - x)/n| \le 1/n$ for any circle mapping f. This estimate shows that the mapping $f \to \rho(f)$ is continuous in the topology of uniform convergence. Moreover, if $\{f_\mu\}$ is a one-parameter family of circle mappings jointly continuous in x and μ, then $\rho(f_\mu)$ is continuous in μ.

We make the convention that when we say p/q is a rational number we mean that p and q are integers with $q \ge 1$.

Proposition: Let f be a circle mapping and let p/q be rational. Then

(1) $\rho(f) > p/q$ if and only if $f^q(x) > x+p$ for all $x \in \mathbb{R}$

(2) $\rho(f) < p/q$ if and only if $f^q(x) < x+p$ for all $x \in \mathbb{R}$

(3) $\rho(f) = p/q$ if and only if there exists $x_0 \in \mathbb{R}$ such that $f^q(x_0) = x_0 + p$.

Proof: We prove only statement (1) since (2) will follow similarly and (3) is deduced from the logical negation of (1) and (2). Clearly, $\rho(f) > p/q$ is equivalent to $q\rho(f) - p > 0$ which, by definition of ρ, is equivalent to $\rho(f^q - p) > 0$. Hence it suffices to show that given a circle mapping f, $\rho(f) > 0$ iff $f(x) > x$ for all x. First suppose $f(x) > x$ for all $x \in \mathbb{R}$. Then, since $f(x) - x$ is continuous and periodic, there is some $\epsilon > 0$ such that $f(x) - x \ge \epsilon$ for all x. By adding and subtracting iterates of f we see that $f^n(x) - x = (f(f^{n-1}(x)) - f^{n-1}(x)) + (f(f^{n-2}(x)) - f^{n-2}(x)) + \ldots + (f(x) - x) \ge n\epsilon$ and hence $\rho(f) \ge \epsilon$. Conversely, suppose $\rho(f) > 0$. By the adding and subtracting trick above, we see that $f(x) - x > 0$ for some $x \in \mathbb{R}$. So, either $f(x) - x > 0$ for all x or there exists x_0 such that $f(x_0) = x_0$. The latter case yields the contradiction $\rho(f) = \lim_{n\to\infty} (f^n(x_0) - x_0)/n = 0$.

We remark that the point x_0 in part (3) of the proposition is a periodic point of the action induced by f on $\mathbb{R}/\mathbb{Z}$ with period dividing q. Indeed, if p/q is in lowest terms the period is q. This remark can be expanded on in the sense of the following proposition.

Proposition: Let f be a circle mapping with rational rotation number $\rho(f) = p/q$. If p/q is in lowest terms, then every orbit of the action induced by f on $\mathbb{R}/\mathbb{Z}$ is asymptotic to a periodic orbit of period q.

Proof: We give a proof only for the trivial case $\rho(f) = 0$, the basic idea of the proof reformulated in terms of $\mathbb{R}/\mathbb{Z}$ extends to the general case. By the previous proposition and periodicity f has a fixed point in every interval of unit length. Let $x \in \mathbb{R}$ be arbitrary. We will show that the sequence of iterates $x_n = f^n(x)$ converges to a fixed point. Let $\bar{x}$ be a fixed point of f such that $\bar{x} \le x \le \bar{x}+1$. Since f is increasing, $\bar{x} \le x_n \le \bar{x}+1$ and moreover the sequence $\{x_n\}$ is monotone. Hence $\{x_n\}$ converges and the limit must be a fixed point of f.

The proposition above gives us a more or less complete understanding of the dynamics of circle mappings with rational rotation number; namely, the orbits under such mappings are asymptotically periodic. For irrational rotation numbers the situation is quite different and we merely remark upon the state of affairs by stating the following result of Denjoy without proof.

Proposition: Let f be a circle mapping with irrational rotation number $\rho(f)=\rho$. Suppose further that f is continuously differentiable with derivative f′ of bounded variation and strictly positive. Then there exists a circle mapping h such that $(h(f(h^{-1}(x)))=x+\rho$.

The proposition says that in some continuous coordinate system, f is simply a rigid irrational rotation by ρ, and we have in some sense completely classified the dynamics of individual circle mappings. We now turn to the question of the variation of the rotation number with respect to a parameter. We have already seen that if the family $\{f_\mu\}$ of circle mappings is continuous in μ then $\rho(f_\mu)$ is continuous in μ. One might reasonably ask if $\rho(f_\mu)$ is analytic in μ if $\{f_\mu\}$ is analytic in μ. The answer to this question is no, and can be seen via the phenomenon of phase locking: the rational rotation numbers have a certain stability in the sense that if $\rho(f_\mu)=p/q$, then $\rho(f_{\mu'})=p/q$ for μ' in some small interval (the phase locking interval) containing μ. We make this statement precise in the following proposition.

Proposition: Let $\{f_\mu\}_{\mu\in[0,1]}$ be a family of circle mappings such that $\mu\to f_\mu$ is continuous in the topology of uniform convergence and $\mu\to\rho(f_\mu)$ is nondecreasing. Let p/q be a rational number such that $\rho(f_0)<p/q<\rho(f_1)$, and suppose that there is no value $\bar{\mu}$ in [0,1] such that $f^q_{\bar{\mu}}(x)=x+p$ for all x. Then, $\rho(f_\mu)=p/q$ for all μ in some interval of non-zero length contained in [0,1].

Proof: Assume the contrary, i.e. suppose that $\sup\{\mu:\rho(f_\mu)<p/q\} = \inf\{\mu:\rho(f_\mu)>p/q\}=\bar{\mu}$. By the first proposition of this section, $\rho(f_\mu)<p/q$ implies $f^q_\mu(x)<x+p$ and hence, by continuity, $f^q_{\bar{\mu}}(x)\le x+p$. Similarly, $\rho(f_\mu)>p/q$ implies $f^q_\mu(x)>x+p$ so $f^q_{\bar{\mu}}(x)\ge x+p$ and hence $f^q_{\bar{\mu}}(x)=x+p$.

We remark that the condition that there is no value of μ such that $f^q_\mu(x)=x+p$ is satisfied by any family for which all members have a critical point (i.e. f'_μ vanishes for some x) or any family for which each member is entire and not linear.

III. RENORMALIZATION GROUP ANALYSIS

We now turn to the main topic of these lectures, namely the renormalization group analysis of the dependence of the rotation number on a parameter. More precisely, we study circle mappings f which are very smooth (say, analytic) but have a critical point. We analyze certain iterates of such mappings in the immediate vicinity of the critical point. For convenience, we take the critical point to be $x=0$ and define a <u>critical circle mapping</u> to be a differentiable circle mapping f such that $f'(0)=0$. We note that , since f is strictly increasing, f" must also vanish at $x=0$. We of course are most interested in one parameter families of such mappings, the most standard example being the sine family $f_\mu(x)=x+\mu-(\sin 2\pi x)/2\pi$. Before discussing some phenomenological aspects of the lengths of various phase locking intervals for the sine family or others like it we must first set some notation for continued fractions. Given a sequence $r_1 r_2,\ldots,r_n$ of positive integers we denote

$$[r_1,r_2,\ldots,r_n] = \cfrac{1}{r_1+\cfrac{1}{r_2+\cfrac{1}{\ddots+\cfrac{1}{r_n}}}}\ .$$

An infinite sequence $r_1,r_2,\ldots$yields an infinite continued fraction by a limiting process. We remark that the continued fractions we are defining are always between 0 and 1.

The first phenomenological observation we make is that there is a particular number δ_1 with numerical value -2.83361...such that the length of the phase locking interval associated with rotation number $\rho=[1,1,\ldots,1]$ $(r_1=r_2=\ldots=r_n=1)$ behaves asymptotically for large n as a constant multiple of $|\delta_1|^{-n}$. What is of interest is that the number δ_1 is independent of the family. Indeed, for any $r=1,2,\ldots$there is a number of δ_r such that the length of the phase locking interval with rotation number $\rho=[r,r,\ldots,r]$ $(r_1=r_2=\ldots=r_n=r)$ behaves for large n as a constant multiple of $|\delta_r|^{-n}$, and, given any fixed sequence $r_1,r_2,\ldots,r_m$ the length of the phase locking interval for rotation number $\rho=[r_1,r_2,\ldots,r_m,r,r,\ldots r]$ $(r_{m+1}=r_{m+2}=\ldots=r_{m+n}=r)$ has the same behavior (with perhaps a different constant multiplier). The numbers δ_r give us quite detailed universal information about how the rotation number changes with the parameter.

For more general rotation numbers the phenomenology is more complicated. For the sine family and others like it there are two universal "structure functions" $\gamma(r_0|r_1,r_2,\ldots)$ and $\sigma(r_0|r_1,r_2,\ldots)$ with exponential decaying dependence on r_j for large j such that, for large n, the length of the phase locking interval with $\rho=[r_1,r_2,\ldots,r_n]$ is given approximately by $\sigma(r_n|r_{n-1},r_{n-2},\ldots)\gamma(r_{n-1}|r_{n-2},r_{n-3},\ldots)\ldots$ $\gamma(r_1|r_0,r_{-1},\ldots)$ in the sense that the ratio of this quantity with the actual length of the interval is bounded and bounded away from zero uniformly in $n,r_1,\ldots,r_n$. The problematical appearance of the sequence $r_0,r_{-1},r_{-2},\ldots$ can be dealt with via the exponentially decaying dependence mentioned above.

Different pieces of the phenomenological picture have different status. The first observation above is in essence a theorem for a nonempty open set of one-parameter families. The proof involves traditional renormalization analysis and requires detailed numerical estimates proved rigorously by computer. This computer assisted proof could almost surely be modified to prove the corresponding statement for values of r greater than 1. We believe that the more comprehenive statement for general rotation number can be derived by a generalized renormalization group analysis in which the renormalization operator has an invariant hyperbolic Cantor set (instead of a fixed point as in the traditional analysis) with a one-dimensional unstable manifold. In the remainder of these lectures we will describe a geometric picture, which, along with assumptions on hyperbolicity of relevant operators, leads to the desired formula for the lengths of phase locking intervals with general rotation number. In order to present this picture and the ideas of the renormalization group analysis we first need some more information about continued fractions.

We denote the integer and fractional parts of a real number x by int(x) and frac(x) respectively. For $\rho\epsilon(0,1)$ let $r_1=\text{int}(1/\rho)$, $\rho_1=\text{frac}(1/\rho)$ and recursively define $r_j=\text{int}(1/\rho_{j-1})$, $j=2,3,\ldots$. The recursion stops after finitely many steps if it happens that one of the ρ_j is equal to zero (this occurs if and only if ρ is rational). At each step we have

$$\rho = \cfrac{1}{r_1 + \cfrac{1}{r_2 + \cfrac{1}{\vdots + \cfrac{1}{r_j+\rho_j}}}}$$

which we denote by $\rho=[r_1,r_2,\ldots,r_j+\rho_j]$. Given such a collection of numbers r_j and ρ_j we define positive integers p_j and q_j via the recursion relations $p_j=r_jp_{j-1}+p_{j-2}$, $q_j = r_jq_{j-1}+q_{j-2}$ with initial conditions $p_0=0, p_1=q_0=1$, $q_1=r_1$. From the recursion relations follow several facts which we leave for the reader to prove. First we note that $p_j/q_j = [r_1,r_2,\ldots,r_j]$, hence, if $\rho_j=0$ then $\rho=p_j/q_j$. In any case, $\rho = (p_j+p_{j-1}\rho_j)/(q_j+q_{j-1}\rho_j)$ which can be inverted to give $\rho_j = (-\rho q_j+p_j)/(\rho q_{j-1}-p_{j-1})$. The p_j and q_j satisfy the determinant condition $q_jp_{j-1}-p_jq_{j-1} = (-1)^j$ and hence the fraction p_j/q_j is in lowest terms. An important piece of intuition is that the numbers p_j/q_j are the best rational approximations to ρ in the sense that q_j is the smallest positive integer q such that $\text{dist}(\rho q,Z) < \text{dist}(\rho q_{j-1},Z)$.

We now begin our discussion of the ideas of the renormalization group analysis. Let f be a critical circle mapping with rotation number $\rho(f)=\rho=[r_1,r_2,\ldots]$ and let ρ_j, p_j, and q_j be constructed as above. We introduce circle mappings $f_j(x)=f^{q_j}(x)-p_j$ and set $x_j=f_j(0)$. The motivation for studying these special iterates of f is that p_j/q_j is a particularly good rational approximation to $\rho(f)$ and hence $\rho(f_j)$ is particularly small, i.e., x_j is particularly close to zero. Indeed, since f is a critical circle mapping we are well advised in our analysis to concentrate on behavior near the critical point, and the idea is that f_j should have simple behavior "on the scale of x_j". Before making this notion of scaling more precise, we make the fundamental observations that the various f_j's commute, satisfy the propagation rule $f_{j+1}=f_j^{r_{j+1}} \circ f_{j-1}$, and are such that $\rho(f_j)/\rho(f_{j-1}) = -\rho_j$. We remark that the last of these facts implies that one can reconstruct ρ_j based on knowledge of only f_j and f_{j-1}. We introduce pairs of rescaled functions $\eta_j^f(z)=(1/x_{j-1})f_j((x_{j-1})z)$, $\xi_j^f(z)=(1/x_{j-1})f_{j-1}((x_{j-1})z)$. From the observations above we see that η_j^f and ξ_j^f commute and a straightforward rescaling leads as to the propagation formulas

$$\eta_{j+1}^f(z) = (1/\lambda_j^f)(\eta_j^f)^{r_{j+1}} \circ \xi_j^f(\lambda_j^f z),\ \xi_{j+1}^f(z) = (1/\lambda_j^f)\eta_j^f(\lambda_j^f z),\ \lambda_j^f \equiv \eta_j^f(0).$$

The renormalization operator is in essence the operator sending the pair (ξ_j^f,η_j^f) to the pair $(\xi_{j+1}^f,\eta_{j+1}^f)$.

We now wish to build the machinery necessary to study, at least asymptotically, iterates of the renormalization operation. We need a suitable space of pairs of functions large enough to contain the pairs (ξ_j^f, η_j^f) coming from critical circle mappings as described above. For the purposes of our analysis, the appropriate space is the space of weakly commuting pairs: We say (ξ, η) is a weakly commuting pair if ξ and η are strictly increasing smooth functions satisfying

1) η is defined on $[0,1]$ and is such that $\eta(x)<x$
2) ξ is defined on $[\eta(0),0]$ and satisfies $\xi(x)>x$, $\xi(0)=1$
3) $\xi \circ \eta(0)=\eta \circ \xi(0)$.

An example of a weakly commuting pair is the pair $\xi_o^f(x)=x+1$, $\eta_o^f(x)=-f(-x)$ where f is a circle mapping with $0<\rho(f)<1$.

We now define the concept of a rotation number for a weakly commuting pair, an invariant notion which in the case of a pair which is a rescaled pair of circle mappings reduces to the ratio of the rotation numbers of the circle mappings. Let (ξ, η) be a weakly commuting pair and let x be an arbitrary point in $[0,1]$. Repeated iteration of η implies that $\eta^n(x)<0$ for some positive integer n. Applying ξ, $\xi^{M_n}(\eta^n(x))\epsilon[0,1]$ for some M_n. The limit of the ratio M_n/n as $n\to\infty$ exists and is independent of the initial x. We say that this limit is the rotation number for the pair (ξ, η) and we denote it by $\rho(\xi, \eta)$.

Theorem: Let (ξ, η) be a weakly commuting pair with rotation number $\rho(\xi, \eta)$ satisfying $1/(r+1) < \rho(\xi, \eta) < 1/r$ for some positive integer r. Let $\lambda=\eta(0)$ and define $\hat{\eta}(x)=(1/\lambda)\eta^r \circ \xi(\lambda x)$, $\hat{\xi}(x)=(1/\lambda)\eta(\lambda x)$. Then $(\hat{\xi}, \hat{\eta})$ is a weakly commuting pair and $\rho(\hat{\xi}, \hat{\eta})=\text{frac}(1/\rho(\xi, \eta))$.

We omit the proof of the theorem but remark that it follows in a perfectly straightforward way. The point of the theorem is that we may now define operators T_r, r=1,2,..., from the domain D_r = {weakly commuting pairs (ξ, η) with $1/(r+1) < \rho(\xi, \eta)<1/r$} into the space of weakly commuting pairs via $T_r(\xi, \eta) = (\hat{\xi}, \hat{\eta})$. We remark that the theorem implies that T_r gives an action on rotation numbers which is simply the traditional Gauss map. We further remark that if f is a circle mapping with $\rho(f)=[r_1, r_2, \ldots]$ then $(\xi_j^f, \eta_j^f) = T_{r_j}T_{r_{j-1}}\cdots T_{r_1}(\xi_o^f, \eta_o^f)$ where ξ_o^f and η_o^f are defined as in the example immediately following the definition of a weakly commuting pair. Since the domains of the various operators T_r are disjoint, we may conveniently view all of the operators as s single operator T.

We now present a useful geometric picture of how the operators act on the space of weakly commuting pairs. We view this infinite dimensional space as being two-dimensional with the vertical direction representing one dimension in which the rotation number ρ increases in the upwards direction and the horizontal direction representing the infinitely many remaining dimensions. As a consequence of the action induced on rotation numbers by the operators T_r the image $Im(T_r) = T_r[D_r]$ contains points having all rotation numbers in (0,1). Roughly speaking, the domain D_r is expanded vertically under the action of T_r (See Figure 1). For the purposes of our analysis, we hope that $T_r[D_r]$ is contracted in all horizontal directions. We refer to this expansivity in the vertical direction and contractivity in the horizonal directions as the <u>global hyperbolicity hypothesis</u>. A possible somewhat more precise version of this hypothesis is as follows: Suppose that coordinates (x,y) can be imposed on the space with x representing all horizontal dimensions and y the vertical dimension. The operator T can then be represented as the pair $T_x,T_y)$ via $T(x,y) = (T_x(x,y),T_y(x,y))$. We require that the inequalities $|\frac{\partial T_y}{\partial y}|\geq a>1$, $||\frac{\partial T_x}{\partial x}||\leq d<1$, $||\frac{\partial T_y}{\partial x}||\leq b$, $||\frac{\partial T_x}{\partial y}||\leq c$ are satisfied uniformly over the whole space for some constants a,b,c,d satisfying $(a-1)(1-d)>bc$. We remark that the traditional renormalization analysis requires a local version of this hypothesis in a neighborhood of a fixed point.

We now concentrate on the consequences of global hyperbolicity. Suppose $r_1,r_2,\ldots$is a fixed sequence of positive integers and consider the sequence of sets $Im(T_{r_1},T_{r_2},\ldots T_{r_n})$. This sequence of images of longer and longer compositions forms a decreasing sequence of thinner and thinner cylinders converging down to a smooth curve $W^u(r_1,r_2,\ldots)$ (See Figure 2). The curve $W^u(r_1,r_2,\ldots)$ is a branch of an unstable manifold which we choose not to construct explicitly. The operators T_r induce an action on the branches $W^u(r_1,r_2,\ldots)$ given by $T_r(W^u(r_1,r_2,\ldots) \cap D_r) = W^u(r,r_1,r_2,\ldots)$.

The next key idea is that the branches admit a special parameterization, indeed there is an essentially unique way of smoothly parameterizing all of them simultaneously in such a way that the action of the operators T_r becomes linear in the respective parameterizations. This is a generalization of the traditional analysis in which coordinates are introduced such that the induced action of the renormalization operator on the (single) unstable manifold is simply linear. We denote by $W^u_\mu(r_1,r_2,\ldots)$ the unstable manifold parameterized by $\mu\epsilon(0,1)$ and normalized such that ρ approaches 0 (respectively,1) as μ approaches 0 (respectively,1).

We now turn our attention to more general one parameter families which are "sufficiently vertical" curves in terms of our geometric picture. Without making this notion of verticality precise, we notice that if $\{\zeta_\mu\}_{\mu\in[0\ 1]}$ is such a family with the parameterization normalized in the sense above then by applying T_r to that part of the family which lies in D_r we generate another such family which, however, is no longer parameterized correctly. (See Figure 3). We construct, for each positive integer r, an operator T_r^* taking one parameter families into one parameter families whose action is to first restrict the parameter μ to the interval for which $\zeta_\mu \epsilon D_r$, then to apply T_r to this piece of the family, finally to make a linear change of parameterization to restore the normalization. For the fixed sequence $r_1, r_2, \ldots$ the sequence of one parameter families $T_{r_1}^* T_{r_2}^* \ldots T_{r_n}^* \zeta_\mu$ converges to $W_\mu^u(r_1, r_2, \ldots)$ not only as point sets but also as parameterized curves. Given this fact we may easily read off information about the lengths of phase locking intervals for the family ζ_μ. Indeed, the length of the phase locking interval for rotation number $\rho=[r_1, r_2, \ldots, r_n]$ is the product of the length of the phase locking interval where $\rho(T_{r_{n-1}}^* T_{r_{n-2}}^* \ldots T_{r_1}^* \zeta_\mu)=1/r_n$ times the rescaling factor in the parameter in going from ζ_μ to $T_{r_{n-1}}^* T_{r_{n-2}}^* \ldots T_{r_1}^* \zeta_\mu$. For large values of n, the first factor is approximately the length of the parameter interval where $\rho(W_\mu^u(r_{n-1}, r_{n-2}, \ldots, r_1, \ldots))=1/r_n$ which we denote by $\sigma(r_n | r_{n-1}, r_{n-2}, \ldots)$. The second factor is itself a product of the length of the parameter interval where $\rho(T_{r_{n-2}}^* T_{r_{n-3}}^* \ldots T_{r_1} \zeta_\mu) \epsilon$ $(1/(r_{n-1}+1), 1/r_{n-1})$ times the rescaling factor in going from ζ_μ to $T_{r_{n-2}}^* T_{r_{n-3}}^* \ldots T_{r_1}^* \zeta_\mu$. The first factor in this product is, for large n, approximately equal to the length of the parameter interval where $\rho(W_\mu^u(r_{n-2}, r_{n-3}, \ldots)) \epsilon (1/(r_{n-1}+1), 1/r_{n-1})$ which we denote by $\gamma(r_{n-1} | r_{n-2}, r_{n-3}, \ldots)$. The second factor has the same structure as the term that we started with, we deal with it in the same manner and continue this process to obtain that the length of the phase locking interval with rotation number $\rho=[r_1, r_2, \ldots, r_n]$ is given, up to bounded corrections, by the product $\sigma(r_n | r_{n-1}, r_{n-2}, \ldots)\gamma(r_{n-1} | r_{n-2}, r_{n-3}, \ldots)\gamma$ $(r_{n-2} | r_{n-3}, r_{n-4}, \ldots) \cdots \gamma(r_1 | r_0, r_{-1}, \ldots)$. The quantities σ and γ are universal in the sense that they represent intrinsic features of the family W_μ^u of unstable manifolds.

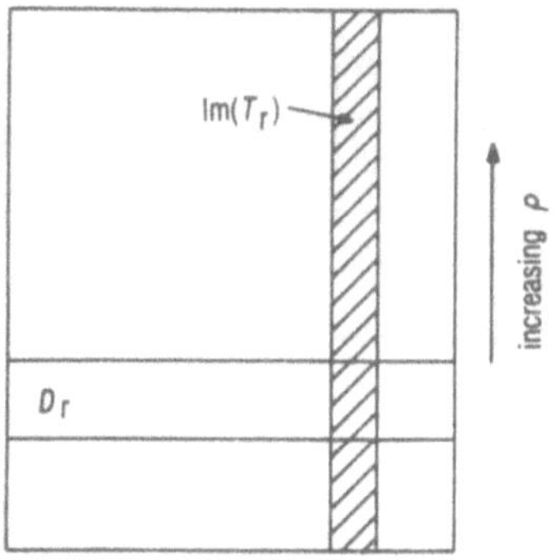

Figure 1.

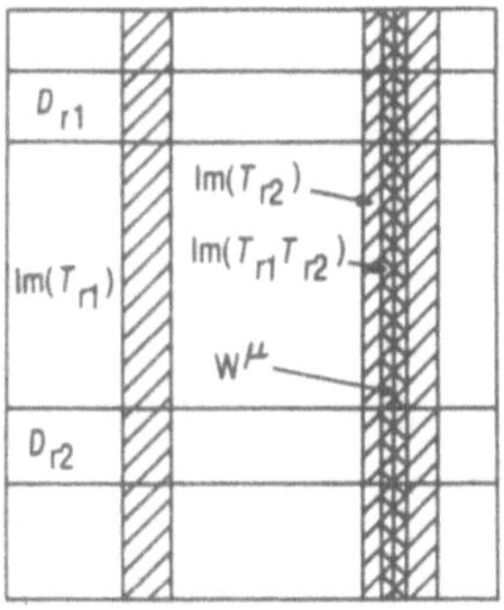

Figure 2.

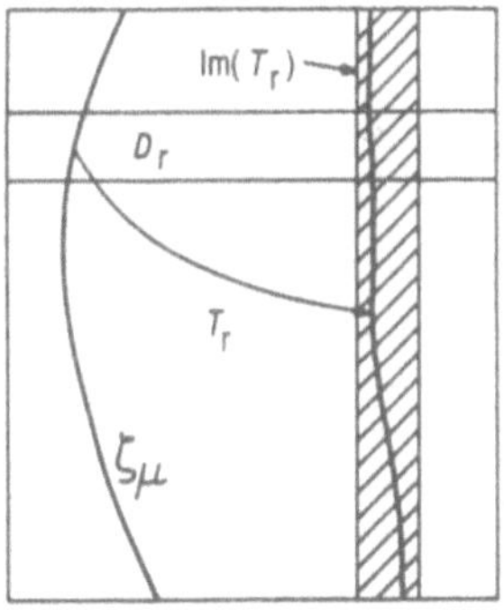

Figure 3.

IV. BIBLIOGRAPHY

For more detailed information on the classical theory of circle mappings, see Chapter 3 of

1. I. P. Corfeld, S. V. Fomin and Ya. G. Sinai, Ergodic Theory, Springer-Verlag, Berlin Heidelberg New York, 1982.

or

2. M. Herman, 'Sur la conjugation différentiable des difféomorphismes du circle a des rotations', Publ. Math. IHES 49 (1979), 5-234.

For proofs and ideas concerning the traditional renormalization analysis, see

3. M. J. Feigenbaum, L. P. Kananoff and S. J. Shenker, 'Quasi-periodicity in dissipative systems: a renormalization group analysis', Physica 5D (1982), 370-386.
4. O. E. Lanford III and R. de la Llave, 'Solution of the functional equation for critical circle mappings with golden ratio rotation number', in preparation.
5. B. D. Mestel, 'A computer assisted proof of universality for cubic critical maps of the circle with golden mean rotation number', Ph.D. Thesis, Mathematics Department, University of Warwick, 1985.
6. S. Ostund, D. Rand, J. Sethna and E. Siggia, 'Universal properties of the transition from quasi-periodicity to chaos in dissipative systems', Physica 8D (1983), 303-342.

For a more complete exposition of the analysis presented in Section III, see

7. O. E. Lanford, 'Renormalization group methods for circle mappings', in Statistical Mechanics and Field Theory: Mathematical Aspects, ed. by T. C. Dorlas, N. M. Hugenholtz and M. Winnik, Springer-Verlag, Berlin Heidelberg New York, 1986, pp. 176-189.

PHASE TRANSITIONS ON STRANGE SETS

Predrag Cvitanović[+]

Niels Bohr Institute
Blegdamsvej 17
DK-2100 Copenhagen

Abstract: The "thermodynamic" formalism provides a very general division of strange (Cantor, fractal) sets into two classes; those which do exhibit phase transitions, and those which do not.

The "thermodynamic formalism"[1,2,3,4)] is based on the observation that the sum used in determining the Hausdorff dimension of a strange (fractal, Cantor) set

$$\sum_{i=1}^{N} \ell_i^{-\tau}$$

resembles a partition sum over "configurations" i, with τ playing the role of "temperature", and covering interval sizes playing the role of "Boltzmann weights".

The "thermodynamic" functions extracted from the above partition sum exhibit a phenomenon that might have gone unnoticed were it not for the thermodynamic formalism. They can undergo "phase transitions". These phase transitions can be visualized in the following way: the exponent τ acts as a "magnifying lens" which blows up some of the covering intervals, and (relatively), shrinks the others. For negative τ, the fat intervals are expanded, and the thin ones are made even thinner. For positive τ, the thin intervals are (relatively) blown up. Below the phase transition, the sum is dominated by a few fat intervals. Visually, the cover consists of a few

---

[+]Carlsberg Fellow

black blocks and vanishing amount of fine "dust". Above the phase transition, the large number of thin intervals overwhelms the few fat ones; visually, the strange set looks gray. All known examples [5-12)] are of first order; at the transition the average segment width jumps discontinuously, by a finite amount.

An important (conjectured) example of such system is the Hénon attractor. In the "hyperbolic" phase, the attractor is well described as a hyperbolic set, composed of sheets whose scale can be determined from the associated unstable cycles[13,14]. In the other phase, the "thermodynamics" is dominated by the "turnbacks".[8)]

In physics existence of a phase transition is a dramatic effect--a liquid freezes into a solid--and a realization that a certain strange set undergoes a phase transition is (potentially) of greater import than knowing that its Hausdorff dimension equals .8701....

In this way the "thermodynamics" offers a broad classification of all strange sets into those which exhibit phase transitions, and those which do not. Examples of the first class are the Hénon attractor and the set of irrational windings for critical circle maps;[5,6)] examples of the second class are the period doubling attractor and hyperbolic (axiom A) sets.

τ in the partition sum is a mathematical invention, not a physical temperature. There is no "heat bath" that an experimentalist can dip a strange set into. If you thus find varying τ objectionable, you should probably think of the above classification as a division of strange sets into those with smooth measures (axiom A) and those with measures that are not smooth. In the "thermodynamic" language, the first class has smooth scaling spectra ($f(\alpha)$ funotions); the second does not. Scaling spectra are accessible directly, by binning the experimental data, but in practice the results are very noisy, and introduction of an "artificial temperature" τ is the preferred technique in searches for phase transitions.

From the experimental point of view, phase transitions are elusive. The problem is that the experiments are limited by noise to extraction of a few hundred intervals ("thermodynamic configurations"), and are very far from the thermodynamic limit. In such small systems phase transitions signal their presence by a further slowdown in convergence, and can be extracted only by finite scaling techniques.

A warning: strange sets, unlike physical systems, often have no unique "thermodynamics"; the weights associated with different pieces of the set might reflect mathematical prejudices[6)] rather than physical imperatives.

1. THERMODYNAMIC DESCRIPTION OF STRANGE SETS

Given a strange set[+] on an interval (for example, the period doubling attractor), and limited patience, one tries to pin down the set by partitioning the interval into a finite number of segments. i-th segment is either "empty (a "hole" Δ_i), or it contains a piece of the strange set (a "cover" ℓ_i). This partition is not unique, and as a sloppy cover can make a strange set appear fatter than it actually is, one needs to insure that the cover is "optimal". The simplest approach is to cut up the interval into equal size segments and discard the holes. This is the blind man's approach to strange sets, oblivious both to the actual layout of the set and its internal systematics. A better cover is obtained by poking holes of maximal size (with a point of the set at both ends), and with labels which reflect the structure of the set.

What is the typical size of a covering interval? Intuitively, if the cover is good, and there are N covering intervals, each covering interval is smaller than 1/N. We can sharpen this intuition by reminding ourselves how Cantor sets are generated by dynamical systems. Typically they arise by dynamical stretching and squeezing, with a "mother" interval at time t giving rise to several "daughter" intervals at time $t+\Delta t$

$$\ell_d = \sigma_{dm} \ell_m \tag{1}$$

The <u>scaling function</u> σ_{dm} is related to the contraction rate of the system. If σ_{dm} is on average of size σ, and the average number of daughters is α, than after k time steps $\ell \simeq \sigma^k$, $N \simeq \alpha^k$, and $\ln \ell \simeq \ln N$. Clearly, instead of using ℓ_i (which

---

[+]A definition: compact, perfect, totally disconnected set.

shrink to zero as we refine the cover), we should define the scaling exponents

$$\mu_i = -\frac{1}{t} \ln \ell_i$$
$$t = \ln N \tag{2}$$

For "good" covers the covering intervals do not vary too widely in size, and μ_i are bounded.

Exercise 1. A two-scale Cantor set is generated by recursively replacing each interval ℓ_m by a "fat" subinterval $\ell_{mo} = \sigma_o \ell_m$ and a "thin" subinterval $\ell_{m1} = \sigma_1 \ell_m$, $o<\sigma_1<\sigma_o<1$, $\sigma_o+\sigma_1<1$. Show that

$$\mu_{min} = -\frac{\ln\sigma_o}{\ln 2}, \quad \mu_{max} = -\frac{\ln\sigma_1}{\ln 2} .$$

A theory of a strange set should yield a good labeling scheme for the covering intervals of the set (the "symbolic dynamics") and predict the scale associated with each interval (the "scaling function"). An experiment yields a finite cover for the set. How are we to efficiently compare the two?

A robust procedure for averaging the experimental data is called for. The first temptation is to extract the "size" of the set from the N experimental covering intervals ℓ_i by forming the sum

$$Z_N = \sum_{i=1}^{N} \ell_i \quad . \tag{3}$$

As the set is dense with holes, this sum tends to vanish, in step with refinement of the cover. Our intuitive estimates (2) of ℓ_i tell us that it vanishes exponentially. We can compensate for the exponential shrinkage by introducing exponent τ:

$$Z_N(\tau) = \sum_{i=1}^{N} \ell_i^{-\tau} \tag{4}$$

For a deftly chosen value $\tau = -D$, the blown-up segments fill up the entire interval, no matter how fine the refinement:

$$\lim_{N\to\infty} Z_N(-D) = 1 \tag{5}$$

For an optimal cover D equals D_H, the Hausdorff dimension: a sloppy cover yields an upper bound on D_H.

Exercise 2. If a set is of dimension D, it can be covered with $n(e) \propto \ell^{-D}$ balls of diameter ℓ. Check that (5) conforms to this notion of dimension by taking equal size covering intervals.

Eq. (4) enables us to extract much more information about the strange set than just its Hausdorff dimension. The parameter τ can be used to extract different scales in the set. For $\tau \to -\infty$ the fattest intervals dominate, and $Z_N(\tau) \to \exp(t\tau\mu_{min})$. For $\tau=0$, $Z_N(o) = N$, and for $\tau \to +\infty$ the thinnest intervals dominate, $Z_N(\tau) \to \exp(t\tau\mu_{max})$. The sum grows exponentially with $t = \ell n\, N$ for the entire range of τ, so the object that has a finite limit as $N \to \infty$ is

$$q_t(\tau) = \frac{1}{t}\,\ell n Z_N(\tau) = \frac{1}{t}\,\ell n \sum_{i=1}^{N} e^{t\tau\mu_i} \to q(\tau) \tag{6}$$

This function grows monotonically with τ. By our normalization convention

$$q_t(o) = 1 \tag{7}$$

The Hausdorff dimension condition (5) can now be restated in terms of a quantity finite in the $N \to \infty$ limit:

$$q(-D_o) = 0 \tag{8}$$

In practice, $q(\tau)$ is a rather boring function, and the distribution of the scales is displayed more effectively by the mean scaling exponent

$$\frac{dq_t(\tau)}{d\tau} = \frac{1}{Z_N(\tau)} \sum_{i=1}^{N} \mu_i e^{t\mu_i\tau} = \mu(\tau) \tag{9}$$

and the higher moments such as

$$\frac{d^2 q_t(\tau)}{d\tau^2} = t(\langle\mu_i^2\rangle_\tau - \mu(\tau)^2) \tag{10}$$

$q''(\tau)$ is strictly positive. $\mu(\tau)$ grows monotonically from $\mu(-\infty) = \mu_{min}$ to $\mu(\infty) = \mu_{max}$.

Exercise 3. Show that for the two-scale Cantor set introduced in exercise 1 the asymptotic $q(\tau)$ is given by

$$q(\tau) = \frac{\ln\left(\sigma_0^{-\tau} + \sigma_1^{-\tau}\right)}{\ln 2}$$

(hint: use binomial theorem). Show that for finite N, $q_t(\tau) = q(\tau) - \tau/t \ln \ell$, where ℓ is the size of the single covering interval at the top level. How can such finite N contributions be eliminated? Plot $q(\tau)$, $\mu(\tau)$ and $q''(\tau)$.

Actually, as we know the $\tau \to \pm \infty$ behaviour of $q(\tau)$, we might just as well subtract it out, and plot the result as the function of μ:

$$S_t(\mu) = q_t(\tau) - \tau\mu(\tau) \tag{11}$$

We recognize (11), together with (9) and

$$\tau = -\frac{dS_t(\mu)}{d\mu} \tag{12}$$

to be a Legendre transformation. In practice it is performed by evaluating (6), (9) and plotting $\mu(\tau)$, $S_t(\mu(\tau))$ pairs for a range of τ.

$S_t(\mu)$ function has almost always compact support $\mu_{min} \le \mu \le \mu_{max}$, is positive (by strict positivity of $q''(\tau)$), is convex by (12), and bounded by 1 (by normalization convention (7)), so it fits nicely on a piece of graph paper.

Exercise 4. Plot $S(\mu)$ for the 2-scale Cantor set. How does $S_t(\mu)$ depend on the finite t?

$S(\mu)$ is called the scaling spectrum and is the most succint summary of all information obtainable from averages over the covering interval lengths. In order to elucidate its physical significance, reconsider the defining sum (4). The sum depends only on the interval sizes and not on the way in which set was generated (reflected in the label i in the sum (4)). Therefore it can be written as

$$Z_N(\tau) = \int_{\ell_{min}}^{\ell_{max}} d\ell N(\ell) \ell^{-\tau} \tag{13}$$

where $N(\ell)d\ell$ is the number of intervals whose size falls into range $[\ell, d\ell]$. $N(\ell)$ grows with, and is bounded by N, the total number of intervals. We extract the rate of growth by defining

$$s_t(\mu) = \frac{1}{t} \ell n(\ell N(\ell)) \tag{14}$$

The sum (4) now takes form

$$Z_N(\tau) = e^{tq_t(\tau)} = e^{t(S_t(\mu)+\tau\mu)} = t \int_{\mu_{min}}^{\mu_{max}} d\mu \; e^{t(s_t(\mu)+\tau\mu)} \tag{15}$$

In the $t \to \infty$ limit a saddle point evaluation yields (11) and (12), with identification $s(\mu) = S(\mu)$, so the scaling spectrum $S(\mu)$ indicates how numerous is scaling exponent μ. The saddle point estimate picks out the global maximum of $s(\mu)+\tau\mu$, and it can jump discontinously in μ. As we show in the next section, this implies that $S(\mu)$ is actually the convex envelope of $s(\mu)$. $S(\mu)$ is smooth and easy to obtain numerically; counting involved in constructing $s(\mu)$ requires a detailed understanding of the strange set.

Exercise 5. Evaluate $N(\ell)$ and $s(\mu)$ for the 2-scale Cantor set. Estimate $q(\tau)$, $S(\mu)$ by a saddle-point evaluation of the integral (15). (Hint: use the Stirling approximation for k!).

Exercise 6. Consider scaling functions (1) which depend on finite memory (the preceeding few symbols in the mother interval label in Exercise 1). Relate $q(\tau)$ to the eigenvalues of the matrix σ_{dm}.

Remark 1: falfas

A strange set is usually highly inhomogenous; some covering intervals can contain most of the set, and others practically nothing. The inhomogeneity of the cover can be taken into account by replacing (2) by a weighted sum[3,15,8]

$$1 = \sum_{i=1}^{N} \frac{p_i{}^q}{\ell_i^\tau} \tag{16}$$

For dynamically built-up strange sets (such as the Hénon attractor) p_i are determined by a unique "natural" measure: p_i is the fraction of the points of the set which land in the i-th covering interval:

$$p_i = \frac{N_i}{N} \tag{17}$$

For many other strange sets (such as the set of all irrational winding parameter values of critical circle maps[6]) there is no "natural" choice of p_i, other than the uniform probability $p_i = 1/N$. If the probability is uniform, (16) reduces to (4).

However, in analysis of experimental data it is sometimes easier to fix $\ell_i = \ell = \text{const}$, and measure the p_i in (16). In this case it is $\tau(q)$, rather than $q(\tau)$, that is extracted from the data;

$$\tau(q) = - \frac{\ell n \sum_{i=1}^{N} p_i{}^q}{\ell n \ell} . \tag{18}$$

Instead of the scaling exponent μ_i, one now defines[3] the "pointwise dimension"

$$\alpha_i = \frac{\ell n p_i}{\ell n \ell_i} \tag{19}$$

Intuitively, the pointwise dimension should be the same regardless of whether we compute it by choosing balls of the same radius or the same probability. If we chose uniform probability, we see from (2) that the pointwise dimension and the scaling exponent are related by

$$\alpha_i = 1/\mu_i \tag{20}$$

Following the same line of reasoning that led to the scaling spectrum (11) one now arrives at the "f of alpha" function[3]

$$f(\alpha) = -\tau(q) + q\alpha(q)$$
$$f(\alpha) = S(\mu)/\mu \ . \tag{21}$$

The scaling spectrum and $f(\alpha)$ are the same function, up to coordinate redefinitions.

<u>Exercise 7</u>. Prove that the scaling spectrum $S(\mu)$ is independent of the choice of covering intervals ℓ_i, as long as the probabilities p_i arise from the same measure.

<u>Remark 2: previously defined dimensions</u>

The function $\tau(q)$ of (18) is related to the generalized dimensions[16-19] by

$$D_q = \frac{\tau}{q-1} \tag{22}$$

D_o is the Hausdorff dimension (8). By (11) and (12), D_o is the slope of the tangent to $S(\mu)$ at $\mu_H = \mu(-D_o)$ which satisfies

$$D_o = \frac{dS(\mu_H)}{d\mu} = \frac{S(\mu_H)}{\mu_H} \tag{23}$$

The information dimension is the dimension of the most numerous interval length, with scaling exponent $\mu_I = \mu(o)$:

$$D_1 = \left.\frac{d\tau(q)}{dq}\right|_{q=1} = \left.\frac{1}{\frac{dq(\tau)}{d\tau}}\right|_{\tau=o} = 1/\mu(o) \ . \tag{24}$$

For $\tau \to \pm\infty$ we obtain

$$D_\infty = 1/\mu_{max}, \ D_{-\infty} = 1/\mu_{min} \tag{25}$$

<u>Remark 3: scale invariance</u>

The sum (4) has an obvious flaw; nothing defines the units in which the lengths ℓ_i are to be measured. Under change of scale $\ell_i \to e^{-\beta}\ell_i$, the value of $Z_N(\tau)$ scales as $Z_N(\tau) \to e^{\tau\beta}Z_N(\tau)$, and the $q_t(\tau)$ function picks up a correction of $O(t^{-1})$:

$$q_t(\tau) \to q_t(\tau) + \beta\tau/t \ , \quad \mu(\tau) \to \mu(\tau) + \beta/t \ .$$

This is a $1/\ell nN$ correction, and in practice large. It is usually eliminated by computing $q(\tau)$ from ratios of Z_N's:

$$q_{t-t'}(\tau) = \frac{1}{t-t'} \ \ell n \frac{Z_N(\tau)}{Z_{N'}(\tau)} \tag{25}$$

The scaling spectrum $S(\mu)$, and $q''(\tau)$ are invariant under the change of scale

Exercise 8. Argue that $q(\tau)$ and the functions derived from it are invariant under the smooth deformations of the strange set.

Remark 4: Scaling spectrum vs. scaling function

Note that a Cantor set with only 2 scales has a continuous spectrum, so infinity of the scaling exponents does not imply an infinity of generating scales.

Remark 5: true thermodynamics

In our evaluation of the partition sum (4) we have assumed something that we can rarely afford in statistical mechanics: that the Boltzmann weight for each and every of the N "configurations" i is explicitely given. The price we pay for this is slow $\ell n N$ convergence of the thermodynamic functions $q_t(\tau)$, $S_t(\mu)$ toward the asymptic $t \to \infty$ limits $q(\tau)$, $S(\mu)$. It would be more in the spirit of statistical mechanics to estimate the $q(\tau)$ by computing a few ℓ_i for high N, and explore the "configuration" space by the Monte Carlo methods.

Exercise 9. Estimate the thermodynamic functions for Hénon attractor from a subset of ℓ_i computed from very long unstable cycles.

Remark 6: ℓ_i for asymptotic N can be computed only by renormalization methods, ie. recursive rescalings of local neighborhoods, such that ℓ_i at n-th level is of order of unity.

2. PHASE TRANSITIONS

Determining that there is a phase transition on basis of a finite covering of a strange set can be tricky. Here we describe three methods: finite size scaling, convexity of the scaling spectrum, and hysteresis.

Approximate the partition sum (15) by two scaling exponents μ_1, μ_2

$$Z_N(\tau) = e^{t(s_1+\mu_1\tau)} + e^{t(s_2+\mu_2\tau)} . \tag{26}$$

In the $N \to \infty$ limit the sum is dominated either by the first or by the second term, and $q(\tau)$ goes through a first order phase transition (here we assume $\mu_1 < \mu_2$)

$$q(\tau) = \begin{cases} s_1+\mu_1\tau & \tau < \tau_c \\ s_2+\mu_2\tau & \tau > \tau_c \end{cases} , \tag{27}$$

where the critical τ is determined by the condition that the two terms in (26) contribute equally

$$\tau_c = - \frac{s_2-s_1}{\mu_2-\mu_1} . \tag{28}$$

This simple model teaches us how to diagnose the existence of a first order phase transition, and estimate the critical τ and the gap in μ from the finite N effects. At the phase transition we have

$$\begin{aligned} q_t(\tau_c) &= q(\tau_c) + \frac{\ell n2}{t} \\ \dot{q}_t(\tau_c) &= \mu_t(\tau_c) = \mu(\tau_c) = \frac{\mu_1+\mu_2}{2} \\ q_t''(\tau_c) &= t \left(\frac{\mu_2-\mu_1}{2}\right)^2 \end{aligned} \tag{29}$$

This is illustrated in Fig. 1

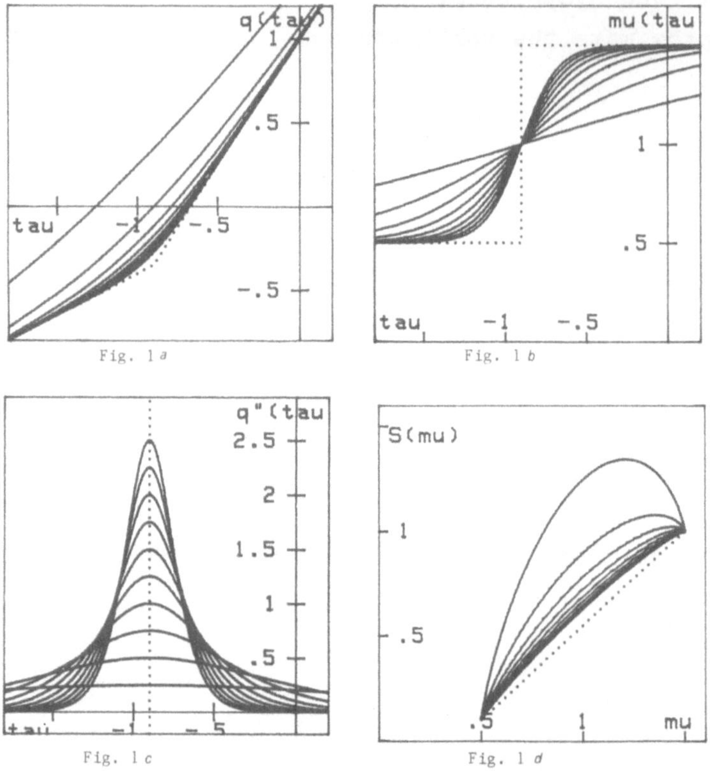

Fig. 1. Thermodynamic functions for the 2-scale model (26), with t = 1,2,...,10 (t=10 corresponds to about 22,000 covering intervals). The asymptotic limits are indicated by the dotted lines.

It is hard to appreciate how slow $1/\ell nN$ convergence is before you actually try to beat it[5]. The problem is not that experimental N is of order hundred; the problem is that no super-computer can bludgen such creeping convergence into submission, unless the nature of convergence is well understood.

If the phase transition is indeed of form (27), than $s_1, s_2, \mu_1, \mu_2, t$ and τ_c can be extracted by fitting the finite N thermodynamic curves. However, in practice both s_i and μ_i are also N-dependent, and one is forced to resort to unreliable numerical acceleration techniques.[6]

The second method for diagnosing existence of a phase transition is based on comparison of $s(\mu)$, the detailed interval counting (14), with $S(\mu)$ from (11), extracted from the global average (4). The two functions differ if $s(\mu)$ is not convex. To see why $S_t(\mu)$ must be convex, consider a three scaling exponent $(\mu_1<\mu_2<\mu_3)$ approximation to the sum (15)

$$Z_N(\tau) = e^{t(s_1+\mu_1\tau)} + e^{t(s_2+\mu_2\tau)} + e^{t(s_3+\mu_3\tau)} \qquad (30)$$

$q(\tau)$ is given by the upper boundary of the union of rays $q = s_i+\mu_i\tau$, see Fig. 2.

In this sum, the second term contributes only if the ray $q = s_2+\mu_2\tau$ crosses the intersection of the two rays from above. The critical s_2 is given by the coalescence condition for the three intersections:

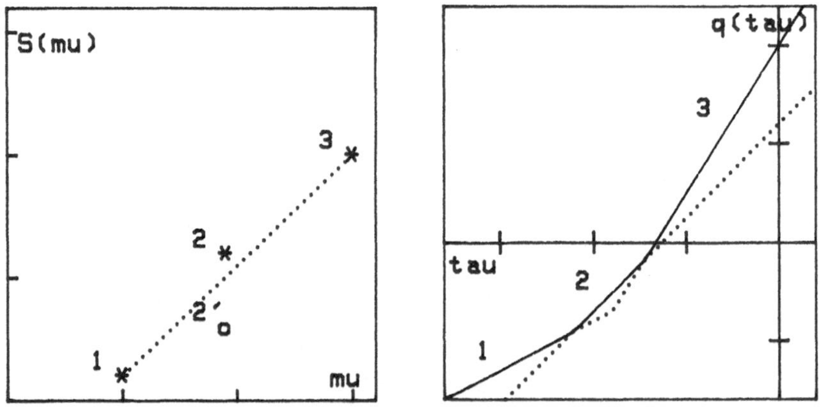

Fig. 2. The second term in (30) contributes to $q(\tau)$, $S(\mu)$ only if $s(\mu_2)$ falls above the dotted line (case 2). In case 2' $S(\mu)$ and $s(\mu)$ differ; $\mu(\tau)$ jumps from μ_1 to μ_2 without detecting the μ_2 scale.

$$\frac{s_2-s_1}{\mu_2-\mu_1} = \frac{s_3-s_2}{\mu_3-\mu_2} \qquad (31)$$

This is the condition that three $s(\mu)$ points lie on a line. If $s(\mu_2)$ for $\mu_1<\mu_2<\mu_3$ falls below this line, $\mu(\tau)$ jumps from μ_1 to μ_3 as τ goes through τ_c given by (28) and $s(\mu_2)$ does not contribute to $S(\mu)$. The critical τ is given by the slope of the straight section of $S(\mu)$ (see 12)). In other words, if the scale counting function $s(\mu)$ has a concave dimple, the system exhibits a first order phase transition. In practice one evaluates $s'(\mu_c)$, which requires counting only in the neighborhood of μ_c. If $s'(\mu_c)$ falls below a convexity bound imposed by known $S(\mu)$ values, $s(\mu)$ is concave.

Exercise 10. Generalize the above arguments to continuous $s(\mu)$. (Hint; estimate (15) by saddle point methods)

Unfortunately, I do not know of anything as simple as the 2-scale Cantor set (exercise 3) with a non-trivial concave $s(\mu)$ segment. The best I can offer here is a rather artificial set which arises in the study of circle maps[6): at n-th level, there are $N = 2^n$ intervals, labelled by all distinct strings of n 0's and 1's. If there are k 1's, the covering interval is given by

$$\ell_k = \sigma(x)^{-2nx}, \quad x = k/n$$
$$\sigma(x) = (1+\sqrt{1+4x^2})/(2x) \qquad (32)$$

$\sigma(x)$ here is in an estimate of the scaling factor associated with an average string of 0's. $n/k = 1/x$ is the average length of 0-string, if there are k 1's. For this set, the thermodynamic functions are plotted in Fig. 3.

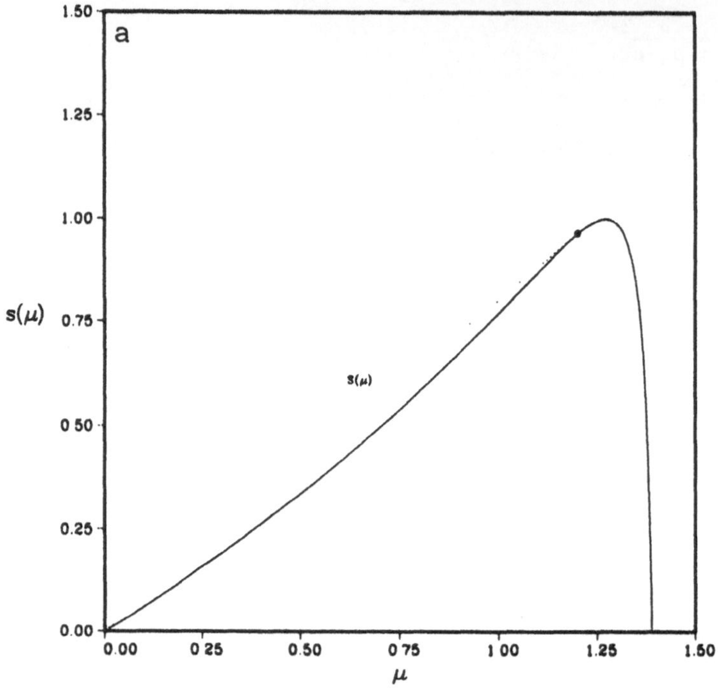

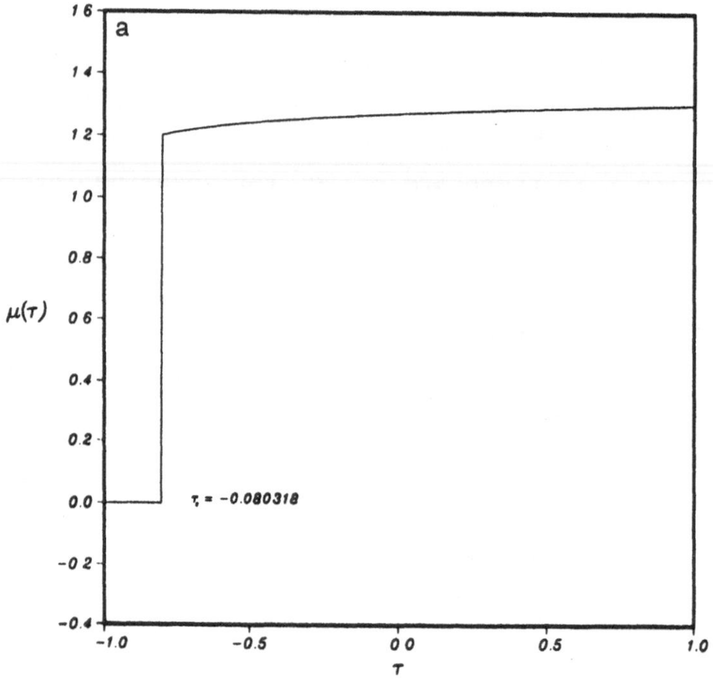

Fig. 3. The first order transition for the covering intervals defined by (32), as seen in the average scaling exponent $\mu(\tau)$, and in the difference between the counting function $s(\mu)$ and its convex envelope, the scaling spectrum $S(\mu)$.

Exercise 11. Show that interval ℓ_k occurs $\binom{n}{k}$ times, and that

$$Z_N(\tau) = \int_0^1 dx\ e^{t(f(x)+\tau\mu(x))}$$

$$f(x) = -\frac{x\ell nx + (1-x)\ell n(1-x)}{\ell n2}, \quad \mu(x) = \frac{2x\ell n\sigma(x)}{\ell n2}$$

Plot $s(\mu)$, $S(\mu)$. Estimate τ_c. (Hint; use the Stirling approximation. $\tau_c = -0.80318\ldots$.)

Remark 7: Hysteresis. The "true thermodynamics" offers a third way to establish a phase transition; metastability around τ_c makes it difficult for Monte Carlo updates to discover the true "ground state". Hence hysteresis is expected.

To summarize: determination of phase transitions on (finite size) strange sets is a non-trivial task. If the transition is of first order, finite size scaling, convexity of $S(\mu)$ and hysteresis can offer the essential clues. Transitions of higher order, to best of my knowledge, are not known in this context.

The thermodynamics formalism is an effective tool for extracting average scale information from experimental data; it is theoretically elegant (through its connection to scaling functions[2,4]); and it leads to the phenomenon of strange set phase transitions.

The thermodynamic formalism views a strange set as a static, given object. It is blind to the dynamics of the system, the detailed layout and ordering of its various parts, to dynamic correlations. It is simply an averaging technique; understanding a strange set requires a detailed study of its symbolic and scaling dynamics.

3. REFERENCES

1. R. Bowen, Lecture notes in Math. 470 (Springer, N.Y. 1975); D. Ruelle, "Statistical Mechanics, Thermodynamic Formalism", Addision-Wesley, Reading (1978).
2. E. B. Vul, YaG. Sinai, and K. M. Khanin, Usp. Mat. Nauk. 39, 3 (1984) [Russ. Mat. Surveys 39, 1 (1984)].

3. T. Halsey, M. H. Jensen, L. P. Kadanoff, I. Procaccia and B. I. Shraiman, Phys. Rev. A33, 1141 (1986).
4. M. J. Feigenbaum, J. Stat. Phys. 46, 919 and 925 (1987).
5. P. Cvitanović, in Proceedings of the Workshop in Condensed Matter Physics", Trieste, Italy 1986. (To be published); in R. Gilmore, ed., "Proceedings of the XV Int. Colloquium on Group Theoretical Methods in Physics" (World Scientific, Singapore, 1987).
6. R. Artuso, P. Cvitanović and B. Kenny (to be published).
7. T. Tel, Phys. Rev. A36 (1987) to appear.
8. R. Badii, P. Grassberger and A. Politi (to be published).
9. T. Bohr and D. Rand, Physica 25D, 387 (1987).
10. T. Bohr and M. H. Jensen (to be published).
11. D. Katzen and I. Procaccia, Phys. Rev. Lett. 58, 1169 (1987).
12. P. Szépfalusy, T. Tél, A. Csordás and Kovács, preprint (1987).
13. P. Cvitanović and G. Gunarathne (to be published).
14. P. Cvitanović, G. Gunarathne and I. Procaccia (to be published).
15. P. Grassberger, Phys. Letters 107A, 101 (1985).
16. J. Balatoni and A. Renyi, Publi. Math. Inst. Hung. Acad. Sci. 1, 9 (1956) (english translation in The selected papers of A. Renyi, Vol. 1, p. 588 (Akademia Budapest, 1976)); A. Renyi, Probability Theory (appendix) (North-Holland, 1970).
17. H. G. Hentchel and I. Procaccia, Physica 8D, 435 (1983).
18. B. B. Madelbrot. J. Fluid Mech. 62, 331 (1974).

STOCHASTIC PERTURBATIONS OF THE INVARIANT MEASURE OF SOME HYPERBOLIC DYNAMICAL SYSTEMS

Pierre Collet
Centre de Physique Théorique
Ecole Polytechnique
F91128 Palaiseau **Cedex**, (France)

I. INTRODUCTION

In this paper we shall consider random perturbations of uniformly hyperbolic dynamical systems. This question has been discussed in the physical literature mostly for flows[6]. A review of some rigorous results is given by Ventsel and Freidlin[5]. Our aim is to give some precise information about the invariant measure in the case of discrete hyperbolic systems[2]. We shall in particular show under some precise hypothesis that the invariant measure can be represented as an asymptotic series in the amplitude of the noise. We shall not aim at the full generality of this result but rather stay within strong hypothesis to make the exposition as simple as possible. Note however that the general hypothesis needed to prove this result are yet unknown. At several points of the proof we shall meet some technical conditions which can probably be lifted at the expense of some work.

The plan of this paper is as follows. In the remainder of this section, we shall give precise hypothesis on the dynamical systems and the random perturbations. We shall also state the main result and present two simple examples which indicate clearly the dominant contribution of the noise to the invariant measure. In section 2 we shall give the geometric constructions of dynamical manifolds in the random case. They are quite analogous to the constructions for the non random case and we shall not give proofs unless they are really different. In section 3 we shall use statistical mechanics to control the perturbation of the transverse measure. Finally in section 4 we will prove the main result.

We now state more precisely our hypothesis and results. We shall consider a C^∞ diffeomorphism T of $\mathbf{R}^2$ with a compact attractor K. We shall also assume that T is Axiom A and mixing[2] on K. For simplicity we shall work with the special case $n = 2$, and $d(.,.)$ will denote the Euclidean distance. We shall also consider a sequence of random variables $(\omega_j)_{j\in\mathbf{Z}}$ with values in $\mathbf{R}^2$ which are independent and identically distributed with a density ϕ which is C^1 and with compact support. Ω will denote the phase space of this process, and $\underline{\omega}$ an element, *i.e.* a realization of a sequence of the independent random variables. $\mathbb{E}$ will denote the expectation with respect to the product measure. For a fixed real number ϵ we now define a Markoff process $(X_j)_{j\in\mathbf{Z}}$ by

$$X_j = T(X_{j-1}) + \epsilon\omega_j \ .$$

Let μ_ϵ be the invariant measure of this process which for small ϵ gives a non zero weight to K. It is known that if ϵ tends to 0, μ_ϵ converges weakly to the Bowen-Ruelle measure μ_0 of T (see Kifer[7] or Young[12]). We shall investigate here the dominant correction in μ_ϵ. In order to be able to state our result, we need some more information about the measure μ_0. It is well known[2] that μ_0 can be disintegrated along the unstable manifolds with absolutely continuous conditional measures. Namely to each local unstable manifold W in K is associated a density ρ_W with respect to the Lebesgue measure $d_W M$ on W. There is also a transverse measure ν on the set $\mathcal{W}^u$of local unstable manifolds, and we have

$$d\mu_0 = d\nu(W)\,\rho_W(M) d_W M \ .$$

Let $\mathbf{n}^{\mathbf{u}}(M)$ denote the normal to the unstable manifold of M. Our main result is

Theorem 1.1. *For ϵ small enough and for each local unstable manifold W of K there is a positive function $\Theta_{\epsilon,M}$ such that the measure μ^0_ϵ defined on continuous functions g of $\mathbf{R}^2$ by*

$$\int g\, d\mu^0_\epsilon = \int_{\mathcal{W}^u} d\nu(W) \int_W g(M + y\mathbf{n}^{\mathbf{u}}(M))\Theta_{\epsilon,M}(y)\rho_W(M) d_W M\, dy$$

has the property

$$\int |d\mu_\epsilon - d\mu^0_\epsilon| = \mathcal{O}(\epsilon(\log\epsilon)^2) \ .$$

It should become clear from the proof that the above result can be extended to higher orders in ϵ, giving an asymptotic series for the invariant measure. We also note that the perturbation is of rather singular nature. The measures μ_ϵ and μ^0_ϵ are absolutely continuous with respect to the Lebesgue measure of $\mathbf{R}^2$ but μ_0 is not since it is supported by K which has a Hausdorff dimension smaller than 2. The above statement is of course stronger than the weak convergence and can be used to obtain explicit estimates. As explained above we have made rather strong hypothesis to keep the exposition reasonably simple. It would be interesting to know an equivalent result under the nice hypothesis of Young[12].

Before starting the proof of the main theorem we shall consider two simple examples which display the main ideas beyond the result. The first example is the map of the circle $x \to 2x \mod 1$. It is an expanding map and the Lebesgue measure is invariant. It is easy to verify that if one uses random iteration with a noise possessing an absolutely continuous density, then the Lebesgue measure is still the invariant measure. This example motivates the idea that the invariant measure will not change very much in the unstable direction. In order to get an example for the perturbation associated to the stable directions, we consider the following map which is a dissipative baker's transformation of the unit square

$$(x,y) \to \begin{cases} (2x, y/6 + 1/6) & \text{if } 0 \leq x \leq 1/2; \\ (2x-1, y/6 + 2/3 & \text{if } 1/2 < x \leq 1. \end{cases}$$

It is easy to verify that under iteration of the above map, almost every point converges to a set which is the product of horizontal segments (the local unstable manifolds) by a (vertical) Cantor set. The invariant S.R.B. measure is the product of the horizontal Lebesgue measure by the vertical Cantor measure. We now add a noise which has only a vertical component (*i.e.* it acts only in the stable direction). One can show easily that for ϵ small, and for any configuration of the noise there is a set varying coherently with the maps and which attracts the trajectories which started in the infinite past. This set is again the product of horizontal segments by a noise dependent Cantor set. There is associated a coherently varying measure which is again the product of the horizontal Lebesgue measure by a Cantor measure. In this example the noise moves the attractor in the stable direction without changing the transverse measure. The averaging on this displacement will produce the contribution of the noise to the invariant measure. More precisely, if we assume that the noise has an absolutely continuous measure, each piece of unstable manifold will give rise to a two dimensional absolutely continuous measure, and the invariant measure for the Markoff process is a superposition of the contributions of all these pieces (which have locally many overlaps). We shall see in this paper that in the general situation this is again the main source of correction to the invariant measure. What is different in the general case is that the transverse measure (here a Cantor measure) depends on the noise in a non trivial way.

II. SOME GEOMETRICAL RESULTS

We shall first recall some results for noisy systems which generalize well known results for deterministic axiom A systems[9,11,12]. In the sequel we shall assume that ϵ is fixed to a small enough value. This means that all our estimates will be valid in a small enough neighborhood of 0. Many quantities will depend on that number ϵ, however we shall not mention explicitly this dependence when it is not crucial to an argument. From the hypothesis on our system, it is easy to verify that there is an open set U, independent of ϵ (small enough) such that for almost any $\underline{\omega}$ in Ω we have

$$T_{\underline{\omega}}(U) \subset \mathrm{Int}\{U\} ,$$

where $\mathrm{Int}\{U\}$ denotes the interior of U. Moreover for almost every $\underline{\omega}$ in Ω and every M in U there are two regular manifolds $W^s_{\underline{\omega}}(M)$ and $W^u_{\underline{\omega}}(M)$ containing M and satisfying

$$T_{\underline{\omega}}(W^s_{\underline{\omega}}(M)) \subset W^s_{S\underline{\omega}}(T_{S\underline{\omega}}(M)) ,$$

and

$$T^{-1}_{\underline{\omega}}(W^u_{S\underline{\omega}}(M)) \subset W^u_{\underline{\omega}}(T^{-1}_{S\underline{\omega}}(M)) .$$

We are now able to show the existence of a (random) Markoff partition.

Theorem 2.1. *For ϵ small enough and for almost any $\underline{\omega}$ there is a finite partition $(A^j_{\underline{\omega}})_{1\le j\le a}$ of K by rectangles, where the integer a does not depend on $\underline{\omega}$, and such that if $\overline{W^s_{\underline{\omega}}}(M)$ is one of the two connected pieces of $\partial^s(A^j_{\underline{\omega}})$ then $T_{\underline{\omega}}(\overline{W^s_{\underline{\omega}}})(M)$ is contained in a connected piece of $\partial^s(A^k_{S\underline{\omega}})$ for some $0 \le k \le p$. A similar statement holds for $\partial^u(A^j_{\underline{\omega}})$. The maximal diameter of the atoms of the partition can be taken arbitrarily small.*

Here $\partial^s(A)$ denotes the part of the boundary of A which is a union of two local stable manifolds, and similarly for ∂^u. As usual we shall use the term Markoff partitions although we are not considering real partitions but a finite covering by sets which overlap only at the boundary. The proof of this theorem relies on the Shadowing lemma[2,8].

Lemma 2.2.(Shadowing Lemma) *If U and ϵ are small enough, for all $\underline{\omega}$ there is a Hölder homeomorphism $g_{\underline{\omega}}$ defined on U together with its inverse such that*

$$T_{\underline{\omega}} = g^{-1}_{\underline{\omega}} \circ T \circ g_{S^{-1}\underline{\omega}} .$$

This map also provides a correspondence between the unstable (stable) local manifolds for the noisy and deterministic dynamical systems. The image by $g^{-1}_{S^{-1}\underline{\omega}}$ of a Markoff partition of T is a random Markoff partition. By a local unstable (stable) manifold we shall mean a connected part of the intersection of an unstable (stable) manifold with an element of the Markoff partition. Note that this definition depends on the Markoff partition. We shall use the same notations $W^u_{\underline{\omega}}$ and $W^s_{\underline{\omega}}$ for the global and local manifolds since it will be clear from the context which one is relevant. Once a Markoff partition is chosen, we shall denote by $\mathcal{W}^u_{\underline{\omega}}$ the set of all local unstable manifolds.

Following Young[12] we can now repeat the standard construction of the S.R.B. measure[10] for almost every $\underline{\omega}$. We have the following theorem

Theorem 2.3. *For ϵ small enough and for almost every $\underline{\omega}$ there is a measure $\mu_{\underline{\omega},\epsilon}$ on $\mathbf{R}^2$ such that*

$$d\mu_\epsilon = \mathbb{E}(d\mu_{\underline{\omega},\epsilon}) \, .$$

Moreover, the structure of the measures $\mu_{\underline{\omega},\epsilon}$ is similar to the structure of μ_0. Namely

Theorem 2.4. *For ϵ small enough and for almost every $\underline{\omega}$ there is a measure $\nu_{\underline{\omega},\epsilon}$ on $\mathcal{W}^u_{\underline{\omega}}$ and for each element W of $\mathcal{W}^u_{\underline{\omega}}$ a density $\rho_{\underline{\omega},\epsilon,W}$ such that*

$$\int g \, d\mu_{\underline{\omega},\epsilon} = \int_{\mathcal{W}^u_{\underline{\omega}}} d\nu_{\underline{\omega},\epsilon} \int_W g(M) \, \rho_{\underline{\omega},\epsilon,W}(M) d_W M \, .$$

We shall need later on some quantitative results on the correspondence between stable (unstable) manifolds which is provided by the Shadowing Lemma. A more precise version of the correspondence is given in the following theorem. Let l denote the minimal length of the local stable (unstable) manifolds. This number can be chosen uniformly for ϵ small enough (including zero). For a point M in U we shall denote by $\mathbf{n}^s(M)$ (respectively $\mathbf{n}^u(M)$) the normal to the stable (unstable) manifold of M with respect to the map T. The notation $\mathbf{t}^s(M)$, $\mathbf{t}^u(M)$ will be used for the tangent. Note that along the corresponding invariant manifolds the fields $\mathbf{n}^s$ and $\mathbf{n}^u$ are differentiable. This is why we shall use them instead of the fields of unstable (stable) directions. The corresponding fields of directions for the random manifolds will be denoted by $\mathbf{n}^u_{\underline{\omega}}$, $\mathbf{t}^u_{\underline{\omega}}$ etc.

Theorem 2.5. *There is a positive number η such that for any M in U and almost every $\underline{\omega}$ there is a diffeomorphism $\sigma_{\underline{\omega}}$ defined on*

$$\widetilde{W}^u(M) = \{M_1 \in W^u(M) | d(M, M_1) \le \eta\}$$

with values in $W^u(M)$, and a real C^∞ function $r_{\underline{\omega}}$ on $\widetilde{W}^u(M)$on such that

1) $\{r_{\underline{\omega}}(M_1)\mathbf{n}^u(M_1) | M_1 \in \widetilde{W}^u(M)\}$ is a piece of unstable manifold for $T_{\underline{\omega}}$.

2) Moreover

$$T_{\underline{\omega}}(T^{-1}(M_1) + r_{S^{-1}\underline{\omega}}(T^{-1}(M_1))\mathbf{n}^u(T^{-1}(M_1)) = \sigma_{\underline{\omega}}(M_1) + r_{\underline{\omega}}(\sigma_{\underline{\omega}}(M_1))\mathbf{n}^u(\sigma_{\underline{\omega}}(M_1)) \, .$$

Note that $r_{\underline{\omega}}$ depends on $W^u(M)$ although for simplicity we have not included this dependence in the notation. A similar result holds for the stable manifolds. The proof of this theorem will use the following lemma.

Lemma 2.6. *In the frames $\mathbf{t}^u(M), \mathbf{n}^u(M)$ and $\mathbf{t}^u(T(M)), \mathbf{n}^u(T(M))$ we have*

$$DT_M = \begin{pmatrix} \lambda^u(M) & a(M) \\ 0 & b(M) \end{pmatrix}$$

where $\lambda^u(.)$ is the dilatation in the unstable direction. $\lambda^u(.)$, $a(.)$, and $b(.)$ are differentiable along $W^u(M)$.

A similar statement holds for the stable direction. Moreover, it follows from the uniform hyperbolicity that the sequence

$$\prod_{j=0}^{j=l} b(T^j(M))$$

converges to 0 exponentially fast in l and uniformly in $M \in U$.

We now start the proof of Theorem 2.5. Let $\underline{\omega}$ be a fixed configuration of the noise. Let M_0 be a point in U. We shall fix later on a uniform (in M_0, ϵ, and $\underline{\omega}$) small positive number ξ. $D_\xi(M)$ will denote the disk in $W^u(M)$ of radius ξ centered at M_0. We shall first prove the existence of two sequences $(\zeta_j)_{j\in\mathbf{N}}$ and $(r_j)_{j\in\mathbf{N}}$,where ζ_j is an injective continuous map from $D_\xi(T^{-j}(M_0))$ onto a neighborhood of $T^{-j}(M_0)$ in $W^u(T^{-j}(M_0))$, and r_j is a continuous map from $D_\xi(T^{-j}(M_0))$ to a neighborhood of 0 in $\mathbf{R}$. Moreover these functions satisfy

$$T(T^{-1}(\zeta_j(M)) + r_{j+1}(T^{-1}(\zeta_j(M))\mathbf{n}^u(T^{-1}(\zeta_j(M))) + \epsilon\omega_{-j} = M + r_j(M)\mathbf{n}^u(M)$$

for all M in $D_\xi(T^{-j}(M_0))$ and for all integer j. Note that if $\zeta_j(M)$ is near enough to M, and if ξ is small enough, the above equation is well defined. We shall also impose that there exist four uniform positive numbers L_1, L_2, L_3, L_4, to be chosen later on, such that if $c_j(M) = \zeta_j(M) - M$, then

i) $\sup_j |r_j| \leq \epsilon L_1$,
ii) $\sup_j \| c_j \| \leq \epsilon L_2$,
iii) $\sup_j |r_j(M) - r_j(M')| \leq \epsilon L_3 d(M, M')$,
iv) $\sup_j \| c_j(M) - c_j(M') \| \leq \epsilon L_4 d(M, M')$.

Projecting on $\mathbf{n}^u(M)$, we get the following equations

$$\begin{aligned} r_j(M) &= < r_{j+1}(T^{-1}(\zeta_j(M)))\mathbf{n}^u(M) | DT_{T^{-1}(M)}\mathbf{n}^u(T^{-1}(M)) > \\ &\quad + R_1 + < \epsilon\omega_{-j} | \mathbf{n}^u(M) > , \qquad (2.1) \\ c_j(M) &= r_j(M)\mathbf{n}^u(M) - r_{j+1}(T^{-1}(\zeta_j(M)))DT_{T^{-1}(M)}\mathbf{n}^u(T^{-1}(M)) - R_2 - \epsilon\omega_{-j} , \end{aligned}$$

where $< \,|\, >$ denotes the Euclidean scalar product. R_1 and R_2 are two quadratic remainders which will give small contributions provided ξ is small enough (although uniform and $\mathcal{O}(1)$). Let a and b be the two positive numbers defined by

$$a = \sup_{M\in U} |a(M)| \quad , \quad b = \sup_{M\in U} |b(M)| ,$$

and let

$$\lambda^u = \inf_{M\in U} |\lambda^u(M)| .$$

We shall now assume that $b < 1$, and $\lambda^u > 1$. This may not be true under our previous assumptions. However it is true if we consider some iterate of the map T, and it is easy to modify the present arguments to deal with that more general case. We now impose that $aL_1 < L_2$ and $aL_3 < L_4$. It is easy to verify that one can choose uniformly the constants L_1, L_2, L_3, L_4, such that the above constraints are satisfied, and also the left hand side of (2.1) defines a map which preserves the assumptions i) to iv). Moreover, this map is a contraction in the C^0 norm. Therefore it has a unique fixed point under the above assumptions. $\sigma_{\underline{\omega}}$ is the inverse of ζ_0, and $r_{\underline{\omega}}$ is equal to $r_{\underline{0}}$. The differentiability properties are proven similarly.

We shall use later on the explicit form of the dominant term in $r_{\underline{\omega}}$. This expression is obtained easily by linearising the expression in formula 2.1.

Corollary 2.7. *Under the hypothesis of Theorem 2.5 there is a C^∞ function $s_{\underline{\omega}}$ on $\widetilde{W}^u$ such that*

$$r_{\underline{\omega}}(.) = \epsilon \sum_{j=0}^{j=\infty} < \omega_{-j} | \mathbf{n}^{\mathbf{u}}(T^{-j}(.)) > \prod_{l=1}^{l=j} b(T^{-j}(.)) + \epsilon^2 s_{\underline{\omega}} \,.$$

In the previous theorem, it is not necessary to assume that the reference map T is constant. The same result holds if we iterate a noisy map. We shall use the notation $r_{\underline{\omega},\underline{\tilde{\omega}}}$ in this more general setting, when referring to the two noise configurations $\underline{\omega}$ and $\underline{\tilde{\omega}}$ (the previous case corresponds to $\underline{\tilde{\omega}} = 0$). We shall need later on an estimate of the variation of the unstable expansion coefficient $\lambda^{\mathbf{u}}_{\underline{\omega}}(.)$ in this situation.

Proposition 2.8. *There are two uniform positive constants L and τ such that if $\underline{\omega}$ and $\underline{\tilde{\omega}}$ are two noise configurations satisfying $\omega_j = \tilde{\omega}_j$ for $-l < j < l$ (l a positive integer) and M and M' are two points in U such that*

$$M = g^{-1}_{S^{-1}\underline{\omega}} \circ g_{S^{-1}\underline{\tilde{\omega}}}(M') \,,$$

then

$$|\log(\lambda^{\mathbf{u}}_{\underline{\omega}}(M)) - \log(\lambda^{\mathbf{u}}_{\underline{\tilde{\omega}}}(M'))| \leq L\epsilon 2^{-\tau l} \,.$$

Proof. Let $c = \max(b, {\lambda^u}^{-1})$, as explained before, we can assume that $c < 1$. It follows easily from the Shadowing Lemma[8] that the distance between M and M' is bounded by $\mathcal{O}(\epsilon)c^l$. Let M'' be defined by

$$M'' = M + r_{\underline{\omega},\underline{\tilde{\omega}}}(M)\mathbf{n}^{\mathbf{u}}_{\underline{\omega}}(M) \,.$$

It follows from Lemma 2.6 that the distance between M and M'' is also $\mathcal{O}(\epsilon)c^l$. From the differentiability of r and of the manifolds it follows that

$$\| \mathbf{t}^{\mathbf{u}}_{\underline{\omega}}(M) - \mathbf{t}^{\mathbf{u}}_{\underline{\tilde{\omega}}}(M') \| \leq \mathcal{O}(\epsilon)c^l \,.$$

The result follows now from the above estimates using the fact that the map T is C^2.

III. PERTURBATIONS OF THE TRANSVERSE MEASURES

In this section we shall analyse the perturbation of the transverse measure. By the correspondence provided by the Shadowing Lemma 2.2, we can transport for ϵ small enough the noisy transverse measure to a measure on the set $\mathcal{W}^u$ of local unstable manifolds of T. The difficulty at this point is that this measure need not be absolutely continuous with respect to the non noisy transverse measure. In order to deal with this problem, we shall use the formalism of statistical mechanics[10]. Let $A = \{\alpha_1, \ldots, \alpha_a\}$ be a finite set of cardinality a. In later applications, A will be a Markoff partition. We shall assume that an $a \times a$ matrix P with entries equal to 0 or 1 is given. The configuration space $\mathcal{A}$ of the system is the set of bi-infinite sequences $\underline{S} = (S_j)_{j\in\mathbf{Z}}$ satisfying

$$P_{S_j, S_{j+1}} = 1 \, .$$

The shift $\mathcal{S}$ is defined on $\mathcal{A}$ in the usual way, and $\dot{d}$ will denote the usual distance[10].

We shall consider sequences of potentials $\underline{V} = (V_j)_{j\in\mathbf{Z}}$ where each element is a continuous function from $\mathcal{A}$ to $\mathbf{R}$ with the property that $V_j(\underline{S})$ depends only of $S_0, S_1, \ldots$. We shall also assume that they are uniformly Hölder continuous. As a consequence, their dependence on S_j will be uniformly exponentially small for large j. We shall denote by $\nu_{\underline{V}}$ the associated (in general non translation invariant) Gibbs states[2,10]. We shall need below estimates relating Gibbs states for different potentials. Let $\underline{V}^1$ and $\underline{V}^2$ be two potentials as above. Let $\Delta\underline{V} = \underline{V}^1 - \underline{V}^2$, and let Δ be the maximum C^0 norm of the elements $\Delta\underline{V}_j$. We have

Lemma 3.1. *There is a positive number B_1 such that if Δ is small enough, we have*

$$\nu_{\underline{V}^1}(S_0, \ldots, S_q) \le e^{B_1 q \Delta} \nu_{\underline{V}^2}(S_0, \ldots, S_q)$$

for any integer q and any configuration $\underline{S}$.

Proof Let τ be a fixed positive number smaller than 1 (τ will be the Hölder exponent of the potentials). For Γ a positive number, let C_Γ denote the cone of continuous positive functions f on $\mathcal{A}$ which satisfy for any $\underline{S}$ and $\underline{S}'$ in $\mathcal{A}$

1) $f(\underline{S}) = f(\underline{S}')$ for all $\underline{S}$ and $\underline{S}'$ such that $S_j = S'_j \, \forall j \ge 0$,

2) $\dfrac{f(\underline{S})}{f(\underline{S}')} \le e^{\Gamma d(\underline{S}, \underline{S}')^\tau}$.

If $(1 - 2^{-\tau})\Gamma$ is larger than the τ Hölder norm of the potential V, the sequence of functions defined for $r \in \mathbf{N}$ by

$$\sum_{S_{-r} \ldots S_{-1}} e^{\sum_{j=1}^{r} V(\mathcal{S}^{-j}\underline{S})}$$

converges exponentially fast in the projective metric[1,4] of the cone C_Γ. The lemma follows now from standard ratios estimates[10].

Note that the above result is still true for non-translation invariant potentials provided the bounds are uniform.

We shall also need a bound for asymptotically equal Gibbs states. As before, the same result will hold if the potentials are not translation invariant but the bounds are uniform. Let p be a fixed integer. We shall denote by $\tilde{\nu}_p$ the Gibbs state which is the thermodynamic limit of the Gibbs state with energy $E_{-N,N}$ in the volume $[-N, N]$ given by

$$E_{-N,N}(\underline{S}) = \sum_{j=-N}^{-p} V_j^1(S^j\underline{S}) + \sum_{j=-p+1}^{p} V_j^2(S^j\underline{S}) + \sum_{j=p+1}^{N} V_j^1(S^j\underline{S}) ,$$

where of course N is bigger than p and the configuration $\underline{S}$ is truncated outside the interval $[-N, N]$.

Lemma 3.2. *The Gibbs measures $\nu_{\underline{V}^1}$ and $\tilde{\nu}_p$ are equivalent, and the Radon Nikodym derivative Φ satisfies*

$$e^{-p\Delta} \le \Phi \le e^{p\Delta} .$$

This is obvious from the definition of Gibbs states.

We now come to the applications of the above results to the transversal measure. We first observe that if $\nu_{\underline{\omega}}$ and $\nu_{\underline{\tilde{\omega}}}$ are associated to Gibbs states for the corresponding random Markoff partitions, their images $\nu^0_{\underline{\omega}}$ and $\nu^0_{\underline{\tilde{\omega}}}$ by the maps $g^{-1}_{S^{-1}\underline{\omega}}$ and $g^{-1}_{S^{-1}\underline{\tilde{\omega}}}$ of the Shadowing Lemma 2.2 are Gibbs states for the partition A. The potentials used to construct the Gibbs states are the logarithms of the expansion coefficients along the unstable directions. We shall denote by $l^u_{\underline{\omega}}(M)$ this quantity at the point M for the configuration ω of the noise. Namely, $l^u_{\underline{\omega}}(M) = \log(\lambda^u(M))$. We shall use the notation $l^u_{\underline{\omega}}(\underline{S})$ when the point M is coded into the symbolic sequence $\underline{S}$. It is a well known fact[2] that $l^u_{\underline{\omega}}$ has an exponentially small dependence on far away spins. Moreover, here we can assume that this exponentially small dependence is uniform in the noise. We shall need an estimate on the dependence of the noise configuration for a fixed spin configuration $\underline{S}$ (recall that from our previous choices, the set of admissible spin configurations does not depend on the noise configuration).

Lemma 3.3. *There are two positive numbers B_1 and σ_1 such that for any admissible configuration of spins $\underline{S}$, and any pair of noise configurations $\underline{\omega}$ and $\underline{\omega}'$ such that $\omega_j = \omega'_j$ for $-l \le j \le l$, we have*

$$|l^u_{\underline{\omega}}(\underline{S}) - l^u_{\underline{\omega}'}(\underline{S})| \le B_1 \epsilon 2^{-\sigma_1 l} .$$

This statement follows at once from proposition 2.6.

We now define our final potential $V_{\underline{\omega}}$. Let

$$\mathcal{R}(\underline{S}) = \begin{cases} S_k & \text{if } k \geq 0, \\ \alpha_1 & \text{otherwise,} \end{cases} \quad \text{and} \quad \mathcal{R}(\underline{\omega}) = \begin{cases} \omega_k & \text{if } k \geq 0, \\ 0 & \text{otherwise.} \end{cases}$$

The potential $V_{\underline{\omega}}$ is given by

$$V_{\underline{\omega}}(\underline{S}) = 1^u_{\mathcal{R}(\underline{\omega})}(\mathcal{R}(\underline{S})) + \sum_{j=0}^{\infty} \left(1^u_{S^{j+1}\mathcal{R}(\underline{\omega})}(S^{j+1}\mathcal{R}(\underline{S})) - 1^u_{S^j\mathcal{R}(S\underline{\omega})}(S^j\mathcal{R}(S\underline{S})) \right) .$$

Note that $V_{\underline{\omega}}$ depends only on the variables S_j for $j \geq 1$, and moreover it is equivalent[10] to $1^u_{\underline{\omega}}$ in the sense that there is a potential Σ such that

$$1^u_{\underline{\omega}} = V_{\underline{\omega}} + \Sigma - \Sigma \circ S^{-1} .$$

It is now easy to check that for a given noise configuration $\underline{\omega}$, the potential $V_{\underline{\omega}}$ satisfies the hypothesis of Lemmas 3.1 and 3.2. It follows from Proposition 2.8 that we also have for $V_{\underline{\omega}}$ the same Hölder dependence with respect to the noise variable.

We shall now show that one can replace the noisy measure by an equivalent one which is independent of the components $\omega_{-2p}\omega_{-2p+1}\cdots\omega_p$ where p is some integer. This will introduce an error which is controlled in term of p. To each element $\underline{\omega}$ of Ω we associate a new element $\underline{\tilde{\omega}}$ of Ω by

$$\tilde{\omega}_j = \begin{cases} \omega_j & \text{if } j \leq -2p \text{ or } j \geq p; \\ 0 & \text{otherwise.} \end{cases}$$

We shall now compare the transverse measures $\nu_{\underline{\omega}}$ and $\nu_{\underline{\tilde{\omega}}}$. More precisely, we shall compare their images under $g^{-1}_{S^{-1}\underline{\omega}}$ and $g^{-1}_{S^{-1}\underline{\tilde{\omega}}}$ respectively (see the Shadowing Lemma 2.2). We shall use the notations $\nu^0_{\underline{\omega}}$ and $\nu^0_{\underline{\tilde{\omega}}}$ for these two measures. Note that they both are measures on the set $\mathcal{W}^u$ of local unstable manifolds of the map T.

Theorem 3.4. *The measures $\nu^0_{\underline{\omega}}$ and $\nu^0_{\underline{\tilde{\omega}}}$ are absolutely continuous and their density is bounded above and below by $e^{p\mathcal{O}(1)\epsilon}$.*

Proof. This theorem is an easy consequence of Lemma 3.2.

IV. PROOF OF THEOREM 1.1.

In this section, g will denote a continuous real function on U, and $\| g \|$ its C^0 norm. We also fix once for all a Markoff partition A with sufficiently small atoms, such that in particular Lemma 2.5 holds on the local unstable manifolds. We shall first estimate the correction to the conditional measure on a noisy unstable manifold. It is controlled in the following lemma

Lemma 4.1. *If g is a continuous function on U and if W is a local unstable manifold for $T_{\underline{\omega}}$, if W_0 is the local unstable manifold of T associated to W by Theorem 2.5, we have*

$$\int_W g(M)\rho_{\underline{\omega},\epsilon,W,}d_W(M) = \int_{W_0} g(M + r_{\underline{\omega}}(M)\mathbf{n}^{\mathrm{u}}(M))\rho_{W_0}(M)\, d_{W_0}M + \mathcal{O}(\epsilon\log\epsilon^{-1}) \parallel g \parallel .$$

Proof. This theorem is proven using Lemma 2.5. There are three contributions to the error term. The first one is coming from the change of variable *i.e.* the change of unstable manifold. The second one is coming from the difference between $\rho_{W_0}(M)$ and $\rho_{\underline{\omega},\epsilon,W}(M)$. It is estimated using Lemma 3.3 and the explicit (infinite product) expression for $\rho_{W,\epsilon}$. The third correction is coming from the end of the manifolds and is due to the fact that the image of a local manifold by the correspondence of Lemma 2.5 may not be a local manifold. This problem is solved by using the estimate on $r_{\underline{\omega}}$ of Lemma 2.5.

From the above result and Lemma 3.4 we now have

$$\int g\, d\mu_{\underline{\omega}} = \int_{\mathcal{W}^{\mathrm{u}}} d\nu^0_{\underline{\omega}}(W) \int_W g(M+r_{\underline{\omega}}(M)\mathbf{n}^{\mathrm{u}}(M))\rho_W(M)\, d_W M + \epsilon\mathcal{O}(p+\log\epsilon^{-1}) \parallel g \parallel .$$

We shall first average over the variables $\omega_0, \cdots, \omega_{-p}$. This will only affect the argument of g. We shall denote by $g_{\underline{\omega}_{\leq -p}}$ the resulting function, *i.e.*

$$g_{\underline{\omega}_{\leq -p}}(W) = \int_W \rho_W(M)\, d_W M \int \cdots \int g(M + r_{\underline{\omega}}(M)\mathbf{n}^{\mathrm{u}}(M)) \prod_{j=0}^{j=p} \phi(\omega_{-j})\, d\omega_{-j} .$$

This is of course a function of W and of the remaining ω variables $\omega_{-p}, \omega_{-p-1}, \cdots$. From now on we shall assume that p is larger than $\log\epsilon^{-1}$. We shall first simplify the expression of $g_{\underline{\omega}_{\leq -p}}$. Let $\overline{g}_{\underline{\omega},p}$ be defined by

$$\overline{g}_{\underline{\omega},p}(W) = \int_W \rho_W(M)\, d_W M \int \cdots \int g(M + S^p_{\underline{\omega}}(M)\mathbf{n}^{\mathrm{u}}(M)) \prod_{j=0}^{j=p} \phi(\omega_{-j})\, d\omega_{-j} ,$$

where

$$S^p_{\underline{\omega}}(M) = \epsilon \sum_{j=0}^{j=p} < \omega_{-j} | \mathbf{n}^{\mathrm{u}}(T^{-j}(M)) > \prod_{l=1}^{l=j} b(T^{-j}(M)) ,$$

and the product is taken equal to 1 when $j = 0$.

We now show that we can replace $g_{\underline{\omega}_{\leq -p}}$ by $\overline{g}_{\underline{\omega},p}$ with a small error.

Lemma 4.2. *There is a positive constant D such that*

$$|g_{\underline{\omega}_{\leq -p}} - \overline{g}_{\underline{\omega},p}| \leq D b^p \epsilon \parallel g \parallel .$$

We recall that $b = \sup_{M \in U} |b(M)|$, and we can assume $b < 1$.

Proof. Let M be a point of the local unstable manifold W. Let $\omega_0^\perp$ be the projection of ω_0 along $\mathbf{n}^u(M)$, $\omega_0^\parallel$ the projection along $\mathbf{t}^u(M)$. Let $y = \omega_0^\perp + \epsilon^{-1} r_{\underline{\omega}}(M) - \epsilon^{-1} S_{\underline{\omega}}^p(M)$. We shall first show that one can change variables from $\omega_0^\perp$ to y, the second component $\omega_0^\parallel$ of ω_0 remaining the same. In other words we have to show that the above equation defining y is invertible in $\omega_0^\perp$ for $\omega_0^\parallel$, $\omega_{-1}, \omega_{-2}, \ldots$ and M fixed. It is easy to verify using Lemma 2.5 that $\partial r_{\underline{\omega}}(M)/\partial \omega_0^\perp$ is uniformly non zero on the domain of variations of M, $\omega_0, \omega_{-1}, \cdots$. It is in fact equal to $\epsilon + \mathcal{O}(\epsilon^2)$. The invertibility follows now from the implicit function theorem[3]. Let $m(M, y, \omega_0^\parallel, \omega_{-1}, \ldots)$ denote the inverse function. It is easy to see that this function is differentiable in all its variable in the range of variations, and in particular the derivative with respect to y is $1 + \mathcal{O}(\epsilon)$. Therefore, up to an error which is uniformly $\mathcal{O}(\epsilon)$, the difference between $\overline{g}_{\underline{\omega},p}(W)$ and $g_{\underline{\omega}_{\leq -p}}$ is equal to

$$\int_W \rho_W d_W M g(M + S_{\underline{\omega}}^p \mathbf{n}^u(M)) \int \cdots \int \prod_{j=1}^{j=p} \phi(\omega_j) d\omega_j$$

$$\times \{\phi(m(M, \omega_0^\perp, \omega_0^\parallel, \omega_{-1}, \ldots), \omega_0^\parallel) - \phi(\omega_0)\} d\omega_0 \, .$$

In the last formula we have also changed the name of the variable y back to $\omega_0^\perp$. Lemma 4.2 follows now easily from the fact that ϕ has a derivative in L^1, and from the estimate on $m(M, \omega_0^\perp, \omega_0^\parallel, \omega_{-1}, \ldots) - \omega_0^\perp$.

We now show that $\overline{g}_{\underline{\omega},p}(W)$ is in fact a rather regular function of W (even though g is only C^0).

Lemma 4.3. *There are positive constants B, p_0, τ, and B', independent of g, $\underline{\omega}$ and ϵ such that if $p \leq -p_0$, if W and W' are two local unstable manifolds of T, belonging to the same atom of the Markoff partition A and such that $d(W, W') \leq B'\epsilon$, then*

$$|\overline{g}_{\underline{\omega},p}(W) - \overline{g}_{\underline{\omega},p}(W')| \leq B(\epsilon^{-1} d(W, W')^\tau + \epsilon) \parallel g \parallel \, .$$

The above constants will depend on the density ϕ of the noise.

Proof. The proof is very similar to that of Lemma 4.2 although here we shall work with two dimensional random variables. We shall first fix $\omega_0^\parallel$, $\omega_{-1}, \cdots, \omega_{-p}$. Let $\mathcal{D}_W$ be the subset of $\mathbf{R}^2$ defined by

$$\mathcal{D}_W = \{M + y\mathbf{n}^u(M) \,|\, M \in W \, , \, y \in \mathbf{R}\} \, .$$

The random variable $M + S_{\underline{\omega}}^p(M)\mathbf{n}^u(M)$ stays in a compact subset $\mathcal{D}_W^\epsilon$ of $\mathcal{D}_W$ when M and $\omega_0^\perp$ vary (in W and the support of $\phi(., \omega_0^\parallel)$ respectively). We observe that $\mathcal{D}_W^\epsilon$ is a slab of width $\mathcal{O}(\epsilon)$ around W and the probability distribution of the random variable $M + S^p_{\underline{\omega}}(M)\mathbf{n}^{\mathrm{u}}(M)$has a density θ_W with respect to the Lebesgue measure $d_W M dy$ which is given by

$$\theta_W(M,y) = \epsilon^{-1}\rho_W(M)\phi(\epsilon^{-1}y - \epsilon^{-1}R_W(M), \omega_0^{\parallel}) ,$$

where $R_W(M) = S^p_{\underline{\omega}}(M) - \epsilon\omega_0^{\perp}$. This density is bounded by $\mathcal{O}(\epsilon^{-1})$. The same observations hold for W'. We now define a map Ψ from $\mathcal{D}^\epsilon_W$ to $\mathcal{D}_{W'}$ by associating to the point $P = M + y\mathbf{n}^{\mathrm{u}}(M)$ the unique point M' on $W^u(W')$ such that P belongs to the normal line to $W^u(W')$ at M'. The number y' is defined to be the (algebraic) distance from M' to P on that normal line. Note that this construction is well defined if the two manifolds W and W' are sufficiently close. It is easy to verify that this map is differentiable, and injective. We have

$$\overline{g}_{\underline{\omega},p}(W') = \int_{\mathcal{D}^\epsilon_W} g(M+y\mathbf{n}^{\mathrm{u}}(M))\theta_{W'}(\Psi(M,y))J_\Psi(M,y)d_W M dy + \mathcal{O}(1)d(W,W') \parallel g \parallel ,$$

where J_Ψ is the Jacobian of Ψ. The correction is coming form the fact that the image of $\mathcal{D}^\epsilon_W$ by Ψ is slightly different from $\mathcal{D}^\epsilon_{W'}$.

It is easy to verify that there is a positive constant τ_1 such that

$$J_\Psi = \mathcal{O}(d(W,W')^{\tau_1}) ,$$

and if $(M',y') = \Psi(y,m)$, we have

$$\frac{\rho_{W'}(M')}{\rho_W(M)} = 1 + \mathcal{O}((W,W')^{\tau_1} + \epsilon) .$$

The final contribution to the difference between $\overline{g}_{\underline{\omega},p}(W)$ and $\overline{g}_{\underline{\omega},p}(W')$ is the average over $\omega_0^{\parallel}$, $\omega_{-1}, \cdots, \omega_{-p}$ of the quantity

$$\int_{\mathcal{D}^\epsilon_W} |g(M+\epsilon y\mathbf{n}^{\mathrm{u}}(M))| \rho_W(M)|\phi(y-\epsilon^{-1}R_W(M),\omega_0^{\parallel}) - \phi(y'-\epsilon^{-1}R_{W'}(M'),\omega_0^{\parallel})| d_W M dy .$$

We also have the following estimates

$$y - y' = \mathcal{O}(d(W,W')) \quad \text{and} \quad R_W(M) - R_{W'}(M') = \mathcal{O}(d(W,W')^{\tau_2}) ,$$

where τ_2 is a positive constant. The lemma follows now easily from the fact that ϕ has derivatives in L^1.

We can now give the proof of our main result.

Proof of Theorem 1.1. Let α be an atom of the Markoff partition A of T. For any positive integer q, let A_{-q} denote the admissible symbols of the inverse dynamic for q steps starting in α. We have

$$\alpha = \bigcup_{S_{-q}\ldots S_{-1}\in A_{-q}} \Delta_{S_{-q}\ldots S_{-1}} .$$

Each set $\Delta_{S_{-q}\ldots S_{-1}}$ is a union of local unstable manifolds of T and their distance is smaller than $c^{q\tau_3}$ where $0 < c < 1$ and τ_3 is a positive constant. We now choose $q = \mathcal{O}(\log \epsilon^{-1})$ such that $c^{q\tau\tau_3} < \epsilon^2$, and $p = 2q$. In each set $\Delta_{S_{-q}\ldots S_{-1}}$ we choose once for all an unstable manifold $W_{S_{-q}\ldots S_{-1}}$. We have

$$\int_\alpha d\nu^0_{\underline{\omega}}(W) g_{\underline{\omega}_{\le -p}}(W) = \sum_{S_{-q}\ldots S_{-1} \in A_{-q}} \int_{W \in \Delta_{S_{-q}\ldots S_{-1}}} d\nu^0_{\underline{\omega}}(W) g_{\underline{\omega}_{\le -p}}(W)$$

$$= \sum_{S_{-q}\ldots S_{-1} \in A_{-q}} \int_{W \in \Delta_{S_{-q}\ldots S_{-1}}} d\nu^0_{\underline{\omega}}(W) \overline{g}_{\underline{\omega},p}(W) + \mathcal{O}(\epsilon) \parallel g \parallel$$

$$= \sum_{S_{-q}\ldots S_{-1} \in A_{-q}} \int_{W \in \Delta_{S_{-q}\ldots S_{-1}}} d\nu^0_{\underline{\omega}}(W) \overline{g}_{\underline{\omega},p}(W_{S_{-q}\ldots S_{-1}}) + \mathcal{O}(p\epsilon) \parallel g \parallel$$

$$= \sum_{S_{-q}\ldots S_{-1} \in A_{-q}} \int_{W \in \Delta_{S_{-q}\ldots S_{-1}}} d\nu(W) \overline{g}_{\underline{\omega},p}(W_{S_{-q}\ldots S_{-1}}) + \mathcal{O}(p\epsilon) \parallel g \parallel$$

$$= \sum_{S_{-q}\ldots S_{-1} \in A_{-q}} \int_{W \in \Delta_{S_{-q}\ldots S_{-1}}} d\nu(W) \overline{g}_{\underline{\omega},p}(W) + \mathcal{O}(p\epsilon) \parallel g \parallel$$

$$= \int_\alpha d\nu(W) \overline{g}_{\underline{\omega},p}(W) + \mathcal{O}(p\epsilon) \parallel g \parallel ,$$

where we have used successively Lemma 4.1, Lemma 4.2, and Lemma 4.3.

Using the same argument as in the proof of Lemma 4.2, we can replace in $\overline{g}_{\underline{\omega},p}$ the quantity $S^p_{\underline{\omega}}$ by $S^\infty_{\underline{\omega}}$. This last real random variable is absolutely continuous with respect to the Lebesgue measure when $\underline{\omega}$ varies. We shall denote by $\Theta_{\epsilon,M}$ it's density. We can now take the expectation with respect to the remaining variables $\omega_{-p-1}, \omega_{-p-2}, \ldots$, and this concludes the proof of Theorem 1.1.

REFERENCES

1. G.Birkhoff. Lattice theory, Amer. Math. Soc., Providence (1967).

2. R.Bowen. Equilibrium States and the Ergodic Theory of Anosov Diffeomorphisms. Lecture Notes in Mathematics 470, Springer, Berlin, Heidelberg, New York (1975).

3. J.Dieudonné. Elements d'Analyse. Gauthier-Villars, Paris (1969).

4. P.Ferrero, B.Schmidt. Produit aléatoire d'opérateurs de transfert, to appear.

5. M.Freidlin, I.Ventsel. Random Perturbations of Dynamical Systems, Springer, Berlin, Heidelberg, New York (1984).

6. R.Graham, D.Rokaerts, T.Tel. Phys. Rev **A 31**, 3364 (1985).

7. Y.Kifer. General random perturbations of hyperbolic and expanding transformations. J. Analyse Math. **47**, 111-149 (1986).

8. O.E.Lanford III. Introduction to Hyperbolic Sets. In Regular and Chaotic Motions in Dynamic Systems. G.Velo, A.Wightman ed., Plenum, New York, London (1985).

9. F.Ledrappier, L.S.Young. Entropy formula for random transformations, preprint (1986). Dimension formula for random transformations, preprint (1987).

10. D.Ruelle. Thermodynamic Formalism. Addison-Wesley, Reading (1978).

11. Y.G.Sinai. Gibbsian measures in ergodic theory. Russian Math. Survey **27**, 21-69 (1972).

12. L.S.Young. Stochastic stability of hyperbolic attractors. Ergod. Theor. & Dynam. Sys. **6**, 311-319 (1986).

FIXED POINTS OF COMPOSITION OPERATORS

Henri Epstein
CNRS and
IHES, Bures sur Yvette, France

1. INTRODUCTION

These lectures will be concerned with the following problem:

Let ν be fixed in $[1,\ 2]$. Find real constants $\lambda \in (0,\ 1)$ and $r > 1$, and a solution ϕ of the functional equation

$$\phi(x) = -\frac{1}{\lambda}\phi\left(\frac{1}{\lambda^{\nu-1}}\phi(\lambda^{\nu}x)\right), \quad \phi(0) = 1, \tag{1.1}$$

with the properties **C1**,**C2** *below, and a further property* **C3** *to be stated later:*

C1. *ϕ is C^1 and strictly decreasing on $[0,\ L]$ for some $L \geq 1$. For all $x \in [0,\ L]$, $\phi(\lambda^{\nu}x)$ is in $[0,\ \lambda^{\nu-1}L]$ and (1.1) holds.*

C2. *There exists a function f, analytic, strictly decreasing, without critical points on $[0,\ L^r]$, with $f(0) = 1$, such that, for all $x \in [0,\ L]$,*

$$\phi(x) = f(x^r). \tag{1.2}$$

We denote $\mathcal{E}_r$ the set of functions ϕ satisfying the condition **C2**, and $\mathcal{E}'_r$ the set of functions ϕ defined on some $[-L,\ L]$, $L \geq 1$, such that $\phi|\ [0,\ 1]$ and $\check{\phi}|\ [0,\ 1]$ are in $\mathcal{E}_r$ (with $\check{\phi}(x) = \phi(-x)$).

As it is well known, the theory of period doubling of Feigenbaum [F] centers on a renormalization, or doubling operator

$$(\mathcal{N}\phi)(x) = -\frac{1}{\lambda}\phi(\phi(-\lambda x)), \quad \lambda = -\phi(1), \tag{1.3}$$

where ϕ is in $\mathcal{E}'_r$ for a certain fixed $r > 1$. Any *even* fixed point of this operator is a solution of (1.1) with $\nu = 1$. (It is then called the Cvitanović-Feigenbaum functional equation. The theory of Coullet and Tresser [CT] uses a very slightly different formalism leading to different functions, but with the same role in interval dynamics.) The theory of Feigenbaum rests on the following (here somewhat redundantly formulated) conjectures (see [CEL], [CE]):

Conj. 1 : *$\mathcal{N}$ has a fixed point g in $\mathcal{E}_r^!$.*

Conj. 2 : *Its derivative $D\mathcal{N}(g)$ at g possesses an invariant subspace of dimension 1 (called unstable) where it reduces to the multiplication by a certain $\delta > 1$, and an invariant subspace of codimension 1 (called stable) where it reduces to a strict contraction.*

Conj. 3 : *There exists an open "ball" $\mathcal{O}$ in $\mathcal{E}_r^!$ around g, where the behavior of $\mathcal{N}$ is hyperbolic and qualitatively the same as that of $D\mathcal{N}(g)$. In particular there are two manifolds $\mathcal{V}_u$ and $\mathcal{V}_s$, invariant under $\mathcal{N}$, of dimension 1 and codimension 1, respectively, tangent to the corresponding invariant subspaces of $D\mathcal{N}(g)$.*

It is further conjectured that $\mathcal{O}$ is large enough to intersect, for some n, the superstability manifold

$$S_n = \{\phi \in \mathcal{E}_r^! : \phi^{2^n}(0) = 0\},$$

and it is assumed that

Conj. 4 : *$\mathcal{V}_u$ and S_n have a unique intersection point in $\mathcal{O}$ and are transversal there.*

I will sometimes refer to Conjectures 2-4 under the global name of "hyperbolicity". Needless to say these conjectures are generally considered a certainty. Conjectures 1-4 have the following important consequence. Let $\mathcal{L} \subset \mathcal{O}$ be a smooth curve, i.e. a one-parameter family of elements of $\mathcal{O}$, with smooth dependence upon the parameter. Assume that $\mathcal{L}$ intersects only once, and transversally, the stable manifold $\mathcal{V}_s$. Then, when n tends to infinity,

$$\mathcal{N}^n \mathcal{L} \cap \mathcal{O} \to \mathcal{V}_u \cap \mathcal{O}.$$

It follows that the elements of $\mathcal{V}_u \cap \mathcal{O}$ will inherit all the properties which

a) are preserved by $\mathcal{N}$,

b) are stable under taking limits,

c) are exhibited by the elements of some family such as $\mathcal{L}$.

This principle leads us to forecast that

- The elements of $\mathcal{V}_u \cap \mathcal{O}$ are even functions,
- The inverse functions of their restrictions to $(0, 1)$ have the anti-Herglotz property,
- These inverse functions are, moreover, univalent (i.e. injective) in $\mathbf{C}_+ \cup \mathbf{C}_-$.

The last two statements appeal to the normality properties of the Herglotz functions (see Lemma 1 of Section 2) and to Hurwitz's theorem. It is therefore natural to look for solutions of the functional equation (i.e. (1.1) with $\nu = 1$) such that the inverse function u of $\phi|\ [0, 1]$ and the inverse function U of $f|\ [0, 1]$ both have the anti-Herglotz property. (The meaning of this is recalled in the next section.)

A theory of circle maps with golden ratio rotation number has been formulated in a related spirit by Feigenbaum, Kadanoff, and Shenker [FKS], and by Ostlund,

Rand, Sethna, and Siggia [ORSS]. It also uses a renormalization operator

$$\mathcal{T}\begin{pmatrix}\xi\\ \eta\end{pmatrix} = \begin{pmatrix}-\frac{1}{\lambda}\eta(-\lambda x)\\ -\frac{1}{\lambda}\eta(\xi(-\lambda x))\end{pmatrix}, \quad \lambda = -\eta(0). \tag{1.4}$$

Here, denoting $\phi(x) = \xi(-x)$, the functions ϕ and $-\eta/\lambda$ are in $\mathcal{E}_r$ for some $r > 1$. Any fixed point of the operator $\mathcal{T}$ corresponds to a ϕ satisfying (1.1) with $\nu = 2$. The operator $\mathcal{T}$ is also conjectured to be hyperbolic, and proved to be so in certain cases ([JR],[M]). Hence the same reasons as above lead us to look, also in the case $\nu = 2$, for solutions of (1.1) whose inverse functions have the anti-Herglotz property. And in the hope of dealing with the two cases at the same time, we let ν be arbitrarily chosen in [1, 2]. (For more details on the functional equations in the circle case, see [L3], [L4], [N].) We now digress to establish some notation and recall the classical facts about Herglotz functions.

2. HERGLOTZ FUNCTIONS

2.1 Definition

Let J be an open, possibly empty interval in $\mathbf{R}$. We denote

$$\mathbf{C}(J) = \{z \in \mathbf{C} : \text{Im } z \neq 0 \text{ or } z \in J\} = \mathbf{C}_+ \cup \mathbf{C}_- \cup J, \tag{2.1}$$

where

$$\mathbf{C}_+ = \{z \in \mathbf{C} : \ \text{Im } z > 0\} = -\mathbf{C}_- \,. \tag{2.2}$$

$\mathcal{F}(J)$ is the real Fréchet space of the functions h holomorphic in $\mathbf{C}(J)$, and such that $h^*(z^*) = h(z)$.

$\mathbf{P}(J)$ is the subset of $\mathcal{F}(J)$ consisting of the functions h such that $h(\mathbf{C}_+) \subset \bar{\mathbf{C}}_+$. An element h of $\mathbf{P}(J)$ is called a Herglotz (or Pick) function (and $-h$ is an anti-Herglotz function).

2.2 Integral Representation

Any element h of $\mathbf{P}(J)$ has a once subtracted Cauchy representation:

$$h(z) - h(z_0) = a(z - z_0) + \int d\mu(t)\left[\frac{1}{t-z} - \frac{1}{t-z_0}\right], \tag{2.3}$$

valid for all z and z_0 in $\mathbf{C}(J)$. Here the positive measure μ has support in $\mathbf{R} \setminus J$ and is the boundary value (in the sense of measures) of Im h/π from the upper half-plane. It satisfies

$$\int \frac{d\mu(t)}{1+t^2} < \infty.$$

The constant a is positive and is called the angular derivative of h at infinity. It is uniquely determined by

$$\lim_{\rho\to\infty} |h(\rho e^{i\theta}) - a\rho e^{i\theta}|/\rho = 0, \qquad \epsilon < \theta < \pi - \epsilon, \tag{2.4}$$

valid uniformly in θ for any fixed ϵ in $(0, \pi)$. For proofs of these facts, see e.g. [D] and [V].

$\mathbf{P}_0(J)$ is the subset of $\mathbf{P}(J)$ consisting of the functions h such that the corresponding a is 0. It is dense in $\mathbf{P}(J)$.

2.3 Positivity Conditions on Derivatives

Suppose that $h \in \mathbf{P}(J)$ with $J \neq \emptyset$. Then, for every $z \in J$ and every finite complex sequence $v_0, \ldots v_N$,

$$\sum_{j,k=0}^{N} \frac{h^{(j+k+1)}(z)}{(j+k+1)!} v_j^* v_k \geq 0. \tag{2.5}$$

In particular $h^{(n)}$ is positive on J for all odd n. Note that if h is not constant and real in $\mathbf{C}_+ \cup \mathbf{C}_-$, then $h(\mathbf{C}_+) \subset \mathbf{C}_+$, and h' is strictly positive on J. Also by (2.5) (case $N = 1$),

$$Sh \equiv \frac{h'''}{h'} - \frac{3}{2}\left(\frac{h''}{h'}\right)^2 \geq 0 \text{ on } J. \tag{2.6}$$

Sh is the Schwarzian of h. Recall that if a smooth function has positive Schwarzian and has no critical point, then its inverse function has negative Schwarzian, and vice versa.

2.4 Iteration of Functions in P(∅)

Denote h_+ the restriction of $h \in \mathbf{P}(\emptyset)$ to $\mathbf{C}_+$. Then (see [V]), either h_+ is an isomorphism of $\mathbf{C}_+$ (i.e. a Moebius transformation) or h_+^n converges, uniformly on any compact subset of $\mathbf{C}_+$, to a constant C. There are three possible cases:

1) $C = \infty$: this can happen only if $a > 1$.

2) $C \in \mathbf{R}$.

3) $C \in \mathbf{C}_+$: then C is an attractive fixed point of h_+, i.e. $|f_+'(C)| < 1$.

2.5 Remarks

Let $h \in \mathbf{P}(J)$ with $J \neq \emptyset$, $z_0 \in J$. If h is not a constant, then ,on J, $(h(z) - h(z_0))/(z - z_0)$ is a strictly positive, convex function of z.

Finally, the following trivial fact is stated as a lemma because it plays a major part in the following:

Lemma 1

Suppose $J \neq \emptyset$ and $J \neq \mathbf{R}$. Then for any $B > 0$,

$$\mathbf{P}(J) \cap \{h \ : |h(J)| \leq B\} \tag{2.7}$$

is compact in $\mathcal{F}(J)$.

Proof: we can assume that $B < 1$. Then any element h of the set (2.7) maps $\mathbf{C}(J)$ into $\mathbf{C}((-1, 1))$ which, in turn can be mapped conformally onto the unit disk by an obvious transformation φ, i.e. $|\varphi \circ h| < 1$. The set (2.7) is thus a normal family, and it is obviously closed in $\mathcal{F}(J)$.

3. NEW STATEMENT OF THE EXISTENCE PROBLEM

We first reexpress the conditions **C1**, **C2** imposed on ϕ in Section 1 into conditions to be satisfied by the inverse functions u of ϕ and U of f. These inverse functions will be the focus of our attention for the remainder of these notes.

The problem is thus (for a fixed $\nu \in [1, 2]$) to find real constants $\lambda \in (0, 1)$ and $r > 1$, and real functions u and U with the properties **C′1**, **C′2** and **C3** below.

C′1. *(i) u is continuous and strictly decreasing on $[-\lambda, 1]$, with*

$$u(1) = 0, \quad u(-\lambda) = \frac{1}{\lambda^{\nu-1}}, \tag{3.1}$$

(ii) and, for all $z \in [-\lambda, 1]$,

$$u(z) = \frac{1}{\lambda^{\nu}} u(\lambda^{\nu-1} u(-\lambda z)), \tag{3.2}$$

C′2. *The function U defined on $[-\lambda, 1]$ by*

$$U(z) = u(z)^r \tag{3.3}$$

is analytic without critical points on $[-\lambda, 1]$. In particular there is a $w_0 > 1$ such that U is analytic and has strictly negative derivative in $[-\lambda, w_0]$. It satisfies in this interval:

$$U(z) = \frac{1}{\tau^{\nu}} U(\lambda^{\nu-1} u(-\lambda z)), \quad \tau = \lambda^r. \tag{3.4}$$

From **C′1** *(ii)* it follows that there exists an $x_0 \in (0, 1)$ such that

$$u(0) = \frac{x_0}{\lambda^{\nu-1}}. \tag{3.5}$$

Note that if $z \in [-\lambda, 1]$, $u(-\lambda z)$ is in $[0, \lambda^{1-\nu}]$ so that both sides of eq.(3.2) make sense. **C′2** implies that u is analytic in $(-\lambda, 1)$ and $u(z)$ behaves near 1 like const.$(1-z)^{1/r}$. The functional equation (3.2) provides an extension of u into an analytic, strictly decreasing function on the open interval $(-\lambda^{-1}, 1)$, continuous at the ends of this interval with

$$u(-\lambda^{-1}) = \frac{x_0}{\lambda^{2\nu-1}}. \tag{3.6}$$

$u(z)$ behaves near $-\lambda^{-1}$ like const. $(z + \lambda^{-1})^{1/r} + x_0 \lambda^{1-2\nu}$. Since u is strictly decreasing, (3.6) and (3.1) imply $x_0 > \lambda^{\nu}$.

C3. *u extends to an anti-Herglotz function (also denoted u):*

$$u \in -\mathbf{P}((-\lambda^{-1}, 1)). \tag{3.7}$$

Let us denote

$$\varphi_0(z) = \lambda^{\nu-1} u(-\lambda z), \tag{3.8}$$

so that equation (3.4) can be rewritten

$$U(z) = \frac{1}{\tau^\nu} U(\varphi_0(z)). \tag{3.9}$$

φ_0 belongs to $\mathbf{P}((-\lambda^{-1}, \lambda^{-2}))$ and

$$\varphi_0(-\lambda^{-1}) = 0, \quad \varphi_0(1) = 1, \quad \varphi_0(\lambda^{-2}) = \lambda^{\nu-1} u(-\lambda^{-1}) = \frac{x_0}{\lambda^\nu} < \lambda^{-2}. \tag{3.10}$$

Thus φ_0 is Herglotz, and it maps $\mathbf{C}((-\lambda^{-1}, \lambda^{-2}))$ into itself. The remark at the end of Section 2 shows that the function $x \to [\varphi_0(x) - 1]/(x-1)$ is convex on $[-\lambda^{-1}, \lambda^{-2}]$, hence bounded above there by the maximum m of its values at the ends of this interval, which is < 1:

$$0 < \varphi_0(z) - 1 \le m(z-1) \quad \text{for all } z \in (1, \lambda^{-2}]. \tag{3.11}$$

There is an N such that $m^N(\lambda^{-2} - 1) + 1 < w_0$. Hence eq. (3.9) can be used to extend U into an anti-Herglotz function over $\mathbf{C}((-\lambda^{-1}, \lambda^{-2}))$, satisfying (3.4) in this cut-plane:

$$U(z) = U'(1) \lim_{n\to\infty} \tau^{-\nu n} [\varphi_0^n(z) - 1].$$

Recall that U is negative on $(1, \lambda^{-2})$, and vanishes at 1. Thus u has a simple branch point at 1, of the type $(1-z)^{1/r}$, and hence the same is true for u at $-\lambda^{-1}$ and for U at $-\lambda^{-1}$ and at λ^{-2}. To summarize:

The conditions **C′1**, **C′2** and **C3** are equivalent to:

$$u \in -\mathbf{P}((-\lambda^{-1}, 1)), \quad u(1) = 0, \quad u(0) > 0, \tag{3.12}$$

$$U \in -\mathbf{P}((-\lambda^{-1}, \lambda^{-2})), \tag{3.13}$$

$$U(z) = u(z)^r \quad \forall\, z \in \mathbf{C}((-\lambda^{-1}, 1)). \tag{3.14}$$

together with the functional equations (3.2) and (3.4) in the cut planes $\mathbf{C}((-\lambda^{-1}, 1))$ and $\mathbf{C}((-\lambda^{-1}, \lambda^{-2}))$, respectively. A final transformation consists in denoting $\psi = U/U(0)$, after which we obtain the form of the problem we shall actually use:

Let ν be fixed in $[1, 2]$. Find real constants r, λ, τ, α and two functions V and ψ satisfying:

$$0 < \lambda < 1, \quad r > 1, \quad \tau = \lambda^r, \quad \alpha > 1, \tag{3.15}$$

$$\psi \in -\mathbf{P}((-\lambda^{-1}, \lambda^{-2})), \quad \psi(0) = 1, \quad \psi(1) = 0, \tag{3.16}$$

$$V \in -\mathbf{P}((0, \alpha\tau^{-2})), \quad V(1) = 1, \quad V'(1) = -\frac{1}{\lambda}, \tag{3.17}$$

and the identities

$$V(\zeta) = \frac{1}{\tau^\nu}\psi\left(\left(\frac{\zeta}{\alpha}\right)^{1/r}\right) \quad \text{for all } \zeta \in \mathbf{C}((0,\ \alpha\tau^{-2})), \tag{3.18}$$

$$\psi(z) = V(\psi(-\lambda z)) \quad \text{for all } z \in \mathbf{C}((-\lambda^{-1},\ \lambda^{-2})). \tag{3.19}$$

Given a solution of this problem, a solution of (1.1) satisfying all our requirements is obtained by setting:

$$\phi = u^{-1}, \quad u(z) = \frac{1}{\lambda^{\nu-1}}[\alpha^{-1}\psi(z)]^{1/r}, \quad x_0 = \alpha^{-1/r}. \tag{3.20}$$

On the basis of much numerical and also rigorous work by many authors, we expect that, for each fixed ν, there will be, for each $r > 1$, a locally (and perhaps globally) unique solution of (1.1) satisfying the above requirements. This solution, and in particular λ should depend analytically on r. Note that if the requirement of the existence and analyticity of f is abandoned, or if it is allowed to have critical points, there appear many other solutions: see e.g. [C], [G]. It must be stressed at this point that a considerable amount of work has been devoted to the theories giving rise to the functional equations (1.1), as other lectures at this school will demonstrate. Only a small portion of this literature, having a direct bearing on the subject of these notes, is included in the list of references.

The remainder of these notes is divided into two parts. Part I, consisting of sections 4 and 5, will give an account of the method used by J.-P. Eckmann and myself ([E], [EE]) to prove the existence of solutions. Several rigorous results on this question appear in the literature, by Collet, Eckmann and Lanford [CEL], Lanford [L1,L2], Campanino, Epstein and Ruelle [CER], Eckmann and Wittwer [EW], Falcolini [Fa], Jonker and Rand [JR], Lanford and de la Llave [LL], Mestel [M], Eckmann and Wittwer [EW2](see also [VKS]). The work of Lanford [L1] was the first proof for the case $\nu = 1$, $r = 2$, and served as a model for other computer-assisted proofs. Besides proving the existence of a solution, it also proved its hyperbolic properties, as described in Section 1. The anti-Herglotz property of U was proved in the case $\nu = 1$, $r = 2$, by Epstein and Lascoux [EL]. The work of Eckmann and Wittwer [EW] showed the existence and nature of the limit $r \to \infty$.

Part II of these notes describes some properties of the solutions of the problem stated in this section, and is independent of Part I. Its main purpose is to provide an ingredient for the ideas of Douady and Hubbard [DH] and Sullivan [S] which seem to me a most promising development in this subject. These notes follow very closely [E], [EE], [EE2], and, in the last sections, [EL]. Sections 5 and 6 hopefully contain some improvements.

4. FIXED λ METHOD

As already mentioned, all the available information indicates that (for fixed ν) the system (3.15-3.19) will have a locally unique solution for any given r with λ a (smooth) function of r. Although we would like to be able to prescribe r, the more successful version of our method works with a fixed λ. This is the version I shall describe now. Until further notice, ν is fixed in $[1,\ 2]$ and λ is fixed in $(0,\ 1)$. A function ψ giving a solution of the system (3.15-3.19) will be looked for as a fixed point of an operator M_λ defined as follows:

1) Start with an element ψ_0 of the set

$$\mathbf{E}_0(\lambda) = \{\psi_0 \in -\mathbf{P}((-\lambda^{-1},\ \lambda^{-2}))\ : \psi_0(0) = 1, \quad \psi_0(1) = 0\}. \tag{4.1}$$

A function V is first defined by

$$V(\zeta) = \frac{1}{\tau^\nu}\psi_0\left(\left(\frac{\zeta}{\alpha}\right)^{1/r}\right), \tag{4.2}$$

where the constants τ, α, and r must be determined so that

$$V(1) = 1, \quad V'(1) = -\frac{1}{\lambda}, \quad \tau = \lambda^r, \tag{4.3}$$

and must satisfy

$$r > 1, \quad 1 < \alpha < \tau^{-\nu}. \tag{4.4}$$

2) Then ψ is the solution of

$$\psi(z) = V(\psi(-\lambda z)), \quad \psi(0) = 1, \quad \psi(1) = 0. \tag{4.5}$$

If all these constructions are possible, and if $\psi \in \mathbf{E}_0(\lambda)$, then ψ_0 is in the domain of M_λ, and $\psi = M_\lambda\psi_0$. We shall now investigate in some detail the conditions under which this is the case.

The following notations will be used:

$$\Omega \equiv \Omega(\lambda) = \mathbf{C}((-\lambda^{-1},\ \lambda^{-2})), \tag{4.6}$$

$$J(\lambda) = (-\lambda^{-1},\ \lambda^{-2}), \quad \Sigma(\lambda) = \mathbf{R} \setminus J(\lambda), \tag{4.7}$$

$$A \equiv A(\lambda) = -1/(\lambda \log \lambda), \quad B \equiv B(\lambda) = (1-\lambda^2)A(\lambda). \tag{4.8}$$

Note that $A \geq e$ and B is a decreasing function of λ tending to 2 when $\lambda \to 1$.

4.1 Remarks on $\mathbf{E}_0(\lambda)$

Let $\psi_0 \in \mathbf{E}_0(\lambda)$. The function $-\log \psi_0$ is herglotzian, and

$$\operatorname{Im} z > 0 \Rightarrow 0 < -\operatorname{Im} \log \psi_0(z) < \pi. \tag{4.9}$$

Hence the representation (2.3) for this function takes the form

$$-\log \psi_0(z) = \int \sigma(t)dt \left[\frac{1}{t-z} - \frac{1}{t}\right] \quad \forall z \in \mathbf{C}((-\lambda^{-1},\ 1)), \tag{4.10}$$

where $\sigma \in L^\infty$ has support in $\mathbf{R} - (-\lambda^{-1},\ 1)$, $0 \leq \sigma \leq 1$, and $\sigma(t) = 1$ for all $t \in [1,\ \lambda^{-2}]$. Conversely, for any such σ, the formula (4.10) yields an element ψ_0 of $\mathbf{E}_0(\lambda)$.

Let $\widetilde{\psi}_0$ be another element of $\mathbf{E}_0(\lambda)$ obtained by substituting $\widetilde{\sigma}$ for σ in (4.10). Since σ and $\widetilde{\sigma}$ coincide with 1 on $[1,\ \lambda^{-2}]$, the formulae

$$\log \frac{\widetilde{\psi}_0(z)}{\psi_0(z)} = \int_{\Sigma(\lambda)} [\sigma(t) - \widetilde{\sigma}(t)] \frac{z}{t(t-z)} dt \tag{4.11}$$

$$\frac{\widetilde{\psi}_0'(z)}{\widetilde{\psi}_0(z)} - \frac{\psi_0'(z)}{\psi_0(z)} = \int_{\Sigma(\lambda)} [\sigma(t) - \widetilde{\sigma}(t)] \frac{1}{(t-z)^2} dt \tag{4.12}$$

hold, by analytic continuation, in the whole domain Ω.

In particular let σ_2 be the characteristic function of $[1,\ \lambda^{-2}]$ and σ_3 that of $\mathbf{R} \setminus (-\lambda^{-1},\ 1)$. Let ψ_2 and ψ_3 be the corresponding elements of $\mathbf{E}_0(\lambda)$:

$$\psi_2(z) = \frac{1-z}{1-\lambda^2 z}, \quad \psi_3(z) = \frac{1-z}{1+\lambda z}, \tag{4.13}$$

Then, with ψ_0 as above,

$$\frac{\psi_0(z)}{\psi_2(z)} \leq 1 \leq \frac{\psi_0(z)}{\psi_3(z)} \quad \text{for all} \quad z \in (0,\ \lambda^{-2}), \tag{4.14}$$

i.e.

$$\frac{1}{1+\lambda z} \leq \frac{\psi_0(z)}{1-z} \leq \frac{1}{1-\lambda^2 z} \quad \text{for all} \quad z \in (0,\ \lambda^{-2}), \tag{4.15}$$

(reversed for $-\lambda^{-1} < z < 0$) and

$$-\frac{\psi_2'(z)}{\psi_2(z)} \leq -\frac{\psi_0'(z)}{\psi_0(z)} \leq -\frac{\psi_3'(z)}{\psi_3(z)} \quad \text{for all} \quad z \in J(\lambda) \setminus \{1\}, \tag{4.16}$$

i.e.

$$\frac{1-\lambda^2}{(1-\lambda^2 z)(1-z)} \leq -\frac{\psi_0'(z)}{\psi_0(z)} \leq \frac{1+\lambda}{(1+\lambda z)(1-z)} \quad \text{for all} \quad z \in J(\lambda) \setminus \{1\}, \tag{4.17}$$

These bounds show, in particular, that $\mathbf{E}_0(\lambda)$ is a compact subset of the Fréchet space $\mathcal{F}(J(\lambda))$. Let indeed $\{\chi_n\}$ be any sequence in $\mathbf{E}_0(\lambda)$. Denote $\mathcal{J}_n = (1 - 1/n)J(\lambda)$,

$n = 1, 2, \ldots$. Applying Lemma 1 to each $\mathbf{C}(\mathcal{J}_n)$, and using the diagonal procedure, it is possible to extract from $\{\chi_n\}$ a subsequence which converges in $\mathcal{F}(J(\lambda))$ to an element of $\mathbf{E}_0(\lambda)$. In particular any real continuous function on $\mathbf{E}_0(\lambda)$ has a maximum and a minimum.

Another bound we shall need later is:

$$\frac{-2\lambda}{1+\lambda z} \leq \frac{\psi_0''(z)}{\psi_0'(z)} \leq \frac{2\lambda^2}{1-\lambda^2 z} \quad \text{for all } z \in J(\lambda). \tag{4.18}$$

It is easily derived from the representation (2.3) for ψ_0.

4.2 Determination of the Constants

Let $\psi_0 \in \mathbf{E}_0(\lambda)$. If a function V can be defined by (4.2) so as to satisfy (4.3), then $z_1 \equiv \alpha^{-1/r}$ must satisfy

$$\psi_0(z_1) = \tau^\nu, \qquad \frac{z_1 \psi_0'(z_1)}{\psi_0(z_1)} = -\frac{r}{\lambda}, \tag{4.19}$$

and since $r = (\log \tau)/(\log \lambda)$, $q(z_1) = 0$, where the function q is defined by

$$q(z) = \frac{\psi_0'(z)}{\psi_0(z)} - \frac{A}{\nu z} \log \psi_0(z), \quad A \equiv A(\lambda) = -1/(\lambda \log \lambda) \geq e, \tag{4.20}$$

for all $z \in \mathbf{C}((-\lambda^{-1}, 1))$.

The function zq vanishes at 0 and tends to $-\infty$ when $z \uparrow 1$. For $z \in (0, 1)$,

$$\frac{d}{dz} zq(z) = z \frac{\psi_0'(z)}{\psi_0(z)} G(z),$$

$$G = \frac{\psi_0''}{\psi_0'} - \frac{\psi_0'}{\psi_0} - \frac{1}{z}\left(\frac{A}{\nu} - 1\right),$$

$$G' = S\psi_0 + \frac{1}{2}\left(\frac{\psi_0''}{\psi_0'} - \frac{\psi_0'}{\psi_0}\right)^2 + \frac{1}{2}\left(\frac{\psi_0'}{\psi_0}\right)^2 + \frac{1}{z^2}\left(\frac{A}{\nu} - 1\right) \geq 0.$$

Since $G(z)$ tends to $-\infty$ when $z \downarrow 0$, and to $+\infty$ when $z \uparrow 1$, it vanishes only once in $(0, 1)$. Hence $(zq(z))'$ also vanishes only once in $(0, 1)$, starts by being positive and ends up negative; $zq(z)$ increases from 0 to a unique maximum, then decreases to $-\infty$. Thus q has a unique zero in $(0, 1)$. We define z_1 as this zero, and:

$$\tau^\nu = \psi_0(z_1) \in (0, 1), \quad r = (\log \tau)/(\log \lambda) > 0, \quad \alpha = z_1^{-r} > 1. \tag{4.21}$$

The function q has the representation

$$q(z) = \int \frac{\sigma(t)dt}{t - z}\left[\frac{A}{\nu t} - \frac{1}{t - z}\right] \quad \forall z \in \mathbf{C}((-\lambda^{-1}, 1)). \tag{4.22}$$

Let $z \in (0,\ 1)$. Then the integrand of (4.22) is easily seen to be positive when t is in $\Sigma(\lambda)$. When $1 < t < \lambda^{-2}$ the integrand is strictly positive if $z \leq 1 - \nu/A$. We conclude

$$z_1 > 1 + \nu\lambda \log \lambda > \lambda^{\nu}. \tag{4.23}$$

From (4.23) we can derive a crude lower bound for r by using (4.17):

$$\frac{r}{\lambda} \geq \frac{z_1(1-\lambda^2)}{(1-\lambda^2 z_1)(1-z_1)},$$

$$r \geq \frac{\lambda^{\nu+1}(1-\lambda^2)}{(1-\lambda^{\nu})(1-\lambda^{2+\nu})} \geq \frac{\lambda^3}{(1-\lambda^4)} \tag{4.24}$$

This shows that $r \to \infty$ when $\lambda \to 1$. To obtain better bounds we use the representation (4.22) for q. Let again (as in subsection 4.1) $\widetilde{\psi}_0$ be another element of $\mathbf{E}_0(\lambda)$, and $\widetilde{\sigma}$, $\widetilde{q}$, $\widetilde{z}_1$, $\widetilde{\tau}$, $\widetilde{r}$ the objects associated with $\widetilde{\psi}_0$ in the same way as σ, q, z_1, τ, r with ψ_0. Suppose $\widetilde{\sigma} \leq \sigma$. Let $z \in (0,\ 1)$. Then, since the bracket in the integrand of the representation (4.22) is positive for $t \in \Sigma(\lambda)$, and the integrand is independent of ψ_0 for $t \in [1,\ \lambda^{-2}]$, we have $\widetilde{q}(z) \leq q(z)$. Hence $q(\widetilde{z}_1) \geq 0$, hence $\widetilde{z}_1 \leq z_1$ and

$$\tau^{\nu} = \psi_0(z_1) \leq \psi_0(\widetilde{z}_1) \leq \widetilde{\psi}_0(\widetilde{z}_1) = \widetilde{\tau}^{\nu}, \quad r \geq \widetilde{r}. \tag{4.25}$$

In particular, for $j = 2,\ 3$, let q_j be the function given by (4.22) with $\sigma = \sigma_j$. Let z_j be the zero of q_j in $(0,\ 1)$, and

$$\tau_j^{\nu} = \psi_j(z_j), \quad r_j \equiv r_j(\lambda) \equiv r_j(\lambda,\ \nu) = \frac{\log \tau_j}{\log \lambda}. \tag{4.26}$$

Then

$$\tau_3 \ \leq \ \tau \ \leq \ \tau_2, \qquad r_2 \ \leq \ r \ \leq \ r_3, \tag{4.27}$$

In other words the minimum and maximum of the continuous function $\psi_0 \ \to r$ on $\mathbf{E}_0(\lambda)$ are respectively reached at ψ_2 and ψ_3.

The function q_2 is given by

$$q_2(z) = \frac{A}{\nu z} \log\left(\frac{1-\lambda^2 z}{1-z}\right) - \frac{1-\lambda^2}{(1-\lambda^2 z)(1-z)}. \tag{4.28}$$

To study its behavior in $(0,\ 1)$, it is convenient to use as a variable $\xi = 1/\psi_2(z)$, i.e.

$$\xi = \frac{1-\lambda^2 z}{1-z}, \quad z = \frac{\xi - 1}{\xi - \lambda^2}. \tag{4.29}$$

This gives

$$\begin{aligned} (1-\lambda^2) z q_2(z) &= \chi(\xi) - \xi, \\ \chi(\xi) \ &= \ \frac{B}{\nu} \log \xi + 1 + \lambda^2 - \frac{\lambda^2}{\xi}. \end{aligned} \tag{4.30}$$

The function χ is increasing and concave on $(0, \infty)$ and $\chi(\xi) - \xi$ vanishes at $\xi = 1$ and at a unique $\widehat{\xi} = \psi_2(z_2) > 1$, and

$$\tau_2^\nu = 1/\widehat{\xi}, \quad r_2(\lambda) = \frac{\log \widehat{\xi}}{\nu \log(1/\lambda)}. \tag{4.31}$$

The necessary and sufficient condition for $r_2(\lambda) > 1$ is therefore $\widehat{\xi} > \lambda^{-\nu}$, i.e. $\chi(\lambda^{-\nu}) - \lambda^{-\nu} > 0$, i.e.

$$\begin{aligned} &\lambda^{\nu-1}(1-\lambda^2) - (1-\lambda^\nu)(1-\lambda^{\nu+2}) \\ &\qquad \equiv (1-\lambda^\nu)(\lambda^{\nu+2} + \lambda^{\nu-1} - 1) + (\lambda^\nu - \lambda^2)\lambda^{\nu-1} \; > \; 0. \end{aligned} \tag{4.32}$$

This is always satisfied for $\nu = 1$, but for $\nu > 1$, it is equivalent to $\lambda > \lambda_0(\nu)$, where $\lambda_0(\nu)$ is the unique zero in $[0, 1)$ of the function of λ in the l.h.s. of (4.32). This condition, while inescapable if one wishes to have M_λ defined on the whole of $\mathbf{E}_0(\lambda)$, is not optimal. Indeed the function $\psi(z) = 1 - z$ is a fixed point of M_λ provided $\lambda^\nu + \lambda^{\nu-1} - 1 = 0$. Furthermore the work of Jonker and Rand [JR] makes it all but certain that the necessary and sufficient condition for the existence of a fixed point of M_λ is $\lambda^\nu + \lambda^{\nu-1} - 1 \geq 0$. It is possible to do better than (4.32) at the cost of rather laboriously extracting from $\mathbf{E}_0(\lambda)$ smaller subsets invariant under M_λ. Here it will be assumed, from now on, that $\lambda > \lambda_0(\nu)$. Under this condition, $r \geq r_2(\lambda, \nu) > 1$ for all $\psi_0 \in \mathbf{E}_0(\lambda)$, and it will be shown that M_λ is indeed defined on the whole of $\mathbf{E}_0(\lambda)$.

The function V can now be defined by the formula (4.2), with τ, r and α as in (4.21). Since $r > 1$, it is anti-Herglotz and thus satisfies the conditions (3.17) with $V(0) = \tau^{-\nu}$, $V(\alpha) = 0$.

It is clear that $W = V \circ V$ is a Herglotz function holomorphic in a neighborhood of 1. In fact we have:

$$\begin{gathered} W \in \mathbf{P}((0, \alpha)), \quad W(1) = 1, \quad W'(1) = \lambda^{-2}, \\ W(\alpha) = \tau^{-\nu}, \quad W(0) = V(\tau^{-\nu}) < 0. \end{gathered} \tag{4.33}$$

The last statement holds because (4.23) implies $\alpha < \tau^{-\nu} < \alpha\tau^{-2}$.

It is also useful to consider

$$\widehat{V}(\zeta) \equiv 1 - V(1-\zeta), \quad \widehat{W} = \widehat{V} \circ \widehat{V}, \quad \widehat{W}(\zeta) \equiv 1 - W(1-\zeta), \tag{4.34}$$

which satisfy

$$\begin{gathered} \widehat{V} \in -\mathbf{P}((1-\alpha\tau^{-2}, 1)), \quad \widehat{W} \in \mathbf{P}((1-\alpha, 1)), \\ \widehat{V}(0) = 0 = \widehat{W}(0), \quad \widehat{V}'(0) = -\lambda^{-1}, \quad \widehat{W}'(0) = \lambda^{-2}, \quad \widehat{W}(1) > 1. \end{gathered} \tag{4.35}$$

Note that quantities such as α, $\widehat{W}(1)$, etc. are continuous functions of ψ_0 defined on $\mathbf{E}_0(\lambda)$, thus, as already remarked, possess maxima and minima there. For example there exist constants $\alpha_{\min}$ and $\alpha_{\max}$ depending only on λ and ν such that $1 < \alpha_{\min} \leq \alpha \leq \alpha_{\max}$, and similarly for $\widehat{W}(1)$. It is also easy to get explicit, cruder bounds. For example using again (4.23) and (4.24), we find:

$$\frac{z_1}{1-z_1} \leq \frac{r(1-\lambda^{2+\nu})}{\lambda(1-\lambda^2)} \leq r\left(\lambda + \frac{1}{\lambda}\right),$$

hence since $r > 1$,

$$\frac{1}{z_1^r} \geq \left(1 + \frac{1}{rY}\right)^r \geq 1 + \frac{1}{Y}, \quad Y = \lambda + \frac{1}{\lambda},$$

$$\alpha \geq 1 + \frac{\lambda}{1+\lambda^2}. \tag{4.36}$$

4.3 Construction of ψ

To obtain ψ satisfying (4.5) we first construct a function Ψ satisfying

$$\Psi(z) = \widehat{V}(\Psi(-\lambda z)) = \widehat{W}(\Psi(\lambda^2 z)), \quad \Psi(0) = 0, \quad \Psi'(0) = 1. \tag{4.37}$$

As it is well known, such a function always exists in a small disk Δ around 0, where it is given by the absolutely and uniformly convergent limit

$$\Psi(z) = \lim_{n\to\infty} \Psi_n(z), \quad \Psi_n(z) = \widehat{V}^n((-\lambda)^n z). \tag{4.38}$$

In fact the size of Δ and the convergence of (4.38) are (for a fixed λ) uniform as ψ_0 varies in $\mathbf{E}_0(\lambda)$. Since each Ψ_n is a Herglotz function, and since the sequence is uniformly bounded in Δ, it also converges, by Vitali's theorem, uniformly on any compact, in the whole of $\mathbf{C}_+ \cup \mathbf{C}_- \cup \Delta$, to a Herglotz function. On $\mathbf{R}_+$, Ψ can be extended, using the second functional equation in (4.37), to a strictly increasing, analytic function on an interval $[0, L)$ in which it must take the value 1. Indeed, if it did not, it could be extended to the whole of $\mathbf{R}_+$, then to the whole of $\mathbf{R}_-$ by the first functional equation in (4.37). It would thus be entire and Herglotz, hence linear, hence it would coincide with z, contradicting the hypothesis. Thus $\Psi(\gamma) = 1$ for some finite $\gamma > 0$. By (4.37), Ψ can be extended to the interval $(-\gamma\lambda^{-1}, \gamma\lambda^{-2})$. We now define

$$\widehat{\psi}(z) = \Psi(\gamma z), \quad \psi(z) = 1 - \widehat{\psi}(z). \tag{4.39}$$

Then ψ is anti-Herglotz, and satisfies (4.5). Moreover

$$\psi(-\lambda^{-1}) = V(0) = \tau^{-\nu}, \quad \psi(\lambda^{-2}) = W(0) = V(\tau^{-\nu}). \tag{4.40}$$

As noted in subection 4.2, these quantities are bounded in modulus by bounds depending only on λ. There is a constant $C_1(\lambda)$ (depending also on ν) such that $|\psi(z)| \leq C_1(\lambda)$ for all $z \in J(\lambda)$. Thus

$$M_\lambda \mathbf{E}_0(\lambda) \subset \mathbf{E}_2(\lambda) = \{\psi \in \mathbf{E}_0(\lambda) \ : \ |\psi(z)| \leq C_1(\lambda) \ \ \forall z \in J(\lambda)\}. \tag{4.41}$$

Note that, as an element of $\mathbf{E}_0(\lambda)$, ψ satisfies all the inequalities mentioned in subsection 4.1; in particular by (4.17),

$$1 - \lambda^2 \leq \gamma = -\psi'(0) \leq 1 + \lambda \tag{4.42}$$

It is now necessary to check that M_λ is continuous in the Fréchet topology. It is clear that V, W, r, α, etc. continuously depend on ψ_0. The restriction of Ψ to Δ depends continuously on $\widehat{V}$. Let K_1, K_2 be compact subsets in its domain of analyticity, with K_1 contained in the interior of K_2. The restriction of Ψ to K_2 can be obtained from its restriction to Δ by iterating the second functional equation in (4.37) a finite number N of times. If $\widetilde{\psi}_0$ is allowed to vary in a sufficiently small neighborhood of ψ_0 in $\mathbf{E}_0(\lambda)$, the same number of iterations will suffice to define the corresponding $\widetilde{\Psi}$ on K_1, and the modulus of $(\widetilde{\Psi} - \Psi)|K_1$ can be made as small as desired.

Since the continuous map M_λ sends the convex compact set $\mathbf{E}_0(\lambda)$ into itself the Schauder-Tikhonov theorem asserts:

Lemma 2

For each $\lambda > \lambda_0(\nu)$ there exist fixed points of M_λ in $\mathbf{E}_0(\lambda)$. Any such fixed point is in $\mathbf{E}_2(\lambda)$.

In fact M_λ maps $\mathbf{E}_0(\lambda)$ into a strictly smaller compact subset of $\mathbf{E}_2(\lambda)$. First it can be shown that if $\psi \in M_\lambda \mathbf{E}_0(\lambda)$,

$$(2z \log \lambda)\psi'(z) \ \geq \ a'[1 - \psi(z)]^3 \text{ for all } z \in [0,\ 1]\}, \tag{4.43}$$

where

$$a' = \frac{a}{1 + 3a}, \qquad a = \frac{1}{12}(1 - r_2(\lambda)^{-2}).$$

This can be used to prove the existence of the Eckmann-Wittwer limit [EW] when $\lambda \to 1$: see [E],[EE]. Second it can be shown (by the same method as in Section 7) that, on the real axis, ψ has continuous boundary values, analytic except at the branch points $(-\lambda)^{-n}$, $n \in \mathbf{N}$. Moreover the limits $\psi(\pm i\infty)$ exist, and $\psi(\Omega)$ is bounded.

4.4 Bounds on $r_2(\lambda,\ 1)$ and $r_3(\lambda,\ 1)$

This subsection states some facts needed in the next section. Recall that $r_2(\lambda,\ \nu)$ is given by $\log \widehat{\xi}/\nu \log(1/\lambda)$, if $\widehat{\xi}$ is the zero > 1 of the function $\xi \to \chi(\xi) - \xi$ defined in eq. (4.30). Hence $r_2(\lambda,\ \nu)$ is the strictly positive zero of the function $x \to Z(\lambda,\ x)$ obtained by substituting $\xi = \lambda^{-\nu x}$ in $[\chi(\xi) - \xi]/(\xi - 1)$, i.e.

$$Z(\lambda,\ x) = x\frac{\lambda^{\nu x-1}(1-\lambda^2)}{1-\lambda^{\nu x}} - (1-\lambda^{\nu x+2}). \tag{4.44}$$

It can easily be verified that (the first term of) this expression has a strictly positive derivative in λ when $\nu x \geq 1$. Since $\chi(\xi) - \xi$ is decreasing at $\widehat{\xi}$, we also have $\partial_x Z(\lambda,\ x) < 0$ at $x = r_2(\lambda,\ \nu)$, so that $r_2(\lambda,\ \nu)$ is increasing in λ.

For $\nu = 1$, it has already been noted that $r_2(\lambda,\ 1) > 1$. Actually it can be verified (straightforwardly but laboriously) that

$$r_2(\lambda,\ 1) > \frac{1+\lambda^2}{1-\lambda^2} \tag{4.45}$$

i.e. that the r.h.s. makes $Z(\lambda,\ x)$ positive (see [E]).

Similarly $r_3(\lambda,\ 1)$ is the strictly positive zero of the function

$$x \to R_3(\lambda,\ x) \equiv (1+\lambda)x + \lambda - \lambda^2(1-\lambda^x) - \lambda^{1-x}. \tag{4.47}$$

It can also be shown to be increasing in λ. For $y > 0$,

$$R_3(\lambda,\ 1+y) < (1+\lambda)(1+y) + \lambda - \exp(-y\log\lambda) < 2\lambda - y(-\log\lambda - 1 - \lambda).$$

If, e.g. $\lambda < e^{-2}$, the above expression is negative for $y \geq 4\lambda/[(1-e^{-2})\log(1/\lambda)]$. To summarize:

$r_2(\lambda,\ 1)$ and $r_3(\lambda,\ 1)$ are both strictly increasing functions of λ, with

$$\frac{1+\lambda^2}{1-\lambda^2} < r_2(\lambda,\ 1) < r_3(\lambda,\ 1), \quad r_3(\lambda,\ 1) \leq \frac{4\lambda}{(1-e^{-2})\log(1/\lambda)} \quad \text{for } \lambda \leq e^{-2}. \tag{4.48}$$

For $x \geq 0$ we define $b(x)$ to be 0 if $0 \leq x \leq 1$, and to coincide with the inverse function of $\lambda \to r_2(\lambda,\ 1)$ for $x > 1$. The function $x \to b(x)$ is continuous, increasing, tends to 1 when $x \to \infty$, and is bounded by

$$b(x) < \sqrt{\frac{x-1}{x+1}} \quad \text{for all } x > 1. \tag{4.49}$$

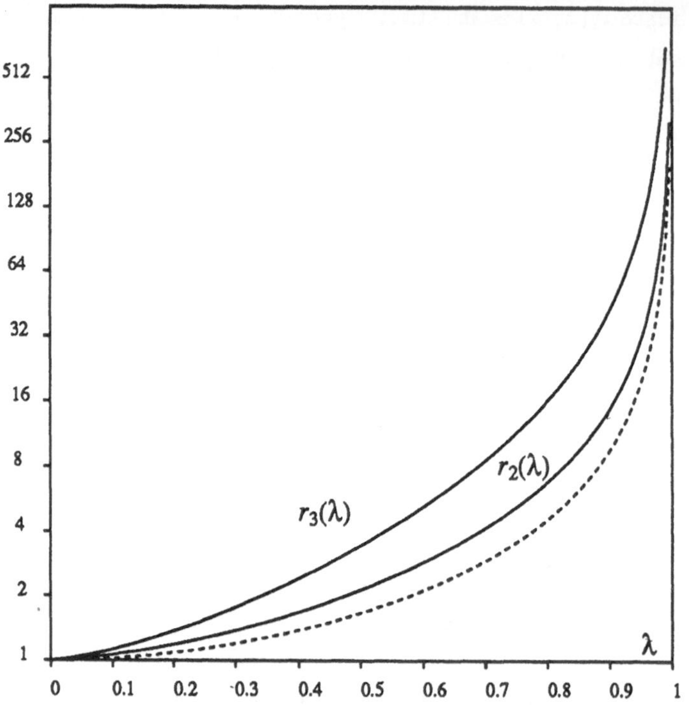

Figure 0. the functions $r_2(., 1)$ and $r_3(., 1)$;
dashed line: $\frac{1+\lambda^2}{1-\lambda^2}$ (logarithmic plot).

5. The FIXED r METHOD (INTERVAL CASE)

The main defect of our method is that, being based on the Schauder-Tikhonov theorem, it does not provide the (local) uniqueness of the solutions and has nothing to say about their continuous dependence upon λ. Thus, given a value of $r > 1$, we cannot assert that there is a corresponding solution. I continue to believe that it will be possible to cure this disease, but, in the meantime, the fixed r method provides the existence of solutions for $\nu = 1$ and all values of $r > 1$.

In this section, ν is fixed and equal to 1, and $r > 1$ is also fixed. A new map S_r is defined as follows.

1) Fix some $b \in (0, 1)$. Start with an element ψ_0 of the set $\mathbf{E}_0(b)$. Define a function V by the same formula (4.2) as in the fixed- λ method, the constants λ, τ, α being determined so that (4.3) holds. They must also satisfy

$$0 < \lambda < 1, \quad 1 < \alpha < \tau^{-1}. \tag{5.1}$$

2) Then ψ is the solution of

$$\psi(z) = V(\psi(-\lambda z)), \quad \psi(0) = 1, \quad \psi(1) = 0. \tag{5.2}$$

If all these constructions are possible, and if $\psi \in \mathbf{E}_0(b)$, then ψ_0 is in the domain of S_r, and $S_r\psi_0 = \psi$. We shall try to chose b so that (for the value of r considered) S_r is defined on the whole of $\mathbf{E}_0(b)$.

Let b be fixed in $(0,\ 1)$ and $\psi_0 \in \mathbf{E}_0(b)$. If a function V can be defined by (4.2) with $\nu = 1$ so as to satisfy (4.3), then $z_1 \equiv \alpha^{-1/r}$ must be such that

$$\lambda^r = \psi_0(z_1) \quad \text{and} \quad \frac{1}{\lambda} = -\frac{z_1 \psi_0'(z_1)}{r\ \psi_0(z_1)}, \tag{5.3}$$

so that z_1 must be a solution of

$$Y(z_1) = 1, \quad Y(z) \equiv -\frac{z\psi_0'(z)}{r\ \psi_0(z)^{1-1/r}}. \tag{5.4}$$

The function Y vanishes at 0 and tends to $+\infty$ as $z \uparrow 1$. Moreover, for $0 < z < 1$,

$$\begin{aligned}\frac{Y'(z)}{Y(z)} &= \frac{1}{z} + \frac{\psi_0''(z)}{\psi_0'(z)} - (1 - 1/r)\frac{\psi_0'(z)}{\psi_0(z)} \\ &\geq \frac{1}{z}\left(\frac{1 - bz}{1 + bz}\right) - (1 - 1/r)\frac{\psi_0'(z)}{\psi_0(z)} > 0.\end{aligned} \tag{5.5}$$

The last inequality uses (4.18). Thus there exists a unique $z_1 \in (0,\ 1)$ such that $Y(z_1) = 1$. We define now

$$\tau = \psi_0(z_1), \quad \lambda = \tau^{1/r} \in (0,\ 1), \quad \alpha = z_1^{-r} > 1, \tag{5.6}$$

and note that (5.3) is satisfied. The function V defined by the formula (4.2) is anti-Herglotz:

$$V \in -\mathbf{P}((0,\ \alpha b^{-2r})), \quad V(0) = 1/\tau, \quad V(\alpha) = 0. \tag{5.7}$$

Note that z_1 is a zero of the function q defined by eq.(4.20) *with the same expression for* A. This function has a representation of the type (4.22) where the support of σ is now $\mathbf{R} \setminus (-b^{-1},\ 1)$, and σ coincides with 1 on $(1,\ b^{-2})$. Despite these differences, the argument leading to

$$z_1 > 1 + \lambda \log \lambda > \lambda, \quad 1 < \alpha < \tau^{-1} = V(0), \tag{5.8}$$

still holds. We define $W = V \circ V$, $\widehat{V}$ and $\widehat{W}$ by (4.34) just as in the preceding section. It is clear that W is defined in a neighborhood of 1, and in $[1,\ \alpha]$ since $W(\alpha) = V(0) = \tau^{-1}$. By (5.8), there is a $\zeta_0 \in (0,\ 1)$ such that $V(\zeta_0) = \alpha$, hence the real interval of analyticity of W contains ζ_0, and $W(\zeta_0) = 0$. Thus $\widehat{W}$ assumes some values > 1. By arguments similar to those of subsection 4.3, Ψ and ψ can be constructed, with

$$\psi(0) = 1, \quad \psi(1) = 0, \quad \psi(-\lambda^{-1}) = V(0) = 1/\tau. \tag{5.9}$$

However we wish to have $\psi \in \mathbf{E}_0(b)$. For this it will suffice that the two following conditions hold:

(i) $\lambda \leq b$: this would imply that ψ is analytic on $(-b^{-1}, 0)$, and moreover

$$-\frac{\lambda}{b^2} \geq -\frac{1}{\lambda} \quad \text{hence} \quad \psi(-\lambda b^{-2}) \leq \psi(-\lambda^{-1}) = \tau^{-1}; \tag{5.10}$$

(ii) $\psi(-\lambda b^{-2}) < \alpha b^{-2r}$: this would imply that $\psi(b^{-2})$ can be defined as $V(\psi(-\lambda b^{-2}))$.

To deal with condition *(i)*, suppose, on the contrary, that $b < \lambda$. Then $\psi_0 \in \mathbf{E}_0(\lambda)$, and the quantities z_1 and r are precisely those which the fixed-λ method would assign to ψ_0 for the same value of λ. Thus (see Section 4)

$$r \geq r_2(\lambda) \quad \text{i.e.} \quad \lambda \leq b(r), \tag{5.11}$$

with $r \to b(r)$ as defined in subsection 4.4. It follows that condition *(i)* will be fulfilled if we suppose, as we shall from now on, that b is chosen equal to $b(r)$.

Condition *(ii)* can be rewritten as

$$\frac{\psi(-\lambda b^{-2})}{\psi(-\lambda)} < b^{-2r}, \tag{5.12}$$

since $0 = \psi(1) = V(\psi(-\lambda))$ implies $\psi(-\lambda) = \alpha$. Note that, in view of (5.10), condition *(ii)* will certainly be satisfied if $\tau^{-1} < \alpha b^{-2r}$, i.e. if $\lambda > z_1 b^2$. We can therefore restrict our attention to the case $\lambda < b^2$.

The function $\log\psi$ has a representation

$$\log\psi(z) = -\int_{X_\lambda} \widehat{\sigma}(t)dt \left[\frac{1}{t-z} - \frac{1}{t}\right] \quad \forall z \in \mathbf{C}((-\lambda^{-1}, 1)), \tag{5.13}$$

where $0 \leq \widehat{\sigma} \leq 1$, and $\widehat{\sigma}$ has support in $X_\lambda = \mathbf{R} \setminus (-\lambda^{-1}, 1)$. Therefore

$$\log\left(\frac{\psi(-\lambda b^{-2})}{\psi(-\lambda)}\right) = \int_{X_\lambda} \widehat{\sigma}(t)\left[\frac{1}{t+\lambda} - \frac{1}{t+\lambda b^{-2}}\right]dt \;\leq\; \int_{X_\lambda}\left[\frac{1}{t+\lambda} - \frac{1}{t+\lambda b^{-2}}\right]dt,$$

$$\frac{\psi(-\lambda b^{-2})}{\psi(-\lambda)} \leq \frac{(1-\lambda)(1+\lambda b^{-2})}{1-\lambda^2 b^{-2}}. \tag{5.14}$$

The r.h.s. is increasing in λ so that (taking into account $\lambda < b^2$) it is bounded above by 2. Thus condition *(ii)* will be satisfied if we show that

$$\forall\, r > 1, \;\; b(r)^{-2r} > 2 \;\text{ or equivalently}: \;\forall\, t \in (0,\, 1), \;\; t^{-r_2(t)} > \sqrt{2}. \tag{5.15}$$

To prove this we recall that $\xi = t^{-r_2(t)}$ is the unique solution > 1 of

$$B\log\xi + 1 + \lambda^2 - \frac{\lambda^2}{\xi} - \xi = 0, \qquad B = \frac{1-t^2}{t\log(1/t)} \geq 2. \tag{5.16}$$

(see subsection 4.2). The l.h.s is positive if we substitute $\xi = 3$ so that its root is $> 3 > \sqrt{2}$.

We conclude that

Lemma 3

For any $r > 1$, and $b = b(r)$, the operator S_r is defined on $\mathbf{E}_0(b)$ and maps it into itself. S_r thus has at least one fixed point in $\mathbf{E}_0(b)$.

Let $\psi_0 = \psi$ be such a fixed point, and z_1 and λ be the corresponding quantities. Denote

$$U(z) = z_1^r \psi(z), \quad u(z) = U(z)^{1/r}, \quad z \in \mathbf{C}((-\lambda^{-1},\ 1)) \tag{5.17}$$

Then u satisfies all the conditions of section 3. In particular U and ψ extend to functions holomorphic in $\Omega(\lambda)$, and define a solution of the problem set in Section 3.

PART II: PROPERTIES OF SOLUTIONS

6. UNIVALENCE

We now suppose given a certain solution of the problem stated in Section 3. Recall that such a solution consists of a pair of functions u and U, both anti-Herglotz,

$$U \in -\mathbf{P}((-\lambda^{-1},\ \lambda^{-2})), \quad u \in -\mathbf{P}((-\lambda^{-1},\ 1)), \tag{6.1}$$

$$U(1) = u(1) = 0, \quad u(0) = \frac{x_0}{\lambda^{\nu-1}} > 0,$$

$$u(z) = U(z)^{1/r} \text{ for all } z \in \mathbf{C}((-\lambda^{-1},\ 1)), \tag{6.2}$$

where ν, λ, r, x_0 are constants such that

$$1 \leq \nu \leq 2, \quad 0 < \lambda < 1, \quad r > 1, \quad \lambda^\nu < x_0 < 1, \tag{6.3}$$

and u, U satisfy, in their respective domains, the functional equations

$$u(z) = \frac{1}{\lambda^\nu} u(\varphi_0(z)), \quad \varphi_0(z) \equiv \lambda^{\nu-1} u(-\lambda z), \tag{6.4}$$

$$U(z) = \frac{1}{\tau^\nu} U(\varphi_0(z)), \quad \tau = \lambda^r. \tag{6.5}$$

As noted in Section 3, φ_0 is a Herglotz function holomorphic in $\Omega \equiv \mathbf{C}((-\lambda^{-1},\ \lambda^{-2}))$, with

$$\varphi_0(-\lambda^{-1}) = 0, \quad \varphi_0(1) = 1, \quad \varphi_0(\lambda^{-2}) = \lambda^{\nu-1} u(-\lambda^{-1}) = \frac{x_0}{\lambda^\nu} < \lambda^{-2}, \tag{6.6}$$

and that $\varphi_0(z) < z$ for all $z \in (1,\ \lambda^{-2})$. We also denote

$$\varphi_1(z) = -\lambda z. \tag{6.7}$$

Both φ_0 and φ_1 map Ω into itself, and the real segment $(-\lambda^{-1},\ \lambda^{-1})$ into the smaller segment $(-\kappa\lambda^{-1},\ \kappa\lambda^{-1})$, with $\kappa = \max\{\lambda\varphi_0(\lambda^{-1}),\ \lambda\} < 1$. This will easily lead to the univalence of U and u.

Let a, b, θ be real with $a < b$, and $\theta \in (0,\ \pi]$. We denote

$$D(a,b,\theta) = \{z \in \mathbf{C}\ :\ 0 < \operatorname{Arg} \frac{z-a}{b-z} < \theta\},$$

$$\widehat{D}(a,b,\theta) = \{z \in \mathbf{C}\ :\ -\theta < \operatorname{Arg} \frac{z-a}{b-z} < \theta\}. \tag{6.8}$$

$D(a,b,\theta)$ is the domain between the real segment $(a,\ b)$ and the circular arc in $\mathbf{C}_+$ with ends a and b whose tangent at a has argument θ. $\widehat{D}(a,b,\theta)$ is the domain between the same arc and its symmetric with respect to the real axis. We start by noting

Lemma 4

Let a, b, a', b' be real with $a < b$, $a' < b'$.

(i) Let φ be a holomorphic map of $\mathbf{C}((a,\ b))$ into $\mathbf{C}((a',\ b'))$ which sends the real segment $(a,\ b)$ into the real segment $(a',\ b')$. Then, for any $\theta \in (0,\ \pi]$, φ maps $\widehat{D}(a,\ b,\ \theta)$ into $\widehat{D}(a',\ b',\ \theta)$.

(ii) For $\theta \in (0,\ \pi)$,

$$\widehat{D}(-1,\ 1,\ \theta) = \left\{z \in \mathbf{C}\ :\ \operatorname{dist}(z,\ (-1,\ 1)) < \log \frac{1+\operatorname{tg}(\theta/4)}{1-\operatorname{tg}(\theta/4)}\right\} \tag{6.9}$$

where dist *denotes the Poincaré distance in* $\mathbf{C}((-1,\ 1))$.

It suffices to prove (i) in the case $a = a' = -1$, $b = b' = 1$. In this form, the statement (i) follows from (ii), since a holomorphic map of a disk into itself never increases Poincaré distances.

To prove (ii), we use the conformal map

$$h(z) = \log \frac{1+z}{1-z} \tag{6.10}$$

of $\mathbf{C}((-1,\ 1))$ onto the strip $S_\pi = \{w \in \mathbf{C}\ :\ |\operatorname{Im} w| < \pi\}$, which maps $D(-1,\ 1,\ \theta)$ onto the strip $\{w \in \mathbf{C}\ :\ 0 < \operatorname{Im} w < \theta\}$ and $\widehat{D}(-1,\ 1,\ \theta)$ onto $S_\theta = \{w \in \mathbf{C}\ :\ |\operatorname{Im} w| < \theta\}$. The latter strip is clearly the set of all points w such that $\operatorname{dist}_{S_\pi}(w, \mathbf{R}) < \operatorname{dist}_{S_\pi}(i\theta,\ 0)$, where $\operatorname{dist}_{S_\pi}$ denotes the Poincaré distance in S_π. This is evaluated by mapping S_π onto the unit disk by $w \to \tanh\ (w/4)$ and yields (ii) (with the normalization of [A]).

Lemma 5

For $\theta \in (0,\ \pi)$, φ_0 and φ_1 map $\widehat{D}(-\lambda^{-1},\ \lambda^{-1},\ \theta)$ into $\widehat{D}(-\lambda^{-1},\ \lambda^{-1},\ \theta')$, with

$$\operatorname{tg} \frac{\theta'}{2} = \kappa \operatorname{tg} \frac{\theta}{2}. \tag{6.11}$$

Since φ_0 and φ_1 map $\mathbf{C}((-\lambda^{-1},\ \lambda^{-1}))$ into $\mathbf{C}((-\kappa\lambda^{-1},\ \kappa\lambda^{-1}))$, by Lemma 4, they map $\widehat{D}(-\lambda^{-1},\ \lambda^{-1},\ \theta)$ into $\widehat{D}(-\kappa\lambda^{-1},\ \kappa\lambda^{-1},\ \theta)$. Lemma 5 is now proved by verifying that, for any $\theta \in (0,\ \pi)$, and any $\kappa \in (0,\ 1)$,

$$D(-\kappa,\ \kappa,\ \theta) \subset D(-1,\ 1,\ \theta'), \quad \operatorname{tg} \frac{\theta'}{2} = \kappa \operatorname{tg} \frac{\theta}{2}. \tag{6.12}$$

(See fig.1).

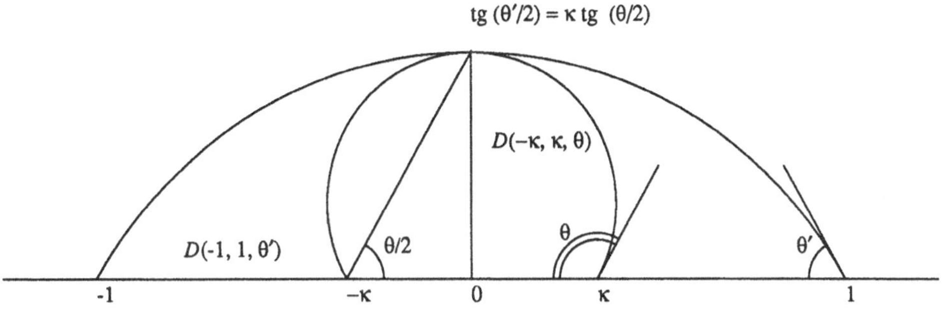

Figure 1

Lemma 6

The function U is injective in $\mathbf{C}((-\lambda^{-1},\ \lambda^{-2}))$.

Suppose that z_0 and w_0 are distinct points in $\mathbf{C}((-\lambda^{-1},\ \lambda^{-2}))$ such that $U(z_0) = U(w_0)$. There exists a $\theta_0 \in (0,\ \pi)$ such that z_0 and w_0 are both in $\widehat{D}(-\lambda^{-1},\ \lambda^{-1},\ \theta_0)$. For $n = 1, 2, \ldots$, we define inductively a pair $(z_n,\ w_n)$ of distinct points, both lying in $\widehat{D}(-\kappa\lambda^{-1},\ \kappa\lambda^{-1},\ \theta_0)$ such that

$$U(z_n) = U(w_n), \tag{6.13}$$

as follows. Assume $z_{n-1},\ w_{n-1}$ constructed. If $u(-\lambda z_{n-1}) = u(-\lambda w_{n-1})$, we set $z_n = -\lambda z_{n-1} = \varphi_1(z_{n-1})$, $w_n = -\lambda w_{n-1} = \varphi_1(w_{n-1})$. Otherwise we must have $\varphi_0(z_{n-1}) \neq \varphi_0(w_{n-1})$ but, by (6.5), $U(\varphi_0(z_{n-1})) = U(\varphi_0(w_{n-1}))$, and we set $z_n = \varphi_0(z_{n-1})$, $w_n = \varphi_0(w_{n-1})$.

By Lemma 5 applied to φ_0 or φ_1, z_n and w_n are both in $\widehat{D}(-\lambda^{-1},\ \lambda^{-1},\ \theta_n)$, with

$$\operatorname{tg} \frac{\theta_n}{2} = \kappa^n \operatorname{tg} \frac{\theta_0}{2}.$$

Moreover, for $n > 1$, z_n and w_n are in $\widehat{D}(-\kappa\lambda^{-1},\ \kappa\lambda^{-1},\ \theta_0)$. Hence z_n and w_n eventually enter a neighborhood of $[-\kappa\lambda^{-1},\ \kappa\lambda^{-1}]$ where U is injective. This contradiction proves Lemma 6.

7. BOUNDARY VALUES OF u

This section closely follows [EL]. For $n = 0, 1, 2, \ldots$, let I_n denote the real segment

$$I_n = (-\lambda)^{-n}[1,\ \lambda^{-2}].$$

Let z follow $I_0 - i0$. Then $u(z)$ follows the segment

$$\tau_0 = e^{i\pi/r}[0,\ |U(\lambda^{-2})|^{1/r}]. \tag{7.1}$$

If z crosses I_0 into $\mathbf{C}_+$, u gets continued by $v_0 \equiv e^{2\pi i/r}u$. If z follows $I_1 - i0$, then $-\lambda z$ follows $I_0 + i0$ and, by (6.4), $u(z)$ follows the analytic arc $\tau_1 = \lambda^{-\nu}u(\lambda^{\nu-1}\tau_0^*)$ which lies entirely in $\mathbf{C}_+$ except for its starting point:

$$u(-\lambda^{-1}) = \frac{x_0}{\lambda^{2\nu-1}}. \tag{7.2}$$

An easy induction shows that when z follows $I_n - i0$, $u(z)$ follows an analytic arc τ_n lying entirely in $\mathbf{C}_+$ for $n > 1$, and u can be continued across I_n by a function v_n, with

$$\tau_n = \lambda^{-\nu}u(\lambda^{\nu-1}\tau_{n-1}^*), \quad v_n(z) = \lambda^{-\nu}u(\lambda^{\nu-1}v_{n-1}(-\lambda z)). \tag{7.3}$$

Thus u is continuous at the boundary of its domain of definition $\mathbf{C}((-\lambda^{-1},\ 1))$, and even analytic except for simple branch points of the type $z^{1/r}$ at $(-\lambda)^{-n}$, $n = 0, 1, 2, \ldots$. Its boundary values are never real except at 1 and $-\lambda^{-1}$. It is also clear by induction that the continuous extension of u to the closure of $\mathbf{C}_+$ (resp. $\mathbf{C}_-$) is injective.

Let Φ be the function defined by

$$\Phi = \frac{1}{\lambda^\nu}u \circ \lambda^{\nu-1}, \quad \Phi^2(z) = \frac{1}{\lambda^\nu}u(\lambda^{-1}u(\lambda^{\nu-1}z)). \tag{7.4}$$

The functional equation (6.4) gives:

$$u(z) = \Phi(u(-\lambda z)) = \Phi^2(u(\lambda^2 z)). \tag{7.5}$$

Applying, as in [EL], the theory of iterations of Herglotz functions [V] to Φ^2 shows that Φ^2 has an attractive fixed point c in $\mathbf{C}_+$. Uniformly on any compact K in $\mathbf{C}_+$, $\Phi^{2n} \to c$ as $n \to \infty$. It easily follows that $u(z) \to c$ when $z \to \infty$ in any closed angle in $\mathbf{C}_-$, i.e. $c = u(-i\infty)$, $c^* = u(+i\infty)$, $\Phi(c) = c^*$. Since

$$\tau_{n+2} = \Phi^2(\tau_n), \tag{7.6}$$

the arcs τ_n converge geometrically to c as $n \to \infty$. Thus the domain $\omega = u(\mathbf{C}_-)$ is bounded, and its boundary is composed of the arcs τ_n, the point c, and the real interval $[0,\ x_0\lambda^{1-2\nu}]$. Recall that ω is contained in the sector $\{z\ :\ 0 < \operatorname{Arg} z < \pi/r\}$.

This implies that the convergence of $\Phi^{2n}|\mathbf{C}_+$ to c as $n \to \infty$ is uniform on compact subsets of $\bar{\mathbf{C}}_+ \setminus \{x_0\}$, and that $u(z) \to c$ when $z \to \infty$ on $\bar{\mathbf{C}}_-$.

8. POLYNOMIAL-LIKE BEHAVIOR FOR $\nu = 1$, r AN EVEN INTEGER

From now on we assume that $\nu = 1$ and r is an even integer, Let $\mathcal{I}(r)$ denote the set of rth roots of unity. Then the Feigenbaum function ϕ (the inverse of u) is holomorphic in

$$\Omega_0 = \text{interior}\left(\bigcup_{\epsilon \in \mathcal{I}(r)} \epsilon(\bar{\omega} \cup \bar{\omega}^*) \right). \tag{8.1}$$

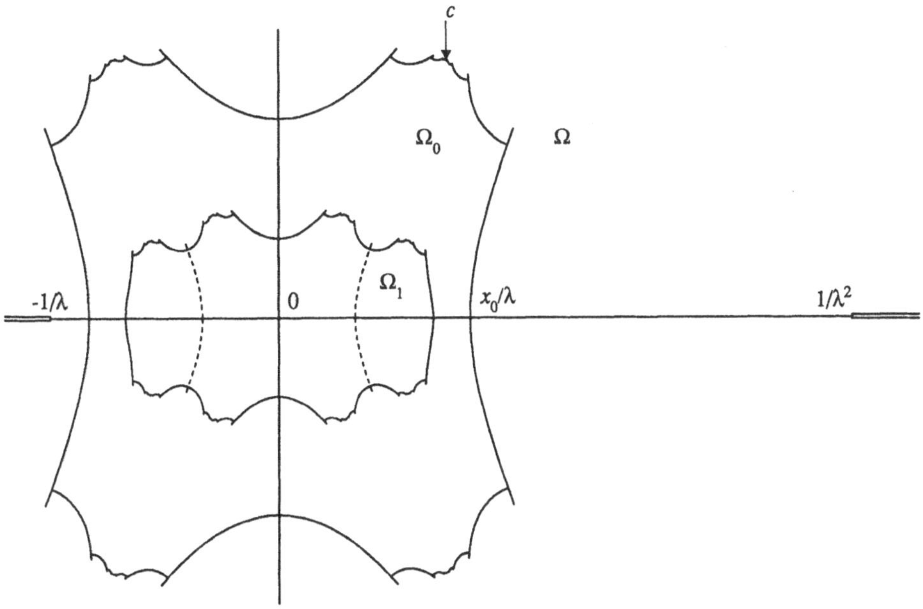

Figure 2
Dashed lines isolate $\lambda\Omega_0$ in Ω_1.

It maps this domain with degree r onto the cut plane $\Omega = \mathbf{C}((-\lambda^{-1},\ \lambda^{-2}))$. The real trace of Ω_0 is $(-x_0\lambda^{-1},\ x_0\lambda^{-1})$. Since the closure of this interval is contained in $(-\lambda^{-1},\ \lambda^{-2})$, the triple $(\Omega,\ \Omega_0,\ \phi)$ is a polynomial-like map of degree r in the sense of Douady and Hubbard [DH].

Remark 1

It follows that $(\Omega,\ (\phi|\ \Omega_0)^{-1}(\Omega_0),\ \phi\circ\phi)$ is a polynomial-like map of degree r^2. On the other hand, since $\phi\circ\phi = -\lambda\circ\phi\circ(-\lambda^{-1})$, the triple $(\mathbf{C}((-\lambda^{-1},\ 1)),\ -\lambda\Omega_0,\ \phi\circ\phi)$ is also polynomial-like with degree r. This can be illustrated by a description of the domain $\Omega_1 = (\phi|\ \Omega_0)^{-1}(\Omega_0)$.

We denote

$$\omega_1 = \omega \cup \omega^* = u(\mathbf{C}_- \cup \mathbf{C}_+). \tag{8.2}$$

The function $(\phi|\ e^{2i\pi n/r}\omega_1\ )^{-1}$ is $e^{2i\pi n/r}u$, with domain $\mathbf{C}_- \cup \mathbf{C}_+$. By (8.1),

$$\Omega_1 = \text{interior}\left(\text{closure} \bigcup_{\epsilon,\eta\in\mathcal{I}(r)} \epsilon u(\eta\omega_1)\right). \tag{8.3}$$

For $n=0$, the function $u\circ e^{2i\pi n/r}u$ coincides with $\lambda u\circ(-\lambda^{-1})$. If $\epsilon_j,\ \eta_j \in \mathcal{I}(r)$, $j = 1,2$, with $\epsilon_1 \neq \eta_1$ or $\epsilon_2 \neq \eta_2$, then $\epsilon_1 u(\epsilon_2\omega_1)$ and $\eta_1 u(\eta_2\omega_1)$ have empty intersection (due to the univalence of u and the fact that $\omega_1 \subset \{z\ :\ |\text{Arg}\ z| < \pi/r\}$).

More generally let

$$\Omega_n = (\phi|\Omega_0)^{-2^n}(\Omega). \tag{8.4}$$

Then, as $n\to\infty$, Ω_n decreases and tends to K_ϕ, the " filled-in Julia-set" [DH]. At the same time, ϕ^{2^n} restricted to $(-\lambda)^n\Omega_0 \subset \Omega_n$ is conjugated to $\phi|\Omega_0$, hence is of degree r. Figures 2 and 3 display, for $r=2$, the domains Ω, Ω_0, Ω_1 and Ω_2.

Remark 2

Keeping $\nu = 1$, suppose that r is not necessarily an integer. Let Q be the open disk with diameter $(-\lambda^{-1},\ \lambda^{-1})$. Let $\psi = U/U(0) = U/(x_0)^r$. Then, by the estimates (4.15),

$$0 < -\psi\left(\frac{1}{\lambda}\right) \leq \frac{1/\lambda - 1}{1-\lambda} = \frac{1}{\lambda}, \tag{8.5}$$

so that

$$0 < -U\left(\frac{1}{\lambda}\right) \leq (x_0)^r/\lambda < (x_0/\lambda)^r, \quad U\left(-\frac{1}{\lambda}\right) = (x_0/\lambda)^r. \tag{8.6}$$

By Lemma 4, $U(Q)$ is contained in the disk centered at 0 with radius $(x_0/\lambda)^r$. Hence

$$\omega_2 = u(Q\cap\mathbf{C}_-) \subset \{z\in\mathbf{C}\ :\ |z| < x_0/\lambda,\ \ 0 < \text{Arg}\ z < \pi/r\},$$

$$u(Q\setminus(1+\mathbf{R}_+)) = \text{interior}\ (\bar{\omega}_2 \cup \bar{\omega}_2^*). \tag{8.7}$$

The boundary of ω_2 is a triangle composed of two straight sides issuing from 0, and one curved analytic arc. If r is an integer, this implies that ϕ has a holomorphic extension in

$$Q' = \text{interior}\left(\bigcup_{n=0}^{r-1} e^{2in\pi/r}(\bar{\omega}_2 \cup \bar{\omega}_2^*)\right) \subset\subset Q. \tag{8.8}$$

and $\phi(Q') = Q$, i.e. $(Q,\ Q',\ \phi)$ is polynomial-like for all integer r.

9. ANALYTICITY OF ϕ FOR $\nu = 1$, r AN EVEN INTEGER

This section adapts in summarized form a part of [EL]. Recall that

$$\phi(x) = f(x^r), \tag{9.1}$$

$$\phi(x) = -\frac{1}{\lambda}\phi(\phi(\lambda x)), \tag{9.2}$$

$$f(t) = -\frac{1}{\lambda}\phi(f(\lambda^2 t)). \tag{9.3}$$

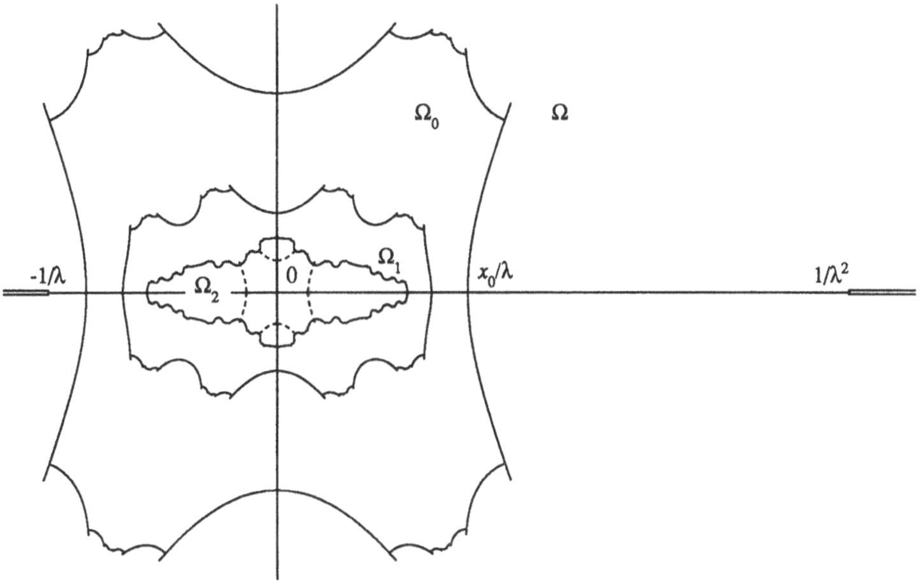

Figure 3

Dashed lines isolate $\lambda^2\Omega_0$ in Ω_2.

As already mentioned, evenness and eq.(9.2) provide an extension of ϕ to the whole of $\mathbf{R}$, and in the present case (r an even integer) this extension is analytic. Since f is analytic and real in a small neighborhood of 0, eq.(9.3) extends it to a function real and analytic on $\mathbf{R}$. Hence ϕ is analytic and real on $\exp(in\pi/r)\mathbf{R}$ for all integer n. The same argument as for $r = 2$ show that the critical points of ϕ on $\mathbf{R}$ are all of the type x^r.

Let $\epsilon = (\epsilon_1, \epsilon_2, ...\epsilon_{2^n})$, $\epsilon_j \in \mathcal{I}(r)$. We denote $|\epsilon| = n$ and

$$u_\epsilon(z) = \frac{1}{\lambda^n}\epsilon_1 u(\epsilon_2 u(...\epsilon_{2^n} u((-\lambda)^n z)...)),$$

$$u_\epsilon(z^*) = u_\epsilon(z)^* \text{ for all } z \in \mathbf{C}_+, \tag{9.4}$$

i.e., on $\mathbf{C}_+$,

$$u_\epsilon = \frac{1}{\lambda^n}\epsilon_1 u \circ \epsilon_2 u \circ ...\epsilon_{2^n} u \circ (-\lambda)^n. \tag{9.5}$$

This defines each u_ϵ as a Herglotz or anti-Herglotz function. Substituting for each occurrence of u in (9.5) the rhs of (6.4), i.e. $\lambda^{-1} u \circ u \circ (-\lambda)$, shows that there is an ϵ' with $|\epsilon'| = |\epsilon| + 1$ such that $u_{\epsilon'} = u_\epsilon$. However the univalence of u and the fact that $0 < \operatorname{Arg} u(\mathbf{C}_-) < \pi/r$ immediately give:

Lemma 7

If $u_\epsilon(z) = u_\eta(w)$ for some z and w in $\mathbf{C}_+ \cup \mathbf{C}_-$ and some ϵ and η, then $u_\epsilon = u_\eta$ and $z = w$. If furthermore $|\epsilon| = |\eta|$, then $\epsilon = \eta$.

Moreover:

Lemma 8

(i) For any pair ϵ, ϵ', there is an η such that

$$u_\eta = \frac{1}{\lambda} u_\epsilon \circ u_{\epsilon'} \circ (-\lambda). \tag{9.6}$$

(ii) For any η with $|\eta| \geq 1$, there is a pair ϵ, ϵ', with $|\epsilon| = |\epsilon'| = |\eta| - 1$, such that (9.6) holds. The pair u_ϵ, $u_{\epsilon'}$ is unique.

Finally it can be shown that:

Lemma 9

(i) For any ϵ, u_ϵ has continuous boundary values on either side of $\mathbf{R}$. It has, on $\mathbf{R}$, a locally finite set of branch points, separated by open intervals of regularity $I_{\epsilon,j}$, $j \in \mathbf{Z}$. The set $u_\epsilon(I_{\epsilon,j} \pm i0)$ is either contained in $\mathbf{R}$, or in $\mathbf{C}_+$, or in $\mathbf{C}_-$. The derivative $u'_\epsilon(z \pm i0)$ never vanishes if z is in one of the regularity intervals. $|u'_\epsilon(z)|$ tends to infinity if z tends to one of the branch points.

(ii) The continuation of u_ϵ across any regularity interval $I_{\epsilon,j}$ is again some u_η.

(iii) Any u_η can be obtained in this way: there exists a path γ_η (resp. γ^*_η), composed of segments along $\mathbf{R} + i0$ or $\mathbf{R} - i0$ connected by suitable crossings (parallel to $i\mathbf{R}$), along which the continuation can be made from u to u_η, and such that the image of γ_η (resp. γ^*_η) during the continuation lies in $\mathbf{C}_+$ (resp. $\mathbf{C}_-$).

Parts (i) and (ii) are easily proved by induction on $|\epsilon|$. Part (iii) is as in [EL]. It can also be proved that all the branch points of any u_ϵ are of the type $z^{1/r}$.

For $k = 1, 2, 3, ...$, let J_k be the kth (open) intercritical interval for ϕ on $\mathbf{R}_+$, and $J_{-k} = -J_k$. Let $u_k = -u_{-k}$ be the inverse function of $\phi|J_k$. For example

$J_1 = (0,\ x_0\lambda^{-1})$ and $u_1 = u$. It is easy to see by induction that for any $k > 1$ there are two integers j and t, with modulus $< k$, such that

$$u_k = \frac{1}{\lambda} u_j \circ u_t \circ (-\lambda),$$

so that the u_k are special cases of the u_ϵ.

The domain of analyticity D of ϕ is given by:

$$D = \bigcup_\epsilon u_\epsilon(\bar{\mathbf{C}}_+) \cup u_\epsilon(\bar{\mathbf{C}}_-), \tag{9.7}$$

where $\epsilon \in \mathcal{I}(r)^{2^n}$ for all $n \in \mathbf{N}$. Using Lemma 8 (ii), it is easy to see that:

Lemma 10

(i) $\lambda D \subset D$ *and* $\phi(\lambda D) \subset D$.

(ii) If $z \in \partial D$ *and* $\lambda z \in D$, *then* $\phi(\lambda z) \in \partial D$.

(iii) $\partial D \subset \lambda \partial D$.

Proof of (iii): Any point $w \in \partial D$ is the limit of a sequence $w_n = u_{\epsilon_n}(x_n \pm i0)$ with x_n real. Since each x_n is in some $\bar{J}_{k_n}$, it can be written as $u_{k_n}(y_n)$ for some real y_n, hence $w_n = \lambda u_{\eta_n}(-\lambda^{-1} y_n \pm i0)$.

We denote

$$c_\epsilon^\pm = u_\epsilon(\pm i\infty), \quad \mathcal{F} = \{c_\epsilon^\pm :\ \epsilon \in \mathcal{I}(r)^{2^n},\ n \in \mathbf{N}\}. \tag{9.8}$$

The set $\mathcal{F}$ is contained in ∂D and:

Lemma 11

(i) $\lambda \mathcal{F} \subset D$ *and* $\phi(\lambda \mathcal{F}) \subset \mathcal{F}$.

(ii) Any point in $\mathcal{F}$ *is an accumulation point of other points in* $\mathcal{F}$.

(iii) There is an infinite sequence in $\mathcal{F}$ *which tends to infinity.*

(iv) $\mathcal{F} \subset \lambda \bar{\mathcal{F}}$.

In fact for every point w of $\mathcal{F}$, there is an N such that $(\phi \circ \lambda)^N(w) = c$ (an immediate consequence of Lemma 8 (ii)).

Remark

We can now give a more precise description of the domains Ω_n mentioned in Section 8, eq.(8.4):

$$\Omega_n = \text{interior} \bigcup_{|\epsilon| \leq n} \lambda^n u_\epsilon(\bar{\mathbf{C}}_+ \cup \bar{\mathbf{C}}_-), \tag{9.9}$$

$$\lambda \Omega_{n-1} \subset \Omega_n \subset \lambda^n D, \quad \text{hence } \lambda K_\phi \subset K_\phi \subset \bigcap_{n \in \mathbf{N}} \lambda^n D. \tag{9.10}$$

Figure 4 shows the subdivision of Ω_2 into the patches $\lambda^2 u_\epsilon(\mathbf{C}_\pm)$, $|\epsilon| \leq 2$ when $r = 2$.

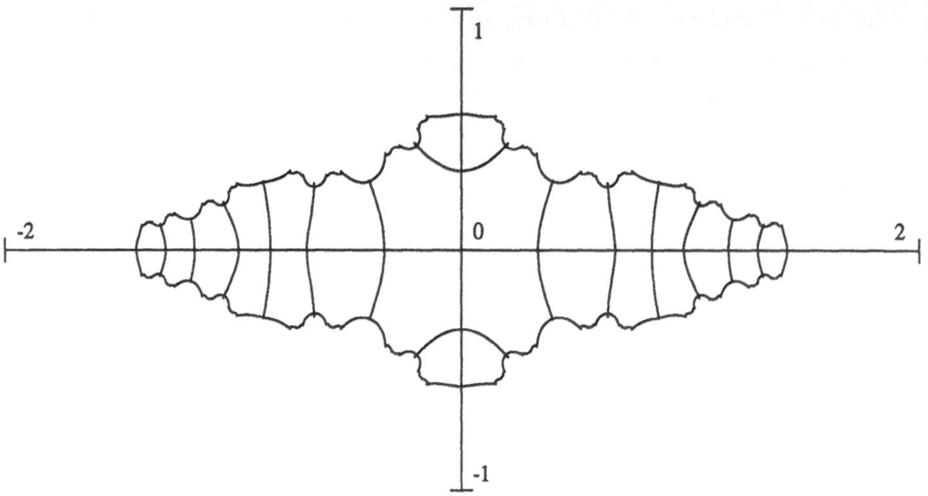

Figure 4

REFERENCES

[A] L.V. Ahlfors: *Conformal Invariants.* McGraw-Hill, New York, 1973.

[C] M. Cosnard : Etude des solutions de l'équation fonctionnelle de Feigenbaum. *Bifurcations, théorie ergodique et applications;* Astérisque, **98-99**, 143-62 (1982).

[CE] P. Collet and J.-P. Eckmann: *Iterated maps of the interval as dynamical systems.* Boston, Birkhaüser 1980.

[CEL] P. Collet, J.-P. Eckmann, and O.E.Lanford III: Universal properties of maps on the interval. Commun. Math. Phys. **76**, 211-54 (1980).

[CER] M. Campanino, H. Epstein, and D. Ruelle: On Feigenbaum's functional equation. Topology **21**, 125-9 (1982). On the existence of Feigenbaum's fixed point: Commun. Math. Phys. **79**, 261-302 (1981).

[CT] P. Coullet and C. Tresser: Itération d'endomorphismes et groupe de renormalisation. J. de Physique Colloque C **539**, C5-25 (1978). CRAS Paris **287 A**, (1978).

[D] W.F. Donoghue, Jr.: *Monotone matrix functions and analytic continuation.* Berlin, Springer Verlag 1974.

[DH] A. Douady and J.H. Hubbard: On the dynamics of polynomial-like mappings. Ann. Scient. Ec. Norm. Sup. 4ème série, **18**, 287-343, 1985.

[E] H. Epstein: New proofs of the existence of the Feigenbaum functions. Commun. Math. Phys., **106**, 395-426 (1986).

[EE] J.-P. Eckmann and H. Epstein: On the existence of fixed points of the composition operator for circle maps. Commun. Math. Phys., **107**, 213-231 (1986).

[EE2] J.-P. Eckmann and H. Epstein: Fixed points of composition operators. VIIIth International Congress on Mathematical Physics (Marseille, 1986), M. Mebkhout and R. Seneor eds. Singapore, World Scientific, 1987.

[EL] H. Epstein and J. Lascoux: Analyticity properties of the Feigenbaum function. Commun. Math. Phys. **81**, 437-53 (1981).

[EW] J.-P. Eckmann and P. Wittwer: *Computer methods and Borel summability applied to Feigenbaum's equation.* Lecture Notes in Physics 227. Berlin, Springer Verlag 1985.

[EW2] J.-P. Eckmann and P. Wittwer: A complete proof of the Feigenbaum conjectures. To appear.

[Fa] C. Falcolini: Some solutions of Feigenbaum's functional equation. Boll. Unione Matematica Italiana (7) **1-A**, 1987, to appear.

[F] M.J. Feigenbaum: Quantitative universality for a class of non-linear transformations. J. Stat. Phys. **19**, 25-52 (1978). Universal metric properties of non-linear transformations. J. Stat. Phys. **21**, 669-706 (1979).

[FKS] M.J. Feigenbaum, L.P. Kadanoff, and S.J. Shenker: Quasi-periodicity in dissipative systems: a renormalization group analysis. Physica **5D**, 370-386 (1982).

[G] J. Groeneveld: On constructing complete solution classes of the Cvitanović-Feigenbaum equation. Physica **138A** 137-166 (1986).

[JR] L. Jonker and D. Rand: Universal properties of maps of the circle with ϵ-singularities. Commun. Math. Phys. **90**, 273-292 (1983).

[L1] O.E. Lanford III: Remarks on the accumulation of period-doubling bifurcations. Mathematical problems in Theoretical Physics, Lecture Notes in Physics vol.116, pp. 340-342. Springer Verlag. Berlin, 1980. A computer-assisted proof of the Feigenbaum conjectures. Bull.Amer.Math.Soc., New Series, **6**, 127 (1984).

[L2] O.E. Lanford III: A shorter proof of the existence of the Feigenbaum fixed point. Commun. Math. Phys. **96**, 521-38 (1984).

[L3] O.E. Lanford III: Functional equations for circle homeomorphisms with golden ratio rotation number. Jour. Stat. Phys. **34**, 57-73 (1984).

[L4] O.E. Lanford III: Renormalization group methods for circle mappings. Proceedings of the Conference on Statistical Mechanics and Field Theory: Mathematical aspects, Groningen 1985 (Springer Lecture Notes in Physics, to appear).

[LL] O.E. Lanford III and R. de la Llave: in preparation.

[M] B. Mestel: Ph. D. Dissertation, Department of Mathematics, Warwick University (1985).

[N] M. Nauenberg: On fixed points for circle maps. Phys. Letters AB **92** 319-320 (1982).

[ORSS] S. Ostlund, D. Rand, J. Sethna, and E. Siggia: Universal properties of the transition from quasi-periodicity to chaos in dissipative systems. Physica **8D**, 303-342 (1983).

[S] D. Sullivan: Quasi-conformal conjugacy classes and the stable manifold of the Feigenbaum operator. Preprint, 1986. Quasiconformal homeomorphisms in dynamics, topology, and geometry. Preprint, 1986.

[V] G. Valiron: *Fonctions Analytiques.* Paris, Presses Universitaires de France 1954.

[VSK] E.B. Vul, Ya.G. Sinai, and K.M. Khanin: Feigenbaum universality and the thermodynamical formalism. Uspekhi Mat. Nauk **39**, 3-37 (1984).

DIFFERENTIABLE STRUCTURES ON FRACTAL LIKE SETS, DETERMINED BY INTRINSIC SCALING FUNCTIONS ON DUAL CANTOR SETS

Dennis Sullivan

City University of New York
The Graduate School and University Center
Graduate Center: 33 West 42nd Street
New York, New York 10036-8099

There is an easy notion of differentiable structure on a topological space. In the case of an embedded Cantor set in the line the differentiable structure records the fine scale geometrical structure. We will discuss two examples from the theory of one dimensional smooth dynamical systems, namely Cantor sets dynamically defined by i) folding maps on the boundary of chaos, and by ii) smooth expanding maps.

In example i) there is a remarkable discovery due to M. Feigenbaum[1] and independently P. Coullet and C. Tresser[2] that there is a universality or rigidity in the fine geometric structure of the Cantor set attractor for folding maps on the boundary of chaos. Feigenbaum expressed this discovery in terms of a universal scaling function for the Cantor set. Both papers offer an explanation motivated by the renormalization group idea of physics. These discoveries were empirical, and even today after much theoretical work they are not well understood. For example, the fine structure is codified by a scaling function defined on a logically distinct perfect set – the dual Cantor set. The main unsolved mystery is why the renormalizations converge. We prove here the rigidity conjecture assuming renormalization converges[1] §5,6. We also prove a converse. The proofs use the theory of the second example and a study of non linearity based on the bounded geometry of the Cantor set.

In the example ii) the Cantor set is the opposite of an attractor. It is the maximal invariant set of a $C(1,\alpha)$ expanding mapping of a 1-dimensional manifold. Now the fine structure of the Cantor set is not rigid but depends on many parameters. A complete set of invariants is again a scaling function but now the scaling function is an arbitrary Holder continuous function on a perfect set. Here the theoretical discussion is complete, straightforward and easy §1,2,3.

---

[1] In earlier unpublished work with Feigenbaum we proved the rigidity differently assuming a definite rate of convergence. Recently, David Rand has also derived a rigidity result.

§1 *Differentiable structures on fractal sets*

Let X be a topological space which is locally compact and can be locally embedded in R^n. If Q denotes some adjective like smooth, real analytic, complex analytic, etc. defining a pseudogroup of local isomorphisms of $\mathbb{R}^n$, we can define *a Q-structure on X of dimension n*. Say that a collection of local embeddings in R^n is *Q-coherent* if whenever i and j are two sucn embeddings defined near $x \in X$ there is a local Q isomorphism φ of R^n so that $\varphi \circ i = j$ near x. Then a Q-structure (of dimension n) on X is a maximal collection of Q-coherent local embeddings whose domains cover all of X.

§2 *Linear differentiable structures on Cantor sets*

For concreteness let C denote the set of one sided infinite sequences of O's and l's with the product topology. Let $C(1,\alpha)$ denote the pseudogroup of smooth local diffeomorphisms of R with α-Holder continuous derivatives, for all α $0 < \alpha \leq 1$. We denote this pseudogroup $C(1,\alpha)$ (instead of the usual symbol) because α is not fixed.

We will consider those $C(1,\alpha)$ structures on the Cantor set C where if C_w = {sequences with initial n-segment = w} then there is a finite coordinate cover so that in a chart containing C_w we have the picture

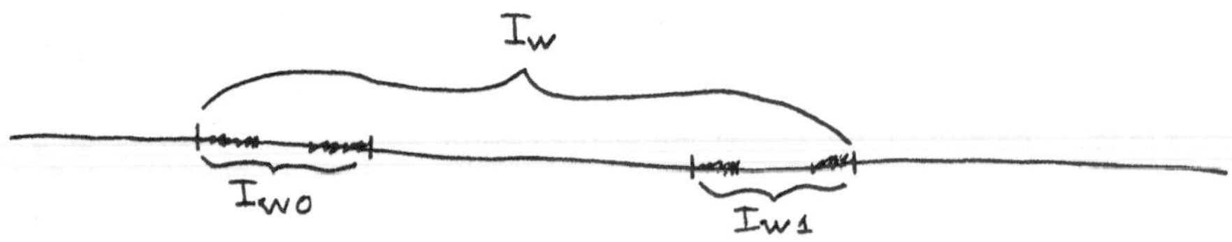

where $I_{w'}$ denotes the smallest interval containing $C_{w'}$. In other words we want I_{w0} and I_{w1} to be disjoint.

We define in terms of the coordinate cover the *ratio geometry of w* to be the 3 ratios length I_{w0}/length I_w, length I_{w1}/length I_w, length w-gap/length I_w.

Definition We say the differential structure has bounded geometry if in addition to the above disjointness property these ratios are bounded away from zero (uniformly in w).

Lemma 1 If length I_w tends exponentially fast to zero in length w, the coordinate ratio function $w \to$ ratio geometry is determined exponentially fast in length w by the differentiable structure.

Proof: Changes of coordinates being $C(1,\alpha)$ have exponentially small non-linearity on intervals of exponentially small size.

Now for some cover of C given by a finite system of charts deform the embeddings into R. Namely, imagine changing the lengths of the I_w and the gaps without changing the local ordering of points of C.

Theorem 2 If C has bounded geometry and if the ratio functions of the deformed charts are only changed by an exponentially small error in length w, then the new charts belong to the original differentiable structure.

Proof: We fill in the diagram locally to construct φ

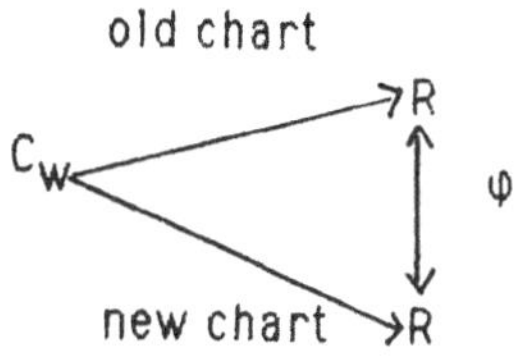

defined between the images of C. The difference quotient $\varphi(x) - \varphi(y)/x-y$ for $x, y \in C_w$ has the form $a_1+a_2+\ldots/b_1+b_2+\ldots$ where a_i and b_i are respective gap lengths in the two different charts and the sums are infinite. These are determined exponentially fast by the respective lengths of I_w and the ratio functions.

One sees the difference quotient for x, y in C_w, is Holder continuous with approximate value (new chart length I_w)/(old chart length I_w).

An elementary extension lemma shows φ has local $C(1,\alpha)$ extensions for some $0 < \alpha \leq 1$. QED

We say two ratio functions are exponentially equivalent if they differ by exponentially small quantities in length w.

Theorem 3 There is a one to one correspondece between $C(1,\alpha)$ differentiable structures on C of bounded geometry with given local order on the one hand and exponential equivalence classes of bounded from zero ratio functions, $\{w\} \to$ ratio geometry, on the other.

Proof: One way is Theorem 2. Conversely, if an abstract ratio function is bounded away from zero one builds the Cantor set C in R directly satisfying i) and ii).

§3 *Differentiable structures with smooth magnification and scaling functions*

Now we ask the question: when is the shift map $(\varepsilon_0\varepsilon_1\varepsilon_2\ldots) \mapsto (\varepsilon_1\varepsilon_2\ldots)$ of C locally a smooth diffeomorphism of class $C(1,\alpha)$ for some given differentiable structure on C.

There is a subtlety we will not deal with here. We will only characterize the situation when one of the two equivalent properties holds:

i) J is smooth and for some smooth metric $J \geq \lambda > 1$ or,

ii) J is smooth and the structure has bounded geometry.

The basic fact for everything is that the non-linearity of the composition

$$I_{w_1} \xrightarrow{J} I_{w_2} \xrightarrow{J} I_{w_3} \xrightarrow{J} \ldots I_{w_n} \quad \text{where } w_{k+1} = {*w_k}, \quad \text{will be controlled by}$$

$\Sigma(\text{length } I_w)^\alpha$ which is part of a geometric series. (See Apendix 1.) This implies the ratio geometry of w stops changing exponentially fast in length w if we add arbitrary symbols to w on the left.

Thus there is a limiting ratio geometry $\sigma(\ldots\varepsilon_2\varepsilon_1\varepsilon_0)$ attached to each left infinite word. These limit ratios are called the scaling function of the differentiable structure. This proves

Theorem 4 If the shift map on the Cantor set of right infinite words is smooth $(C(1,\alpha))$ in a structure of bounded geometry, the coordinate dependent ratio function $w \to$ ratio geometry defines a limiting scaling function which is coordinate cover independent and attached intrinsically to the differentiable structure. The scaling function assigns to each left infinite word a triple of positive ratios adding up to one.

Remark: The proof shows this scaling function is exp-continuous on $(\ldots\varepsilon_2\varepsilon_1\varepsilon_0)$, namely there is exponentially fast determination of the value of σ by knowledge of initial n-segment of $(\ldots\varepsilon_2\varepsilon_1\varepsilon_0)$. We call this property Holder continuity of the scaling function σ.

Theorem 5: Conversely, if there is a Holder continuous limiting scaling function for the differentiable structure (as in the remark) the shift is a smooth $C(1,\alpha)$ expanding map (in some smooth metric).

Remark: All Holder continuous scaling functions on $\{\ldots\varepsilon_2\varepsilon_1\varepsilon_0\}$ occur in this discussion.

The proof of theorem 5 involves exactly the same consideration as that of theorem 2. One sees the relevant difference quotient is Holder using the scaling function. A standard argument shows the shift is expanding in some smooth metric because the bounded geometry implies all the derivatives at period points are greater than unity.

Summary Differentiable structures on C where the shift is a $C(1,\alpha)$ expanding map are precisely those structures which have bounded geometry and whose asymptotic ratio geometry is described by a scaling function.

$$\{\ldots\varepsilon_2\varepsilon_1\varepsilon_0\} \xrightarrow{\sigma} \text{ratio geometry.}$$

All Holder continuous σ occur in this discussion. *There is a one to one correspondence between these $C(1,\alpha)$ structures and exponentially continuous scaling functions Theorems 3,4,5.*

Furthermore if the structure admits a $C(k,\alpha)$ refinement so that the shift is $C(k,\alpha)$, this structure is also determined uniquely by the same scaling function $k=0,1,\ldots$; $k=\infty$, or $k=\omega$. In fact, a shift commuting homeomorphism between structures which has a non zero derivative at one point, already is the restriction of a $C(k,\alpha)$ equivalence. (Appendix, part ii) of corollary)

An unsolved problem here is to determine what further properties of the scaling function σ allows higher smoothness. From earlier work we also know that if the structure is at least C^2 and for any smooth metric the second derivative of the shift is non zero at some point of C, the scaling function itself is determined by the thermodynamics of C which we know to be determined by the underlying Lipschitz structure. By thermodynamics we mean a certain mathematical discussion whose input is the sizes of the I_w, the set of numbers obtained by taking k-fold products of σ over k-fold shifts of k-periodic sequences.

§4 *The period doubling attractor* (Informal discussion)

Let us consider the simplest class of maps which allows a transition from very simple dynamics to complicated dynamics with exponential effects.

These are the folding maps of an interval $I \xrightarrow{f} I$ which have a turning point c in I so that f is increasing before c and decreasing after c.

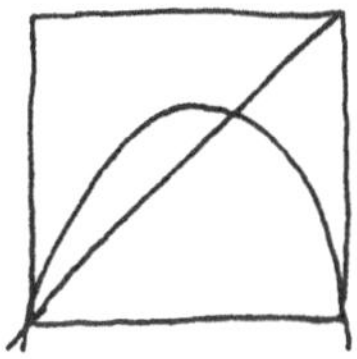

If there is a parameter t in the formula for f which raises the graph enough

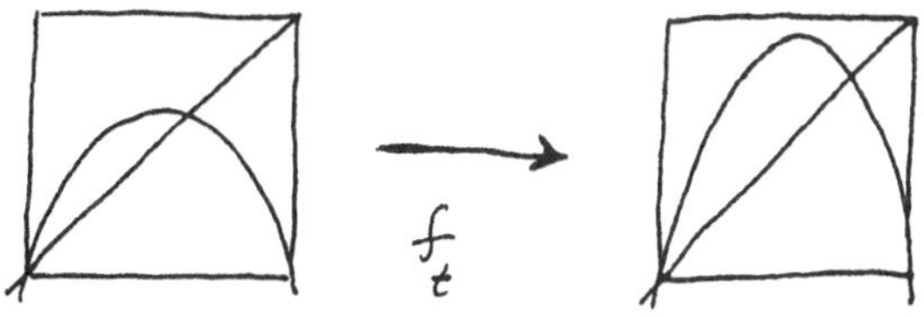

and the family has appropriate smoothness there will be a parameter value a where f_a has an attractive Cantor set (all but a countable sequence of points are asymptotic to C).

This Cantor set is the closure of the forward orbit of the turning point and is created by an infinite sequence of period doublings bifurcations of known form:

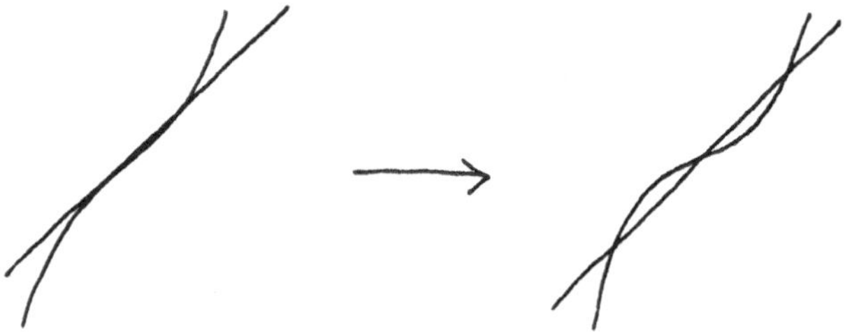

The forward orbit of the critical point denoted $\{1,2,3,...\}$ increases its complexity as t increases to a. At a sequence of values a_1, a_2, a_3, ... tending to a, the critical point has 2^n - periodicity:

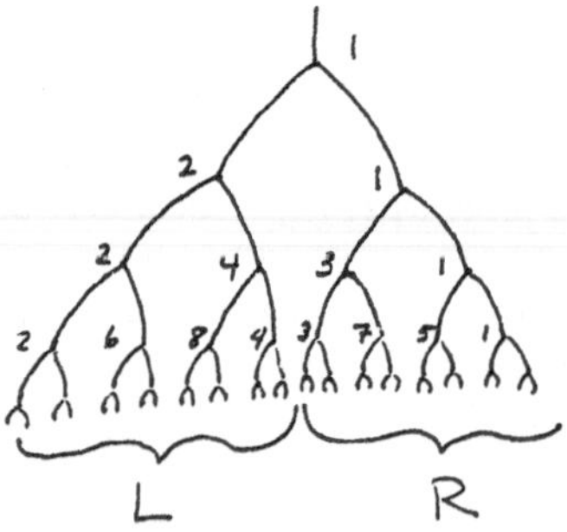

The numbers in L at stage n are just those at stage (n-1) doubled and reversed in order. The numbers in R at stage n are obtained from those in L at stage n by subtracting 1 and reversing the order.

In the limiting map f_a (at the "boundary of chaos") there is a 2-adic Cantor type structure on which f acts by "adding 1". The more precise statement is that the closure of the orbit of the critical point is a Cantor set on which f is equivalent to $(\varepsilon_0\varepsilon_1...) \mapsto g(\varepsilon_0\varepsilon_1...)$ where $g(\varepsilon_0\varepsilon_1...)$ is, change the first zero to a one and all previous ones to zeroes (adding "1" in the 2-adic integers.)

The identification can be chosen so that the critical point is 111 and the critical value is 000··· .

Feigenbaum made the remarkable discovery that for many examples deep ratios in the Cantor set have asymptotic limits independent of the family f_t. His calculations involved smooth functions with quadratic turning points. The only way to change the fine scaling in practice was to change the nature of the critical point or to introduce other critical points.

§5 *The Feigenbaum Rigidity Conjecture*

Let us formulate a precise rigidity statement corresponding to the Feigenbaum discovery. We assume f is a folding map $f:I \to I$ satisfying

i) f has a Lipschitz first derivative i.e. $f \in C(1,1)$.

ii) f has exactly one critical point c, namely $f'c = 0$, and $f'x \neq 0$ for $x \neq c$.

iii) f in some $C(1,1)$ coordinate system near c is just $(x-c)^2 + f(c)$.

iv) $fc, f^2c, f^3c, \ldots$ is deployed in the interval in terms of order as described in §4.

Rigidity Conjecture: The closure of the forward orbit of the critical point $\{fc, f^2c, \ldots\}$ is the 2-adic Cantor set C of one sided sequences of 0's and 1's with f acting on C by adding "1". The $C(1,\alpha)$ differentiable structure on C induced from its embedding in R is unique and described by a universal scaling function (§3).

A corollary of the mere existence of the scaling function for C is that the shift map J of the Cantor set is a $C(1,\alpha)$ expanding map (§3). In the 2-adic notation $x = \varepsilon_0 + \ldots + \varepsilon_i 2^i + \ldots$, $f(x) = x+1$ on C, and the shift map J is $x \to$ greatest integer in $1/2x = [1/2x]$. The calculation $1/2[x+1+1] = [1/2x] + 1$ shows $\boxed{Jff = fJ}$. This is the topological form of celebrated Cvitanovic-Feigenbaum functional equation[1].$\boxed{1}$

This equation can be iterated to obtain $J^n f^{2^n} = fJ^n$. Now J^n provides 2^n diffeomeorphisms between I_C, the interval subtending C, and the 2^n I_w's, the intervals subtending the Cw's, length $w = n$. The I_w are each invariant by f^{2^n} and the branches of $J^n{=}n$ provides smooth conjugacies between f^{2^n}/I_w and f/I_C.

For example, let $I_{w(n)}$, where $w(n)$ has with n 1's, converge down to the critical point $111111\cdots$ fixed by J. If $\alpha = J'(11111\cdots)$, then by calculus $\alpha^n J^{-n}_{w(n)}$ has a limit. Thus if $J^{-n}_n = J_{w(n)}$ and $f_n = f^{2^n}/i_{w(n)}$, then $\alpha^n J_n f J_n^{-1} \alpha^{-n} = \alpha^n f_n \alpha^{-n}$ has a limit. (These equations only hold on the Cantor set, a detail we will clarify in the next paper. The limit g satisfies the Cvitanovic- Feigenbaum functional equation $\boxed{\alpha g^2 \alpha^{-1} = g}$, where $g = \lim_{n\to\infty} \alpha^n f^{2^n} \alpha^{-n}$, $\alpha = J'(111\cdots)$. This proves the first part of Theorem 6. The rest is explained by the argument of the next section.

Theorem 6 If the period doubling Cantor set has a scaling function, than the nth renormalization of f, $\alpha^n f^2 \alpha^{-n}$ converges to g, a solution of the CF functional equation $\alpha g^2 \alpha^{-1} = g$. The limit g only depends on the scaling function.

§6 *The Rigidity Conjecture and Renormalization*

The folding maps f we are considering satisfy (by hypothesis) that $\{1,2,3,\ldots,2k\}$ denoting the first 2k forward iterates of the critical point $k=2^n$ are deployed

$$\underset{2 \quad k+2}{\overset{I_2}{|\!\!-\!\!-\!\!|}} \;\cdots\; \underset{2^n \quad 2^{n+1}}{\overset{I_{2^n}}{|\!\!-\!\!-\!\!|}} \;\cdots\; \underset{j \quad k+j}{\overset{I_j}{|\!\!-\!\!-\!\!|}} \;\cdots\; \underset{k+1 \quad 1}{\overset{I_1}{|\!\!-\!\!-\!\!|}} \,.$$

Consequently, $f^{2^n} = f^k$ preserves each of the indicated intervals and is a folding map of the same form. Each of these is called an nth renormalization of f.

As we observed in §5 one of these renormalizations after linear rescaling by powers of $J'(111\cdots) = \sigma(\cdots 111)$ converge assuming the part of the conjecture about the existence of the scaling function σ.

There is a converse.

Theorem 7 If the renormalization of f about the critical point converges (in the C^0 topology to a folding map with a quadratic critical point), then the Cantor set of f has a scaling function only dependent on this renormalization limit.

Corollary: If two folding maps have the same renormalization limits there is a $C(1,\alpha)$ diffeomorphism between their Cantor sets conjugating the dynamics on the Cantor sets.

Proof of Corollary: Theorem 7 and Summary of §3.

We make the proof of Theorem 7 assuming the Cantor sets have bounded geometry. We will expose our general result (valid for all maps described at the beginning of the section) on bounded geometry and general a priori estimates on the non-linearity of renormalization in the next paper.

Now consider the measure $|dx/x|$ restricted to all the intervals at the nth level except I_{2^n} containing the critical point. Here x=0 is the critical point. By induction on n we prove two properties:

i) the density of the measure is quasi-constant on each interval

ii) the total mass is controlled independent of n.

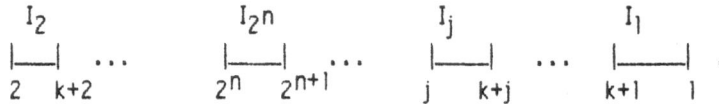

Passing from level n to n+1 we cut away a middle piece from each interval which by i) and the bounded geometry reduces the total mass by a definite factor (and keeps property i). We also add a new interval near the critical point. This only adds a new term of bounded mass and quasi constant density because the interval is nicely situated with respect to the critical point by the bounded geometry assumption on C. This completes the induction.

Now the non linearity of f (the measure $(f''/f')dx$) is controlled by a bounded density measure away from the critical point and the measure $|dx/x|$ near the critical point. Thus $|f''/f'||dx|$ is controlled by a measure satisfying i) and ii) above since $|dx/x|$ and bounded measures satisfy i) and ii).

Now consider the ratio geometry associated a long word w (length r say). Fix j. We can keep the j-segment on the right fixed and change the other digits to l's by applying f no more than 2^{r-j} times. The ratio geometry of w is that of something at level j inside some interval at depth r-j. We transform this over to the critical point interval by applying f no more than 2^{r-j} times.

These iterates all have bounded non linearity by ii) above (applied to level r-j). We care about the distortion of an object j levels deeper. This object is exponentially smaller in j, relatively (by bounded geometry). The non linearity measure (by i) at level (r-j)) we see is exponentially smaller in j. Thus the distortion of the appropriate iterate of f restricted to the smaller object is exponentially small in j. Thus the ratio geometry of w of length r is the same as that of the word beginning with r-j ones and ending with the same last j segment as w with an exponentially small in j error. This much follows from the just bounded geometry assumption on the Cantor set.

Now let r increase still keeping the final j segment of w fixed. The structure of the ratio geometry of 1111111 ... 111(final j segment of w) only depends on the 2^j forward orbit of the renormalized map which is converging in C^0 to a folding map f_∞. Thus we can define the scaling function at arguments ...1111111 (word of length j) in terms of the first 2^j iterates of f_∞. Since j was unrestricted in the argument we have defined the scaling function at all left infinite words which are all l's eventually. Moreover, by the first part of the argument these values are determined exponentially fast by the initial segments on the right. This proves theorem 7 assuming bounded geometry of C.

Appendix (Composition of Contractions in $C(k,\alpha)$).

Consider a composition g of diffemorphisms $I_1 \xrightarrow{f_1} I_2 \xrightarrow{f_2} \ldots \xrightarrow{f_n} I_{n+1}$ where

$|f_i'| \leq \lambda < 1$.

If x_1 and y_1 are 2 points in I_1 let $(x_{i+1}, y_{+i}) = (f_i(x_i),\ f_i(y_i))$.

Note $|x_i - y_i| \leq \lambda^{i-1} |x_1 - y_1|$. If φ_i is a function on I_i satisfying $|\varphi_i(x) - \varphi_i y| \leq C |x_i - y_i|^\alpha$ $0<\alpha\leq 1$, then $\varphi_1(x_1) + \varphi_2(x_2) + \ldots$ is also Holder continuous with constant $C(1/1-\lambda^\alpha)$ and same exponent α.

We apply this to $\varphi_i = D^k \log f_i'$ $k = 0,1,2,\ldots$ to see that if the f_i's satisfy $|D^k \log f_i'(x) - D^k \log f_i(y)| \leq C|x - y|^\alpha$ then so does $D^k \log g'(x)$ for the same α and the new C as above.

Corollary 1) A compositon of uniforms contractions which are individually bounded in $C(k,\alpha)$ (as diffeomorphisms) is also in $C(k,\alpha)$. (same α and new constant).

2) If a sequence of such compositons is renormalized by post composition with linear maps to obtain mappings between unit intervals the sequence is precompact in $C(k,\alpha)$.

References

1. M. Feigenbaum, *The universal metric properties of non linear transformations*, J. Statis. Phys. **19**(1978), 25-52; **21**(1979), 669-706.

2. P. Coullet, J. Tresser, *Iterations d'endomorphisms et groupe de renormalisation*, C.R. Acad.Sc., Paris **287**(1978), 577; Journal de Physique **C5**(1978) 25.

PAINLEVÉ PROPERTY AND INTEGRABILITY

Eric D. Siggia

Laboratory of Atomic & Solid State Physics
Clark Hall, Cornell University
Ithaca, NY 14853-2501

INTRODUCTION

A brief account of this work has appeared in Ref. (1) and an expanded version will follow.[2] These lectures are merely intended to illustrate, in the simplest possible terms, our approach to the Painlevé problem. Precise statements of theorems, in the few instances where we have them, as well as nontrivial examples are all reserved for Ref. (2).

We will consider only polynomial systems of equations, ($\cdot \equiv d/dt$),

$$\dot{x}_i = f_i(\{x_j\}) \qquad i,j = 1,\cdots n\ , \tag{1}$$

so that the right hand side of (1) does not introduce any singularities into the time flows. In fact, the differential equations define the continuation of $x_i(t)$ from real to complex times. All variables will henceforth be understood as complex. The Painlevé conjecture then asserts that if the singularities of $x_i(t)$ are no worse than poles for all t, (i.e., no branch points, logarithms, essential singularities etc.) then (1) is "integrable". We define integrable below. In other terms, a system whose solutions are globally meromorphic is said to possess the Painlevé property.

MOTIVATION

There are several reasons for pursuing this subject that go beyond a specific interest in integrable systems. However integrability is defined, it is clearly a global property of the flows in phase space. Integrals are smooth functions that are defined globally. Any analytic system of equations however, is locally integrable but the local patches of level sets do not in general fit together.

Singularity analysis, is also purely local. Polynomial equations only have singularities when some variable blows up. The leading singularity can be guessed by assuming a form e.g., bt^{-a} and calculating a and b. When the leading singularity is a pole, it is fairly simple to continue the series. Logarithms may then appear which means the system is not Painlevé. If it is on the other hand, then formal Laurent series are obtained. With additional labor their convergence may be established. This is all local analysis, and can be made very explicit.

The problem then is to understand why local analysis should have global, gemetric consequences.

A second general context into which the Painlevé conjecture falls, is that of singularities in nonlinear systems. Poles are a particularly mild singularity, and for Painlevé systems their occurence is merely an indication that one is working on too small a phase space. Most of these lectures will be devoted to constructing an augmented phase space on which the flows are analytic for all times. Other types of singularities may not admit such an interpretation, but it is interesting to understand the Painlevé case first which is not altogether trivial.

A third intriguing aspect of the Painlevé conjecture is that it might be made into a tool for computing integrals. Currently, to prove a system integrable means displaying the integrals or solving the initial value problem. This requires ingenuity and insight. Any constructive algorithm would be useful. Partial results along these lines comprise the last section.

Lastly, a thorough understanding of the relationship between sigularities and integrability could lead to an alternative to the K.A.M. notion of being "near" to integrable.

HISTORY

Kowalevska was the first to use singularity analysis to screen Hamiltonian systems for integrability. Her reasoning was that all other integrable systems known at that time had meromorphic solutions. It was relatively simple to find a previously unknown set of parameter values for which the equations for a top in a gravitational field had only poles. She then explicity solved the initial value problem in terms of hyperelliptic functions,[3] thereby demonstrating integrability.

Painlevé some forty years later considered all equations of the form

$$\ddot{x} = f(t, x, \dot{x})$$

which are analytic in t, and rational in x and $\dot{x}$, and for which the only moveable singularities, (those whose location depends on initial date), are poles. In addition to the known transcendental functions, six new ones were found, the so-called Painlevé transcendents.[4] Success in integrating several of these by inverse scattering methods lead Ablowitz et al. to conjecture a connection between the Painlevé property and integrability.[5] Many examples have been worked out since then.[6]

INTEGRABILITY

There is no single definition of integrability that seems appropriate in all circumstances. We mention two common ones (for autonomous systems).

Liouville integrability[7] follows for an n-degree of freedom Hamilton system if there are n-dependent integrals $K_i(q,p)$ in involution. If the intersection of their level sets is compact, then it is topologically a complex n-torus.

For the system (1), we define O.D.E. integrability to be the existence of $n-1$ independent analytic integrals. We note several pecularities of these definitions.

An integrable Hamiltonian system is generally not O.D.E. integrable. If we insist on functional independence of the (real) integrals at infinity formed from the real and imaginary parts of $\mathcal{H}$, then $\mathcal{H} = p^2 - q^5 - 1$ is not Liouville integrable (i.e., the level surface is of genus 2, not a torus, and any vector field on it must vanish somewhere). Thus although a solution exists for real and finite (q, p), this example is not Liouville integrable over the complexes. The converse pathology occurs for any repulsive and short ranged potential between pairs of particles in an n-particle system on the real line. These examples are always integrable since the momenta at $t = +\infty$ are constant and probably analytic in the initial data.

RICCATI EXAMPLE

The Riccati equation,[4] though ancient, and not Hamiltonian, is an exceedingly instructive illustration of how the singularities serve to augment the original phase space in such a way that on the augmented manifold the flow is analytic. The equation reads,

$$\dot{x} = a_0(t) + a_1(t)x + a_2(t)x^2 , \tag{2}$$

where $a_i(t)$ are entire in t. Clearly $x(t)$ is analytic whenever it is defined (i.e., finite). The only singularities are poles, near which $\dot{x} \sim a_2 x^2$ or $x \sim cst/(t - t_o)$. Consider the substitution, $\tilde{x} = 1/x$, under which (2) becomes,

$$-\dot{\tilde{x}} = a_0 \tilde{x}^2 + a_1 \tilde{x} + a_2 . \tag{3}$$

Clearly (3) is also an analytic equation, and $\tilde{x}(t)$ is analytic whenever it exists, particularly around $\tilde{x} = 0$, where $\tilde{x}$ vanishes as $\sim (t - t_o)$. Therefore we have proven that the Laurent series for x, which could be found by formally expanding (2), actually converges.

Define an augmented phase space as the set,

$$M = \{x \epsilon C\} \cup \{\tilde{x} = 0\} . \tag{4}$$

Cover $\tilde{x} = 0$ with the open set (patch) $\{\tilde{x} \epsilon C\}$ and identify this path with the other one, $\{x \epsilon C\}$, by the transition function $\tilde{x} = x^{-1}$ for x and $\tilde{x} \neq 0$. This makes M into a manifold.

In this example M is compact, and is just the Riemann sphere. Consider the finite time map φ_t. It takes M 1:1 and onto itself and is biholomorphic since equations (2) and (3) are analytic. Hence it must be fractional linear i.e.,

$$y(t) = \frac{\alpha y(t_o) + \beta}{\gamma y(t_o) + \delta} \tag{5}$$

where α, β, γ, and δ are entire functions of t and t_o. (Proof: Assume φ_t maps $y(t_o) = \Delta$ to infinity, write a Laurent series $y(t) = \sum a_n (y(t_o) - \Delta)^n$, then $a_{n>0} = 0$ since otherwise infinity would also map to infinity. Furthermore, $a_{n \le -2} = 0$ since the map is uniquely invertible Q.E.D.).

One could imagine a multivariable generalization of (5) for which there would be no invariant integrals even though one has an explicit formula for the dependence on inital data. Such a system should surely be called solvable though it is not integrable in the technical sense defined above.

The main body of these lectures will be devoted to illustrating how the steps which were required to construct M apply to an arbitrary Hamiltonian and Painlevé system.

BALANCES

The information that one can obtain through a singularity analysis of the equations of motion for a polynomial Hamiltonian will now be considered. Solutions of entire analytic differential equations will only fail to be locally analytic in time and initial data when they blow up. A balance is defined to be the leading term in a formal asymptotic expansion about such a singularity. We say "formal" since the convergence of the series is not obvious. All balances for a Painlevé system must in fact be the first term in a Laurent expansion. Testing for the Painlevé property usually means showing that all formal solutions around any singular point are Laurent.

A principal balance will be a formal Laurent (or equivalently pole) series with the maximum number of free constants allowed by the dimension of the phase space. Lower balances can be ordered by the number of free constants. For instance, for

$$\mathcal{H} = \frac{1}{2}(p_1^2 + p_2^2) + q_1^2 q_2 + 2q_2^2 \tag{6}$$

one finds a principal balance,

$$\begin{aligned} q_1 &= c_1 t^{-1} + \frac{5}{12}c_1 t + c_2 t^2 - \frac{1}{72}c_1^5 t^3 - \frac{5}{6}c_2 t^4 - \frac{1}{9}c_1 c_3 t^5 + \sigma(t^5) \\ q_2 &= -t^{-2} + \frac{1}{12}c_1^2 + \frac{1}{48}c_1^4 t^2 + \frac{1}{3}c_1 c_2 t^3 + c_3 t^4 + \sigma(t^5) \end{aligned} \tag{7}$$

In addition to c_1, c_2, and c_3, a fourth constant, t_o has been hidden in $t =$ time $-t_o$.

For any balance we define a resonance ρ to be the lowest power of t at which a new constant enters the series measured from the leading one i.e., $q_i = t^{-f_i}(cst + \cdots t^\rho)$. Thus in (7), the resonances occur for $\rho = 0, 3, 6$, and -1 for t_o. Equation (6) also has two lower balances

$$q_1 = \pm 6it^{-2}(1 + \cdots), \quad q_2 = -3t^{-2}(1 + \cdots)$$

with resonances $\rho = -1, 6, 8$.

The principal balance series can be used as a formal variable change from $\{q, p\}$ to t_o and the constants. Thus we find that the value of $\mathcal{H}$ may be expressed as

$$E = 14c_3 + \frac{35}{432}c_1^6$$

The 2-form $\omega^{(2)}$ can be rewritten and of course must be $t-$ independent;

$$\omega^{(2)} = \sum_i dp_i \wedge dq_i = dt_o \wedge dE + 3dc_2 \wedge dc_1 \tag{8}$$

Note that (8) establishes a conjugate pairing between the constants and leads to relations among the associated ρ.[2]

ELLIPTIC EXAMPLE

We now consider how the construction of a manifold on which the flows exist, analytically, for all times may be extended to a Hamiltonian system. The general argument plus nontrivial examples is given in Ref. 2. Consider as an example

$$2\mathcal{H}(q,p) = p^2 - 4q^3 - 2q$$

The principal, and only balance reads

$$q = (t - t_o)^{-2}(1 + \cdots), \quad p = -2(t - t_o)^{-2}(1 + \cdots)$$

We will solve the Hamilton-Jacobi equation perturbatively around infinity for a canonical variable change analogous to $\tilde{x} = x^{-1}$ in the Riccati example. Thus

$$\mathcal{H}(q, \partial S/\partial q) = E \tag{9}$$

has the approximate solution,

$$\tilde{S}(q,v) = 4/5q^{5/2} + q^{1/2} - vq^{-1/2} , \tag{10}$$

which satisfies (9) to $\mathcal{O} \mid q^{-1} \mid$. We have replaced E by v in (10) since we want to use v as a variable in the coordinate patch at infinity. Since $\tilde{S}$ is only approximate, v will not be constant and it would be misleading to call it E.

Define a variable change by

$$u = \partial\tilde{S}/\partial v, \qquad p = \partial\tilde{S}/\partial q ,$$

which implies,

$$\begin{aligned} q &= u^{-2} \\ p &= -2u^{-3} - \frac{1}{2}u - \frac{1}{2}vu^3 \end{aligned} \tag{11}$$

Since the variable change is canonical, the equation of motion in (u,v) variables is derived from a Hamiltonian,

$$\mathcal{H}(u,v) = v + \frac{1}{8}u^2 + \frac{1}{4}vu^4 + \frac{1}{8}v^2u^6 .$$

In particular for u small (i.e., p,q near infinity)

$$\dot{u} = 1 + \mathcal{O}(u^4), \qquad \dot{v} = \mathcal{O}(u) .$$

Hence v is approximately constant and $u \sim (t - t_o)$. This could be seen equally well by comparing the transition functions (11) with the time series. Since $\mathcal{H}(u,v)$ is polynomial, the flows are analytic around infinity.

Define an augmented manifold by

$$M = \{q, p\epsilon C^2\} \cup \{u = 0, v\epsilon C\} .$$

Note that we have resolved the singularity, and "infinity" is nothing but $u = 0$, and v an arbitrary complex number.

If we were to take this curve and integrate "backwards" by $-t$ we would obtain another analytic curve consisting of all initial data that hits infinity in a time t. Finally, to make M into a manifold we have to cover infinity by a patch consisting of a tube around $u = 0$ which is narrow enough that the transition functions are uniquely invertible for $u, v(q,p)$.

Note that M is not compact. It ignores points such as $p =$ infinity, $q =$ finite which are never reached by the flow.

AUGMENTED MANIFOLD

In Ref. 2 a general algorithm is presented for constructing an augmented manifold M for any polynomial Hamiltonian system with the properties:

1. $\{p,q\} \epsilon C^{2n}$ is a dense subset of M,
2. $M - C^{2n}$ is a finite union of analytic hypersurfaces,
3. the time flows extend to M, exist for all times, and are analytic, and
4. the transition functions between the patches of M are canonical.

The principal balance(s) in local coordinates always correspond to $u = 0$, $\{v_i \epsilon C, i = 1,2,...2n-1\}$; that is they are codimension one. If one initial point $\{p,q\}$ blows up in a time t_o then an open set of neighboring points blow up in a time near to t_o (i.e., integrate the u equation backward from $u = 0$). The lower balances add points to the boundary of the principal balance (i.e., certain v_i infinite). In appropriate local coordinates, the added sets are just given by $u_i = 0$ with the number of variables u_i equalling the codimension. The codimension 2 lower balances may be realized as a singularity in the principal balance equations in which certain $v_i \to$ infinity as $u \to 0$. (Clearly any singularity in these equations for $u \neq 0$ is due to bad coordinates and disappears when transferred back to q,p variables.)

The Hamiltonian local variables is always analytic and there is a coordinate patch for each balance, implying conditions 2, 3. Lastly all computations are done with the Hamilton-Jacobi equation, guaranteeing that M is symplectic.

ARGUMENT FOR INTEGRABILITY

Given a manifold M with the properties just described, there is a heuristic argument as to why the flow is simple. The assertion is not that it is integrable in the technical sense but rather that either the finite time map is birational as in the Riccati example or a level surface exists in the form of a time dependent entire function, $F(t,q,p)$ whose total time derivative is zero.

The argument which is no better than an intuition exploits the characterization of "entire" functions on a manifold by rate of growth. The simplest example of this reasoning is Liouville's theorm which says that if the maximum modulus of an entire function grows algebraically then it is a polynomial. There is an extension of the reasoning that lead to (5) to several variables which applies when M is either compact, or the rate of growth of the finite time map ϕ_t as its arguments tend towards the omitted regions, (i.e., $p \to \infty$, q finite, for the elliptic example), is algebraic. In the latter case, M admits a formal compactification. Bianalytic maps such as ϕ_t between compact spaces all can be given explicit functional forms as in the Riccati example. We consider all such examples to be solved.

Hence the only problematic case is when ϕ_t behaves essentially as its arguments tend toward the points required to compactify. Consider a trivial example,

$$\mathcal{H} = \frac{1}{2}(pq)^2$$

$$p_t = p_o e^{-p_o q_o t}, \quad q_t = q_o e^{p_o q_o t} .$$

The map ϕ_t is clearly essential but the exponent depends only on an invariant $p_t q_t = p_o q_o$. We believe something like this must happen if composition is not to result in an explosion of essentialness as in exp(exp(...)). Thus we would like to claim that an invariant surface results from the group property of ϕ_t plus essential growth.

The basic argument may be repeated if the first level set does not result in a compactifiable submanifold.

We also mention in this section a refinement of the Painlevé test which permits one to detect O.D.E. integrability.

Conjecture If there are no lower balances, then a Painlevé system is O.D.E. integrable.

The converse is clearly trivial.

To argue in the direction stated, consider the set of all complex t poles for given initial conditions. If there is one pole, there must be an infinite number by Picard's Theorm. The assumption of no lower balances implies that these poles cannot coalesce or disappear as initial conditions are changed. Consider any two and use them to define a map from the data at infinity to itself by integrating from one to the other. This generates a bi-entire map from C^{2n-1} to C^{2n-1}. We would like to assert that integral invariants exist.

SEPARABILITY

We first illustrate how to solve an integrable system by separating variables in the Hamilton-Jacobi equation. There are separable (and hence integrable) equations which are not Painlevé but this occurs in the known examples because one is working in too small a phase space. If a system is separable in a technical sense to be defined, then the local analysis embodied in the Painlevé test yields a good deal of information about how to perform the separation and the form of the other integrals in involution. First as an example, we separate (6).

Let

$$q_1 = i\xi_1\xi_2, \qquad q_2 = \frac{1}{2}(\xi_1^2 + \xi_2^2) \tag{12}$$

Reexpress $\partial S/\partial q$ interms of $\partial S/\partial \xi_i = \eta_i$ in the equation,

$$\mathcal{H}(q, \frac{\partial S}{\partial q}) = h_1$$

and one finds a hyperelliptic curve γ.

$$\eta_i^2 = -1/2\xi_i^8 + 2h_1\xi_i^2 + h_2 \tag{13}$$

The action is

$$S = \sum_1^2 \int^{\xi_i} \eta_i d\xi_i .$$

The numbers h_i are the values of the two integrals in involution. Equation (13) written for $i = 1, 2$ can be simultaneously inverted to find $h_i(\xi, \eta)$ or $H_i(q, p)$. The "times" conjugate to h_i are given by $t_i = \partial S/\partial h_i$. If we differentiate this pair of equations we find an expression for the flows $\xi_i(t_j)$ on the level surfaces.

$$\delta_{ij} = [\frac{\partial^2 S}{\partial h_i \partial \xi_k}][\frac{\partial \xi_k}{\partial t_j}]$$

Definition

For a <u>hyperelliptically</u> <u>separable</u> system there is;

a. a good variable change $q_i = q_i(\xi_j)$ where q_i is a symmetric function of $\{\xi_j\}$,

b. under which the Hamilton-Jacobi equation separates into n copies of a hyperelliptic curve γ with the equation,

$$\begin{aligned} \eta^2 &= \xi^d + \cdots \\ d &\geq 2n + 1 \quad , \end{aligned} \tag{14}$$

c. the integrals h_i occur as the coefficients of ξ^{α_i} with $\alpha_i + 1 - d/2 < 0$,

d. the set $\{t_i = \partial S/\partial h_i, i = 1, 2, \cdots n\}$ modulo the periods of γ is a torus.

The fourth condition guarantees that the Hamiltonian in question is Painlevé. When $d = 2n + 1$, condition (d) is automatic and for larger d some symmetry is required of γ in order to factor the period lattice which would otherwise be of higher dimension than the phase space. The above example is hyperellipticly separable.

The following facts then follow

a. The principal balance corresponds to $\xi_1 \sim t^{-1}$ or t^{-2} (depending on whether d is even or odd) and $\xi_{i>1} \sim cst$. The leading exponents f_i, g_i for $q \sim t^{-f_i}$, $p \sim t^{-g_i}$ are the same for all the n-flows in involution;

b. $\mathcal{H}_i(q, p)$ is polynomial in q, p;

c. there is a lowest balance with just n free constants, the h_i, plus t_o;

d. from an expansion of the Hamilton-Jacobi equation at a principal balance, there follows the degree d of γ and bounds on the weighted degrees of $H_i(q, p)$. Here each q, p is given the weight f_i, g_i defined in (a).

By comparing the series (7) with the separating variable change (12) it will be observed that property (a) reduces the calculation of (12) to checking only a few possibilities. In (d), the degree of $\eta^2(\xi)$ follows from expanding $S \sim \int^{\xi_1(t)} \eta_1 d\xi_1$ and comparing with the first (largest) term in $S(q(t), v)$. The degrees of H_i basically reflect the order of their occurence in η^2, but their calculation from the Hamilton-Jacobi expansion is laborious.[2]

We believe similar results can be formulated when γ is replaced by a rational function in ξ or something higher order in η.

REFERENCES

1. N. Ercolani and E.D. Siggia, Phys. Lett. A **119**, 112 (1986).
2. N. Ercolani and E.D. Siggia (to appear).
3. V.V. Golubov, Lectures on integration of the equations of motion of a rigid body about a fixed point (State Publishing House, Moscow, 1953).
4. E.L. Ince, Ordinary differential equations (Dover, New York, 1947).

5. M.J. Ablowitz, A. Ramani, and H. Segur, J. Math. Phys. 21 (1980) 715.
6. Y.F. Chang, M. Tabor and J. Weiss, J. Math. Phys. 23 (1982) 531; T. Bountis, H. Segur, and F. Vivaldi, Phys. Rev. A 25 (1982) 1257; T. Bountis, in: Dynamical systems and chaos, ed. L. Garrrido (Springer, Berlin, 1983) p. 227.
7. V.I. Arnold, Mathematical methods of classical mechanics (Springer, Berlin, 1980).

NEKHOROSHEV-LIKE RESULTS FOR HAMILTONIAN DYNAMICAL SYSTEMS

Giancarlo Benettin

Dipartimento di Fisica "G. Galilei"
Università di Padova
Via F. Marzolo, 8 — 35131 Padova, Italy

1. Introduction

1.1 Nearly-integrable systems. The purpose of these lectures is to revisit some topics in classical perturbation theory, in particular the celebrated Nekhoroshev theorem[1–6] on the stability of motions in nearly-integrable Hamiltonian systems, and to show how the basic ideas and techniques entering this theorem can be extended to study some other dynamical systems, which are quite relevant for physics, but are not close to integrable ones.

Let us consider a nearly-integrable Hamiltonian system with n degrees of freedom, written in the form

$$\begin{aligned} H(I,\varphi,\varepsilon) &= h(I) + \varepsilon f(I,\varphi,\varepsilon) \\ I &= (I_1,\dots,I_n) \in B \\ \varphi &= (\varphi_1,\dots,\varphi_n) \in \mathbb{T}^n \ , \end{aligned} \tag{1.1}$$

where B is a convenient domain contained in $\mathbb{R}^n$, and $\mathbb{T}^n$ is the n-dimensional torus; all functions are here assumed to be regular in all variables, including the perturbation parameter ε. For $\varepsilon = 0$ the system is integrable, and the equations of motion are trivially solved by $I(t) = I(0)$, $\varphi(t) = \varphi(0) + \omega\big(I(0)\big)t$, where $\omega = \frac{\partial h}{\partial I}$ is the unperturbed angular velocity. Thus, the phase space $B \times \mathbb{T}^n$ is foliated into invariant tori, of the form $\{I\} \times \mathbb{T}^n$, on which the motion is quasi-periodic. As is well known, according to the celebrated Liouville–Arnold theorem, this very special structure can be achieved whenever one is given a regular Hamiltonian $H(p,q)$, $(p,q) \in \mathbb{R}^{2n}$, which admits n regular integrals of motion $F_1,\dots,F_n = H$ "in involution", i.e. with all Poisson brackets $\{F_i,F_j\}$ vanishing.† Typical examples of integrable systems are the Kepler problem,

† More precisely, if the level set S given by $F_j(p,q) = c_j$, $j = 1,\dots,n$, is a n-dimensional compact manifold, on which the gradients ∇F_j, $j = 1,\dots,n$, are linearly independent, then S is diffeomorphic to a n-torus, and in a neighborhood U of S one can introduce a canonical transformation $(p,q) = \mathcal{C}(I,\varphi)$, $\mathcal{C}\colon B \times \mathbb{T}^n \longrightarrow U$, such that one has $H(\mathcal{C}(I,\varphi)) = h(I)$. For details see for example ref.[7]

or the free rigid body, where the energy, the squared angular momentum and any of its components are integrals in involution. A very special class of integrable dynamical systems is given by the exactly harmonic systems, for example systems with n point masses interacting via linear forces: as is well known, in this case one can introduce the normal coordinates, which give the Hamiltonian the form $H(p,q) = \frac{1}{2}\sum_{j=1}^{n}(p_j^2+\omega_j^2 q_j^2)$, with constant $\omega_1,\ldots,\omega_n$; the action-angle variables (I,φ) are then easily introduced by the substitution $p_j = \sqrt{2\omega_j I_j}\cos\varphi_j$, $q_j = \sqrt{2I_j/\omega_j}\sin\varphi_j$, which gives the Hamiltonian the form $H(I,\varphi) = \sum_{j=1}^{n}\omega_j I_j \equiv h(I)$. These harmonic systems are called *isochronous*, because the angular velocity $\omega = \frac{\partial h}{\partial I}$ is independent of the actions; on the opposite side one has the strictly non-isochronous systems, characterized by $\det\left(\frac{\partial^2 h}{\partial I \partial I}\right) \neq 0$: this is the case, for example, of the free rigid body, while in the Kepler problem one has an intermediate situation.

As a matter of fact, in several physical problems one deals with Hamiltonian systems which are close to integrable ones, and can be given the form (1.1). This is the case, for example, of a rigid body in a small gravitational field, or of a planetary system with small planet masses, compared to the primary mass. Almost harmonic systems are close to integrable isochronous ones, and can be given the special form

$$H(I,\varphi) = \omega\cdot I + \varepsilon f(I,\varphi,\varepsilon)\ , \tag{1.2}$$

where the dot denotes the usual scalar product in $\mathbf{R}^n$. The above form can be obtained, in particular, for any smooth Hamiltonian system restricted to a neighborhood of a stable equilibrium position: indeed, taking this point as origin, and using the normal-modes coordinates, by a Taylor expansion one can write $H(p,q) = \frac{1}{2}\sum_{j=1}^{n}(p_j^2+\omega_j^2 q_j^2)+ H_3 + H_4 + \cdots$, where H_l, $l \geq 3$, is a homogeneous polynomial of degree l in p,q. Then, in a ε-neighborhood of the origin given by $p_j^2 + \omega_j^2 q_j^2 \leq \varepsilon^2 E$, $j = 1,\ldots,n$, one can introduce the rescaling ("blowing up") $p_j = \varepsilon p'_j$, $q_j = \varepsilon q'_j$, which gives

$$H'(p',q') = \varepsilon^{-2}H(\varepsilon p',\varepsilon q') = \frac{1}{2}\sum_{j=1}^{n}({p'_j}^2 + \omega_j^2 {q'_j}^2) + \varepsilon H'_3 + \varepsilon^2 H'_4 + \cdots\ , \tag{1.3}$$

the domain of definition being now given by an ε-independent condition, precisely ${p'_j}^2 + \omega_j^2 {q'_j}^2 \leq E$, $j = 1,\ldots,n$. Finally, one introduces the action-angle variables, thus achieving an expression of the form (1.2). The celebrated Fermi-Pasta-Ulam[8] Hamiltonian and the Hénon-Heiles[9] Hamiltonian have this form.

1.2 Infinite-time stability. Given a Hamiltonian system of the form (1.1), with small ε, the first basic question one would like to answer is on which time-scale the effects of the perturbation εf become important: in particular, on which time-scale do the actions $I_j(t)$ significantly evolve, and deviate from the initial value $I_j(0)$. As a matter of fact, this question was considered to be among the most important of Physics at the turn of the century: for example, Poincarè[10] refers to it as to the "general problem of dynamics", and in fact, most of the great difficulties of classical physics, which led to the birth of quantum mechanics, are more or less directly related with the behavior of Hamiltonian systems of the form (1.1).

From the equations of motion corresponding to the Hamiltonian (1.1), i.e.,

$$\begin{aligned}\dot\varphi &= \omega(I) + \varepsilon\frac{\partial f}{\partial I}(I,\varphi)\\ \dot I &= \quad -\varepsilon\frac{\partial f}{\partial \varphi}(I,\varphi)\ ,\end{aligned} \tag{1.4}$$

one directly deduces only a very poor zero-order estimate of the form

$$|I_j(t) - I_j(0)| = \varepsilon\left|\int_0^t \frac{\partial f}{\partial \varphi_j}(I(t),\varphi(t))dt\right| \le \varepsilon|t|\left\|\frac{\partial f}{\partial \varphi_j}\right\| , \tag{1.5}$$

where $\|\cdot\|$ denotes the supremum norm in the phase space. The above estimate is certainly good for short t, but could be pessimistic for large t: indeed, $\frac{\partial f}{\partial \varphi_j}$ has zero φ-average, so that on long times one expects compensations in the above integral.

A very nice situation would be the existence of a canonical transformation $(I,\varphi) = \mathcal{C}(I',\varphi',\varepsilon)$, ε-close to the identity, i.e.

$$|I_j' - I_j| = \mathcal{O}(\varepsilon) , \qquad |\varphi_j' - \varphi_j| = \mathcal{O}(\varepsilon) , \qquad j = 1,\ldots,n, \tag{1.6}$$

such that the new Hamiltonian $H' = H \circ \mathcal{C}$ has the integrable form $H'(I',\varphi',\varepsilon) = h'(I',\varepsilon)$. In this case one would have $I'(t) = I'(0)$, and thus, taking into account (1.6),

$$|I_j(t) - I_j(0)| = \mathcal{O}(\varepsilon) , \qquad t \in \mathbb{R}, \tag{1.7}$$

i.e. stability for all times. Unfortunately, it is not difficult to find counterexamples to this too optimistic perspective: for example, if

$$H(I,\varphi) = h(I_{r+1},\ldots,I_n) + \varepsilon f(I_1,\ldots,I_n,\varphi_1,\ldots,\varphi_r) , \tag{1.8}$$

then the motion of $I_1,\ldots,I_r,\varphi_1,\ldots,\varphi_r$ is completely governed by the perturbation (where $I_{r+1},\ldots,I_n$ behave as parameters), and by suitably choosing f it can be of any kind.† This example is not as trivial as it appears: indeed, as discussed for example in ref. [4,6], Hamiltonians of the above form naturally appear in perturbation theory, in the the study of the motion inside the so-called "resonant regions".

If one is interested in infinite-time stability, at small but finite ε, then the best result is the so-called Kolmogorov-Arnol'd-Moser, or KAM theorem; a possible formulation[11–12] is the following:

Theorem (KAM): *Consider a nearly-integrable Hamiltonian system of the form (1.1), and assume:*

i. H is analytic in a complex neighborhood of $B \times \mathbb{T}^n$;

ii. h is strictly non-isochronous, i.e., one has $\det\left(\frac{\partial^2 h}{\partial I \partial I}\right) \ge d > 0$ for $I \in B$;

iii. ε is sufficiently small, say $\varepsilon \le \varepsilon_0$, where ε_0 depends on all of the parameters entering the Hamiltonian.

Then there exist a canonical transformation $(I,\varphi) = \mathcal{C}(I',\varphi',\varepsilon)$, an integrable Hamiltonian $h'(I',\varepsilon)$, both $\mathcal{C}$ and h' being of class C^∞ in $B \times \mathbb{T}^n$, and a set $B_\varepsilon \subset B$ of large Lebesgue measure, precisely $\mathrm{mes}(B \setminus B_\varepsilon) \to 0$ for $\varepsilon \to 0$, such that the new Hamiltonian $H' = H \circ \mathcal{C}$ satisfies the relation

† The trivial rescaling $t = \varepsilon^{-1}t'$, $H' = \varepsilon^{-1}H$, completely eliminates the ε-dependence in the equations of motion for $I_1,\ldots,I_r,\varphi_1,\ldots,\varphi_r$, so that the above Hamiltonian, although having the form (1.1), is in no sense close to an integrable Hamiltonian.

$$H'(I',\varphi',\varepsilon) \overset{B_\varepsilon}{=} h(I',\varepsilon) , \tag{1.9}$$

where $\overset{B_\varepsilon}{=}$ *denotes equality restricted to* B_ε.

Remark: Unfortunately, the set $B \setminus B_\varepsilon$, as constructed in the proof of the theorem, is open and dense, so that B_ε has empty interior; for $n > 2$, $B \setminus B_\varepsilon$ is also connected.

Let us make a comment about the relevance of KAM theorem in our discussion. For initial data $I'(0) \in B_\varepsilon$, or equivalently $(I(0),\varphi(0)) \in \mathcal{C}(B_\varepsilon \times \mathbb{T}^n)$, the system behaves as if it were integrable: $I'(t)$ turns out to be constant, and one has perpetual stability. The measure of B_ε is large, so that (for a random choice of the initial data) the condition $I'(0) \in B_\varepsilon$ has large probability; however, B_ε having empty interior, this condition can never be decided physically, and stability cannot be guaranteed. In particular, for $n > 2$, one cannot exclude a diffusion of $I(t)$ throughout the whole space ("Arnol'd diffusion").

1.3 Long-time stability. Let us now introduce in a very informal way the basic ideas entering Nekhoroshev theorem. Here one does not ask for perpetual stability, but "only" for stability for very long times: precisely, one looks for results of the form

$$I(t) - I(0) \sim \varepsilon^a , \qquad a > 0, \tag{1.10}$$

(the symbol $\sim$ means "of the order of") for a time-scale growing exponentially with ε,

$$t \sim \exp\left(\frac{\varepsilon_0}{\varepsilon}\right)^b , \qquad \varepsilon_0, b > 0 . \tag{1.11}$$

Results of this form, although weaker than perpetual stability, can be very relevant for physics: for example, one may want to discuss the stability of the Hamiltonian system representing, say, the Solar system, "only" on a time-scale not greater than the Universe lifetime, or even on a shorter time-scale, before the dissipative effects become important and the Hamiltonian model is no more realistic.

On the other hand, Nekhoroshev theorem provides stability results of the above form in the whole phase space — no strange sets like B_ε there appear — so that, in particular, this theorem provides an upper bound to Arnol'd diffusion. In addition, Nekhoroshev theorem covers some situations where KAM theorem cannot be applied, in particular the case of almost-harmonic systems (1.2), and moreover, the basic ideas entering this theorem can be extended to systems which are even not close to integrable ones. For example:

i. A point mass is "physically" constrained to a surface; this means that (in addition to a possible external potential) there is a strong confining potential of the form $\omega^2 W$, where ω is a large parameter, and W vanishes on the surface and is positive outside. For example, one can try to physically realize a spherical pendulum by using, in place of an ideally rigid bar, a spring with large elastic constant ω^2, which produces an approximate constraint to a spherical surface. The basic question one would like to answer is on which time-scale does the fast vibration transversal to the surface disturb the motion on the surface[13–16]. By an extension of Nekhoroshev methods one proves[17] that (under suitable assumptions) there is no significant energy exchange between the two movements, for a time-scale of order $\exp(\omega/\omega_0)$, ω_0 being a suitable positive constant. The result extends to the realization of a rather wide class of holonomic constraints, and there is no need at all that the constrained system one wants to realize is integrable.

ii. A closely related problem: a (classical) vibrating diatomic molecule, represented by two atoms connected by a spring, collides with the wall of a container. If the molecule were perfectly rigid, then one would have a perfectly elastic collision, but if the spring has a finite elastic constant ω^2, then the collision is no more elastic, and an energy exchange between the internal vibrational degree of freedom and the translational (or rotational) ones is possible. Here one is interested to a short time-scale, of the order of the collision time; one can prove[17] that at the end of the collision the energy exchange is, at most, of order $\exp(-\omega/\omega_0)$.

iii. A parameter is slowly varied in an oscillating system: for example, one has the equation of motion of a pendulum with slowly varying length,

$$\ddot{\theta} + \omega(\varepsilon t)^2 \sin\theta = 0 , \tag{1.12}$$

$\omega(\tau)$ being any regular function. The question is here on which time-scale does the adiabatic invariant (i.e. the action) appreciably change; the answer[18] is that, for small enough ε, one needs an exponentially large time–scale, of order $\exp(\varepsilon_0/\varepsilon)$.

iv. An integrable dynamical system with two degrees of freedom is weakly perturbed, and in the Poincaré section some thin chaotic regions are produced. The "islands" produced by the perturbation extend in general over a large region, of order $\varepsilon^{\frac{1}{2}}$, but for small ε chaotic (homoclinic) phenomena are confined to very thin strips, at the border of the islands, around split separatrices. One proves[19] (see also ref.[20] and the lecture by A. Giorgilli in this volume) that the angle between split separatrices, and consequently the size of these strips, is exponentially small with ε, of order $\exp(\varepsilon_0/\varepsilon)$.

Nekhoroshev-like perturbation theory more or less easily extends to all of these examples; on the contrary, it is not even conceivable to work out for them KAM-like results, as none of that systems is close to an integrable system.

These lectures are organized as follows: In the next Section we will introduce some basic elements of classical perturbation theory, and state formally Nekhoroshev theorem in the simplest possible framework, i.e. for weakly coupled Diophantine harmonic oscillators. Section 3 reports the proof of Nekhoroshev theorem for harmonic oscillators. Finally, Section 4 is devoted to some of the above quoted Nekhoroshev-like results, essentially to examples i. and ii.

2. Classical perturbation theory and Nekhoroshev theorem

In the previous section we described, so to speak, the aim of Nekhoroshev theorem and of related results; here we start with some technical work: precisely, we consider finite order perturbation theory, illustrate its basic difficulties, and explain how one can overcome them; this will lead us in a quite natural way to Nekhoroshev exponential estimates.

2.1 Poincaré non-existence theorem. As already remarked, if one is given a Hamiltonian system of the form (1.1), then from the equation of motion (1.4) one only deduces the zero–order estimate $I(t) - I(0) \sim \varepsilon t$. A natural procedure to go beyond this naive result, and obtain higher order estimates, is to look for a canonical transformation $(I,\varphi)=\mathcal{C}_r(I',\varphi',\varepsilon)$, such that the new Hamiltonian $H_r = H\circ\mathcal{C}_r$ has the following "normal form up to order r":

$$H_r(I',\varphi',\varepsilon) = h(I') + \varepsilon g_r(I',\varepsilon) + \varepsilon^{r+1} f_r(I',\varphi',\varepsilon)\,; \tag{2.1}$$

$\mathcal{C}_r(I',\varphi',\varepsilon)$ must be close to the identity for small ε, and satisfy, in particular, $I'-I\sim\varepsilon$. If one succeeds in this program, then one has $I'(t)-I'(0)\sim\varepsilon^{r+1}t$, so that $I'(t)-I'(0)$ is of order ε at time $t\sim\varepsilon^{-r}$; on the other hand, $I(t)$ can only oscillate around $I'(t)$, so that $I(t)-I(0)$ is also of order ε on the same time-scale $t\sim\varepsilon^{-r}$.

Let us try to construct $\mathcal{C}_r$ for $r=1$. We look for a generating function $S(I',\varphi) = I'\cdot\varphi + \varepsilon W(I',\varphi)$, and write

$$I = I' + \varepsilon\frac{\partial W}{\partial\varphi}(I',\varphi)\,, \qquad \varphi' = \varphi + \varepsilon\frac{\partial W'}{\partial I}(I',\varphi)\,; \tag{2.2}$$

one has then

$$\begin{aligned} H' &= h(I' + \varepsilon\frac{\partial W}{\partial\varphi}(I',\varphi)) + \varepsilon f(I',\varphi,\varepsilon) \\ &= h(I') + \varepsilon\left[\omega(I')\cdot\frac{\partial W}{\partial\varphi}(I',\varphi') + f(I',\varphi',0)\right] + \mathcal{O}(\varepsilon^2)\,. \end{aligned} \tag{2.3}$$

In order to accomplish our program, we must impose that the term of order ε is independent of the angles: thus we must solve an equation of the form

$$\omega(I')\cdot\frac{\partial W}{\partial\varphi}(I',\varphi') + f(I',\varphi',0) = \psi(I') \tag{2.4}$$

for the unknowns W and ψ. As is well known since Poincaré's work, this equation is the basic equation of classical perturbation theory, and is met within all perturbative schemes; in a sense, different branches of perturbation theory differ precisely in the way they approach the above equation.†

The difficulties one meets in solving equation (2.4) were particularly stressed by Poincaré. A very weak form of its celebrated theorem on the non-existence of integrals of motion is given by the following

Theorem (Poincaré): *Consider the above equation (2.4), and assume:*

i. $h(I)$ is non–isochronous, i.e. one has $\det\left(\frac{\partial^2 h}{\partial I\partial I}(I)\right)\neq 0$ for $I\in B$;

ii. $f(I,\varphi,0)$ has sufficiently many non-vanishing Fourier components: precisely, for each $k\in\mathbb{Z}^n$ there exists at least one $k'\in\mathbb{Z}^n$, k' parallel to k, such that the k'-th Fourier component $f_{k'}(I)$ of $f(I,\varphi,0)$ does not vanish in B.

Then the equation (2.4) cannot be solved in any domain of the form $B'\times\mathbb{T}^n$, B' being any open subset of B.

Sketch of the proof: Let us write

$$\begin{aligned} f(I,\varphi,0) &= \sum_{k\in\mathbb{Z}^n} f_k(I)e^{ik\cdot\varphi} \\ W(I,\varphi) &= \sum_{k\in\mathbb{Z}^n} W_k(I)e^{ik\cdot\varphi}\,, \end{aligned} \tag{2.5}$$

---

† For a short review on this subject, see for example ref. [21].

and project equation (2.4) on Fourier components. For $k = 0$ one has $f_0(I) = \psi(I)$, which gives ψ, while for $k \neq 0$ one gets

$$\big(ik \cdot \omega(I)\big) W_k(I) + f_k(I) = 0 \,. \tag{2.6}$$

This equation cannot be solved by $W_k(I) = \frac{f_k(I)}{ik \cdot \omega(I)}$. Indeed, on one hand, in virtue of assumption *i.*, the "resonant set" $\{I \in B;\ \exists k \in \mathbb{Z}^n \backslash \{0\}\colon\ \omega(I) \cdot k = 0\}$, is dense in B; on the other hand, in virtue of assumption *ii.*, $f_k(I)$ cannot vanish for all vectors k such that $k \cdot \omega(I)$ vanishes; thus, in any open set $B' \subset B$ there are points where, for some $k \in \mathbb{Z}^n$, (2.6) cannot be satisfied.

2.2 Birkhoff series. The theorem considered in the previous subsection tells us, in a sense, what is not possible to do in perturbation theory. It may appear strong (it appeared strong, and perhaps discouraged the research activity in perturbation theory for some time), nevertheless, as we now shortly see, it leaves enough ways out to beautiful developments of classical perturbation theory.

The first of them is due to Birkhoff, who succeeded in constructing perturbation theory for Hamiltonian systems in the vicinity of an equilibrium position ("Birkhoff series"). As already remarked, the Hamiltonian of a generic system in the vicinity of a stable equilibrium position can be given the form (1.3), which in turn, after the substitution to action–angle variables, assumes the form (1.2), more precisely

$$H(I, \varphi, \varepsilon) = \omega \cdot I + \sum_{s=1}^{\infty} \varepsilon^s H_s(I, \varphi) \,; \tag{2.7}$$

here H_s is a polynomial of order s in $\cos\varphi$ and $\sin\varphi$, and thus has finitely many non-vanishing Fourier components. Both of the Poincaré's assumptions are here violated. Birkhoff also assumed that ω were non-resonant, i.e., that $\omega \cdot k = 0$ has no solution $k \in \mathbb{Z}^n$ but the trivial one $k = 0$. Within these assumptions it is obvious that equation (2.4) is solved by

$$W(I, \varphi) = \sum_k \frac{f_k(I)}{ik \cdot \omega} \,, \tag{2.8}$$

where the sum extends only to those k's for which $f_k(I)$ does not vanish: indeed, the sum is finite, and all denominators are different from zero. Moreover, for ε sufficiently small, the equations (2.2) defining the canonical transformation can be inverted, so that W really defines a canonical transformation, and the new Hamiltonian has the form (2.1), with $r = 1$. It could be seen that, for ε sufficiently small, the procedure can be iterated any finite number r of times, giving Hamiltonians of the form (2.1).

2.3 The Diophantine condition. As a matter of fact, one can solve an equation of the form (2.4), with constant ω, if $f(I, \varphi, 0)$ is analytic in φ, and ω satisfies a non-resonance condition stronger than Birkhoff's one, called "Diophantine condition", precisely

$$|k \cdot \omega| > \gamma\, |k|^{-n} \qquad \forall k \in \mathbb{Z}^n \setminus \{0\} \,, \tag{2.9}$$

where γ is a positive constant and $|k| = |k_1| + \ldots + |k_n|$; one easily proves that the set of non-Diophantine frequencies has measure vanishing with γ. The Diophantine condition is used in the following way: the solution to (2.4) can be written in the form (2.8) and, as in Birkhoff's case, denominators never vanish. However, the sum is now extended to the whole $\mathbb{Z}^n \backslash \{0\}$, so one must prove its convergence. The idea is quite simple: if $f(I,\varphi,0)$ is analytic and bounded in a complex strip $|\mathrm{Im}\,\varphi_j| < \rho$, then, denoting $F = \sup_{I,\varphi} |f(I,\varphi,0)|$, one has:

i. $f_k(I)$ decays exponentially with $|k|$: precisely, one has $|f_k(I)| < Fe^{-|k|\rho}$, and consequently, in any reduced strip $|\mathrm{Im}\,\varphi_j| \le \rho - \delta$, $|f_k(I)e^{ik\cdot\varphi}| < Fe^{-|k|\delta}$;

ii. on the other hand, the denominators decay only as a power of $|k|$, so that each term of the sum still decays exponentially, and the sum is easily seen to converge to a function which is analytic and bounded in the strip $|\mathrm{Im}\,\varphi_j| \le \rho - \delta$.

A precise lemma will be stated in the next section, within the proof of Nekhoroshev theorem. From the above considerations it should be clear that here too, as in the Birkhoff case, for small ε one can reach any finite perturbative order r, thus giving the Hamiltonian the normal form (2.1).

2.4 Nekhoroshev theorem for harmonic oscillators. Once the Hamiltonian is in form (2.1), then one has $\dot{I} \sim \varepsilon^{r+1}$, say

$$|\dot{I}_j| \le C_r \varepsilon^{r+1}, \qquad j = 1,\ldots,n, \tag{2.10}$$

C_r being a suitable positive constant. It is now crucial to control the r–dependence of C_r. This requires a lot of technical estimates, which are quite annoying, and even hard if one looks for optimal results: however, it is worthwhile to notice that it is enough to obtain apparently poor estimates for C_r, of the form

$$C_r < AB^r r^{rm}, \qquad A, B, m > 0\,, \tag{2.11}$$

to get highly non-trivial results, precisely the celebrated exponential estimates of Nekhoroshev theorem. Indeed, from the above estimate one deduces $|\dot{I}_j| < \varepsilon A (\varepsilon B r^m)^r$; then (this is the basic idea) one can take r depending on ε, for example in such a way that $\varepsilon B r^m = e^{-1}$, i.e., $r(\varepsilon) = (\varepsilon_0/\varepsilon)^{1/m}$, with $\varepsilon_0 = (eB)^{-1}$ (let us here forget about the fact that $r(\varepsilon)$ is not an integer). From this choice one immediately deduces the exponential estimate $|\dot{I}| < A \exp -(\varepsilon_0/\varepsilon)^{1/m}$.

This simple procedure is really the heart of Nekhoroshev theorem for harmonic oscillators (and also of the analytic part of the non-isochronous case). We shall refer to this theorem as to the Nekhoroshev–Gallavotti theorem, because the isochronous case, with Diophantine ω, was first studied by Gallavotti[4] (see also ref.[6]). In order to produce a formal statement, we need some notations:

i. (Concerning domains). We are interested in the Hamiltonian (1.2) for $(I,\varphi) \in \mathcal{B} \times \mathbb{T}^n$; however, we find convenient to think of φ as belonging to $\mathbb{R}^n$, f being periodic of period 2π. Denote $D_0 = \mathcal{B} \times \mathbb{R}^n$, and given $\rho = (\rho_I, \rho_\varphi)$, with $\rho_I > 0$ and $1 \ge \rho_\varphi > 0$, consider, for each $(I,\varphi) \in D_0$, the polydisc $\Delta_\rho(I,\varphi) = \{(I',\varphi') \in \mathbb{C}^{2n};\ |I'_j - I_j| < \rho_I,\ |\varphi'_j - \varphi_j| < \rho_\varphi,\ j = 1,\ldots,n\}$. Then define $D_\rho = \bigcup_{(I,\varphi)\in D_0} \Delta_\rho(I,\varphi)$. Concerning the parameter ε, we assume it belongs to some real interval containing the origin; by a possible redefinition of functions, one can always assume $|\varepsilon| \le 1$.

ii. (Concerning norms): For any $\rho = (\rho_I, \rho_\varphi)$ we denote by $\|\cdot\|_\rho$ the supremum norm in D_ρ, including, for vector–valued functions, the maximum over components; for ε–dependent functions, the supremum for ε in the interval of definition is also included. For integer vectors $k \in \mathbb{Z}^n$, we denote by $|k|$ the norm $|k_1| + \ldots + |k_n|$.

We can now formulate the following

Theorem (Nekhoroshev–Gallavotti): *Consider a nearly-isochronous Hamiltonian system of the form (1.2) and assume:*

i. H is analytic for $(I,\varphi) \in D_\rho$ and $|\varepsilon| \leq 1$.

ii. ω satisfies the Diophantine condition

$$|k \cdot \omega| < \gamma |k|^{-n} \qquad \forall k \in \mathbb{Z}^n \setminus \{0\}, \tag{2.12}$$

for a suitable $\gamma > 0$.

Then there exists a constant C, which can be taken

$$C < n^{7n+23} n^{2n+3}, \tag{2.13}$$

such that, denoting

$$F = \|f\|_\rho \qquad \varepsilon_0 = C^{-1} \gamma \rho_\varphi^{n+2} \rho_I F^{-1}, \tag{2.14}$$

and assuming $\varepsilon \leq \varepsilon_0$,

a) there exists a real analytic canonical transformation $(I,\varphi) = \mathcal{C}_\varepsilon(I',\varphi')$, $\mathcal{C}_\varepsilon : D_{\frac{1}{2}\rho} \to D_\rho$, which gives the Hamiltonian $H' = H \circ \mathcal{C}_\varepsilon$ the form

$$H'(I',\varphi',\varepsilon) = \omega \cdot I' + \varepsilon g(I',\varepsilon) + \left(\frac{\varepsilon}{\varepsilon_0}\right) e^{-m(\varepsilon_0/\varepsilon)^{1/m}} f'(I',\varphi',\varepsilon), \tag{2.15}$$

with $m = n+3$, and

$$\|g\|_{\frac{1}{2}\rho} \leq 2F, \qquad \|f'\|_{\frac{1}{2}\rho} \leq \frac{1}{16} \gamma \rho_I \rho_\varphi. \tag{2.16}$$

b) The canonical transformation is small with ε, precisely, denoting $I = I' + \alpha(I',\varphi')$, $\varphi = \varphi' + \beta(I',\varphi')$, one has

$$\|\alpha\|_{\frac{1}{4}\rho} \leq \frac{1}{8} \frac{\varepsilon}{\varepsilon_0} \rho_I, \qquad \|\beta\|_{\frac{1}{4}\rho} \leq \frac{1}{8} \frac{\varepsilon}{\varepsilon_0} \rho_\varphi. \tag{2.17}$$

From this theorem one immediately deduces the following

Corollary: *In the above assumptions, denoting by $(I(t),\varphi(t))$ a possible movement for the Hamiltonian (1.2), with initial datum $(I(0),\varphi(0))$ in D_0, and by $(I'(t),\varphi'(t))$ its inverse image $\mathcal{C}_\varepsilon^{-1}(I(t),\varphi(t))$, one has*

$$|I'_j(t) - I'_j(0)| \leq \frac{1}{8} \frac{\varepsilon}{\varepsilon_0} \rho_I \tag{2.18a}$$

$$|I_j(t) - I_j(0)| \leq \frac{1}{2} \frac{\varepsilon}{\varepsilon_0} \rho_I \tag{2.18b}$$

for

$$|t| \leq \gamma^{-1} e^{m(\varepsilon_0/\varepsilon)^{1/m}}. \tag{2.19}$$

In order to prove the corollary, we preliminarly recall a very elementary but important inequality, known as

Cauchy inequality: *given $\sigma > 0$ and $z_0 \in \mathbb{C}$, consider the disc $\Delta = \{z \in \mathbb{C}, |z - z_0| < \sigma\}$. If a function $\mathcal{U} : \Delta \to \mathbb{C}$ is analytic and bounded in Δ, then one has $\left|\frac{d\mathcal{U}}{dz}(z_0)\right| \leq \sigma^{-1} \sup_{z\in\Delta} |\mathcal{U}(z)|$.*

Proof: The proof is a trivial application of Cauchy integral representation.

We come now to the

Proof of the corollary: consider real initial data $(I(0), \varphi(0)) \in D_0$; then, according to (2.17), $(I'(0), \varphi'(0))$ belongs to the real part of $D_{\frac{1}{8}\rho}$, while the whole orbit $(I'(t), \varphi'(t))$ is obviously real. By Cauchy inequality applied to f', thinking of all coordinates but the angle φ'_j as fixed parameters, as far as $(I'(t), \varphi'(t))$ remains in $D_{\frac{1}{2}\rho}$ one has

$$\left|\frac{\partial f'}{\partial \varphi'_j}(I'(t), \varphi'(t))\right| \leq \frac{2}{\rho_\varphi}\|f'\|_{\frac{1}{2}\rho} \leq \frac{1}{8}\gamma\rho_I \tag{2.20}$$

(here it is essential to have $\operatorname{Im}\varphi'_j(t) = 0$), and thus

$$|\dot{I}_j(t)| \leq \frac{1}{4}\frac{\varepsilon}{\varepsilon_0} e^{-m(\varepsilon_0/\varepsilon)^{1/m}} \gamma\rho_I \,.$$

For t satisfying (2.19), one immediately obtains the inequality (2.18a).

Concerning (2.18b), it is sufficient to observe that, because of (2.18a), $(I'(t), \varphi'(t))$ cannot escape $D_{\frac{1}{4}\rho}$, so that one can use (2.17), and (2.18b) is immediately obtained.

3. Proof of Nekhoroshev–Gallavotti theorem

3.1 The Lie Method. Following the strategy outlined in Section 2.4, we introduce a canonical transformation $\mathcal{C} : D_{\frac{1}{2}\rho} \to D_\rho$, of the form $\mathcal{C} = \mathcal{C}_1 \circ \cdots \circ \mathcal{C}_r$, with

$$\begin{aligned} \mathcal{C}_s &: D_{\rho - s\delta} \to D_{\rho-(s-1)\delta} \\ \delta &= (\delta_I, \delta_\varphi) = \left(\frac{\rho_I}{2r}, \frac{\rho_\varphi}{2r}\right). \end{aligned} \tag{3.1}$$

Let us denote $H_0 = H$, and $H_s = H_{s-1} \circ \mathcal{C}_s$, $s = 1, \ldots, r$; we must determine $\mathcal{C}_1, \ldots, \mathcal{C}_r$ in order that H_s, $s = 1, \ldots, r$, is in normal form up to order s, precisely

$$H_s(I, \varphi, \varepsilon) = \omega \cdot I + \varepsilon g_s(I, \varepsilon) + \varepsilon^{s+1} f_s(I, \varphi, \varepsilon)\,. \tag{3.2}$$

A simple and convenient way to define $\mathcal{C}_s$ is given by the so–called Lie method.

Let χ be any real analytic function: $D_{\rho-(s-1)\delta} \to \mathbb{C}$; consider the Hamilton equations associated to the Hamiltonian χ, and denote by $\Phi^\tau_\chi(I', \varphi')$ their solution at time τ, with initial datum $(I', \varphi') \in D_{\rho - s\delta}$. Let $\tau = \varepsilon^s$; for sufficiently small ε, $\Phi^{\varepsilon^s}_\chi$ is certainly a real analytic canonical mapping: $D_{\rho-s\delta} \to D_{\rho-(s-1)\delta}$, so we can set $\mathcal{C}_s = \Phi^{\varepsilon^s}_\chi$, and try to determine χ in order that $H_s = H_{s-1} \circ \Phi^{\varepsilon^s}_\chi$ has the form (3.2), assuming that H_{s-1} already has the form (3.2), with $s - 1$ in place of s.

Let $\mathcal{U}$ be any regular function: $D_{\rho-(s-1)\delta} \to \mathbb{C}$. As is well known, one has

$$\frac{d}{d\tau}(\mathcal{U} \circ \Phi^\tau_\chi) = \{\chi, \mathcal{U}\} \circ \Phi^\tau_\chi\,, \tag{3.3}$$

where $\{.\}$ is the Poisson bracket. Consider then the first– and second–order remainders $R_1^\tau[\chi,\mathcal{U}]$ and $R_2^\tau[\chi,\mathcal{U}]$, defined by

$$\begin{aligned} \mathcal{U}\circ\Phi_\chi^\tau &= \mathcal{U} + R_1^\tau[\chi,\mathcal{U}] \\ \mathcal{U}\circ\Phi_\chi^\tau &= \mathcal{U} + \tau\{\chi,\mathcal{U}\} + R_2^\tau[\chi,\mathcal{U}]\,; \end{aligned} \tag{3.4}$$

one has immediately

$$\begin{aligned} R_1^\tau[\chi,\mathcal{U}] &= \int_0^\tau \{\chi,\mathcal{U}\}\circ\Phi_\chi^{\tau'}\,d\tau' = \mathcal{O}(\tau) \\ R_2^\tau[\chi,\mathcal{U}] &= \int_0^\tau d\tau'\int_0^{\tau'} \{\chi,\{\chi,\mathcal{U}\}\}\circ\Phi_\chi^{\tau''}\,d\tau'' = \mathcal{O}(\tau^2)\,. \end{aligned} \tag{3.5}$$

Let us now write $H_s = H_{s-1}\circ C_s = (\omega\cdot I)\circ\Phi_\chi^\tau + \varepsilon g_{s-1}\circ\Phi_\chi^\tau + \varepsilon^s\circ\Phi_\chi^\tau$; using (3.4) and $\tau = \varepsilon^s$, one obtains

$$\begin{aligned} H_s =& \,\omega\cdot I + \varepsilon g_{s-1} \\ &+ \varepsilon^s(f_{s-1} + \{\chi,\omega\cdot I\}) \\ &+ R_2^{\varepsilon^s}[\chi,\omega\cdot I] + R_1^{\varepsilon^s}[\chi,\varepsilon g_s + \varepsilon^s f_s]\,. \end{aligned} \tag{3.6}$$

Clearly, the first line contains only φ–independent terms, while the third line contains terms which are, at least, of order ε^{s+1}. In order that H_s is in normal form, one then requires that the second line is independent of φ; this directly leads to the equation

$$-\omega\cdot\frac{\partial\chi}{\partial\varphi} + f_{s-1}(I,\varphi,\varepsilon) = \psi(I,\varepsilon) \tag{3.7}$$

for the unknowns χ and ψ. Here we recognize the basic equation of perturbation theory, already met in section 2.1 (with $-\chi$ in place of W; ε is here a parameter). As we know, if ω satisfies a Diophantine condition, then this equation can be solved by

$$\begin{aligned} \psi(I,\varepsilon) &= f_{s-1,0}(I,\varepsilon) \\ \chi(I,\varphi,\varepsilon) &= \sum_{k\in\mathbb{Z}^n\setminus\{0\}} \frac{f_{s-1,k}(I,\varepsilon)}{ik\cdot\omega}\,, \end{aligned} \tag{3.8}$$

where $f_{s-1,k}$ is the k-th Fourier component of f_{s-1}. Once (3.7) is satisfied, H_s has the required form (3.2), with

$$\begin{aligned} g_s &= g_{s-1} + \varepsilon^{s-1}\psi(I,\varepsilon) \\ \varepsilon^{s+1} f_s &= R_2^{\varepsilon^s}[\chi,\omega\cdot I] + R_1^{\varepsilon^s}[\chi,\varepsilon g_{s-1} + \varepsilon^s f_{s-1}]\,. \end{aligned} \tag{3.9}$$

3.2 Estimates. Our purpose is here to work recursively, and estimate g_s and f_s in $D_{\rho-s\delta}$, assuming g_{s-1} and f_{s-1} have been already estimated in $D_{\rho-(s-1)\delta}$. We proceed in five steps:

i) from (3.8) we estimate $\frac{\partial\chi}{\partial I}$ and $\frac{\partial\chi}{\partial\varphi}$ in $D_{\rho-(s-\frac{1}{2})\delta}$;

ii) we estimate a generic Poisson bracket $\{\chi,\mathcal{U}\}$ in $D_{\rho-(s-\frac{1}{2})\delta}$;

iii) we establish a sufficient condition on ε, which guarantees $\Phi_\chi^{\varepsilon^s}(D_{\rho-s\delta}) \subset D_{\rho-(s-\frac{1}{2})\delta} \subset D_{\rho-(s-1)\delta}$;

iv) we estimate g_s and f_s in $D_{\rho-s\delta}$;

v) we estimate the canonical transformation *in a smaller set*, precisely, denoting $\mathcal{C}_s(I',\varphi',\varepsilon) = (I' + \alpha_s(I',\varphi',\varepsilon), \varphi' + \beta_s(I',\varphi',\varepsilon))$, we estimate α_s and β_s in $D_{\frac{1}{4}\rho}$.

Throughout these steps, we will make use of a recurrent assumption: denoting $F_s = \|f_s\|_{\rho-s\delta}$, $s = 0,\dots,r$, and $F = F_0 = \|f\|_\rho$, we assume

$$F_l = \Lambda^l F \tag{3.10}$$

for $l = 0,\dots,s-1$, and try to determine Λ in such a way that the same inequality holds for $l = s$. Λ will depend on all constants entering our problem and on r; its expression will be

$$\begin{aligned}\Lambda &= 2^{4n+14} n^{2n+3} \gamma^{-1} \delta_\varphi^{-n-2} \delta_I^{-1} \\ &= 2^{5n+17} n^{2n+3} \gamma^{-1} \rho_\varphi^{-n-2} \rho_I^{-1} r^{n+3}\,.\end{aligned} \tag{3.11}$$

Let us anticipate the result: if

$$\varepsilon \le \frac{1}{2}\Lambda^{-1}\,, \tag{3.12}$$

then steps i) — iv) can be carried on, and one has

$$\|f_s\|_{\rho-s\delta} \le \Lambda^s F \tag{3.13a}$$

$$\|g_s\|_{\rho-s\delta} \le 2F \tag{3.13b}$$

$$\|\alpha_s\|_{\frac{1}{4}\rho} \le \frac{1}{4} r^{-n-3} \varepsilon^s \Lambda^s \rho_I \tag{3.13c}$$

$$\|\beta_s\|_{\frac{1}{4}\rho} \le \frac{1}{4} r^{-n-3} \varepsilon^s \Lambda^s \rho_\varphi\,. \tag{3.13d}$$

Step i): our estimates are provided by the following

Lemma (on the basic equation): *the function χ defined by (3.8) is analytic in the domain $D_{\rho-(s-1)\delta}$ where f_{s-1} is analytic, and in any reduced set $D_{\rho-(s-1)\delta-\tilde\delta}$ it satisfies the estimates*

$$\begin{aligned}\left\|\frac{\partial\chi}{\partial I}\right\|_{\rho-(s-1)\delta-\tilde\delta} &\le A\gamma^{-1}\tilde\delta_\varphi^{-\eta-1}\|f_{s-1}\|_{\rho-(s-1)\delta} \\ \left\|\frac{\partial\chi}{\partial\varphi}\right\|_{\rho-(s-1)\delta-\tilde\delta} &\le A\gamma^{-1}\tilde\delta_\varphi^{-\eta}\tilde\delta_I^{-1}\|f_{s-1}\|_{\rho-(s-1)\delta}\,,\end{aligned} \tag{3.14}$$

with $A = 2^{3n+10} n^{2n+3}$, $\eta = n+1$.

By comparison with (3.11), we see that one has

$$\Lambda = 2^{n+4} A\gamma^{-1}\delta_\varphi^{-n-2}\delta_I^{-1}\,. \tag{3.15}$$

Proof: we will not prove here this lemma, which is a very classical achievement of perturbation theory. Several proofs can be found in the literature, which differ, in particular, for the value of the constants A and η; as will be more clear later, η is the more relevant one, because it directly determines the exponent $1/m$ in (2.15) and (2.19): indeed, one finds $m = \eta + 2$. Elementary proofs (see for example ref. [15]) give $\eta = 2n$; an improvement by Moser[22] leads to $\eta = n+1$, while some non trivial additional work by Russman[23,24] gives $\eta = n$. We did follow Moser's estimates, with minor modifications necessary to adapt them to our context.

Step ii): To estimate a Poisson bracket $\{\chi,\mathcal{U}\}$, $\mathcal{U}$ being any analytic function $D_{\rho-(s-1)\delta}\to\mathbb{C}$, let us write the identity

$$\{\chi,\mathcal{U}\}(I,\varphi,\varepsilon)=\frac{d}{d\tau}\mathcal{U}\left(I-\tau\frac{\partial\chi}{\partial\varphi}(I,\varphi,\varepsilon),\varphi+\tau\frac{\partial\chi}{\partial I}(I,\varphi,\varepsilon)\right)\bigg|_{\tau=0}. \tag{3.16}$$

Clearly, if for given (I,φ,ε) one can guarantee

$$(I-\tau\frac{\partial\chi}{\partial\varphi}(I,\varphi,\varepsilon),\varphi+\tau\frac{\partial\chi}{\partial I}(I,\varphi,\varepsilon))\in D_{\rho-(s-1)\delta} \tag{3.17}$$

for $|\tau|\le T$, then by Cauchy inequality applied to $\mathcal{U}(I-\tau\frac{\partial\chi}{\partial\varphi}(I,\varphi,\varepsilon),\,\varphi+\tau\frac{\partial\chi}{\partial I}(I,\varphi,\varepsilon))$, thought of as a function of τ in the disc $|\tau|\le T$, one has $|\{\chi,\mathcal{U}\}(I,\varphi,\varepsilon)|\le T^{-1}\|\mathcal{U}\|_{\rho-(s-1)\delta}$. Now, (3.17) is guaranteed if $(I,\varphi)\in D_{\rho-(s-\frac{1}{2})\delta}$ and

$$T\left|\frac{\partial\chi}{\partial\varphi}(I,\varphi,\varepsilon)\right|\le\frac{1}{2}\delta_I,\qquad T\left|\frac{\partial\chi}{\partial I}(I,\varphi,\varepsilon)\right|\le\frac{1}{2}\delta_\varphi; \tag{3.18}$$

using (3.14), with $\tilde{\delta}=\frac{1}{2}\delta$, and the recurrent assumption (3.10), one finds the sufficient condition

$$T\le r^{-n-3}A^{-1}\delta_\varphi^{\eta+1}\delta_I\Lambda^{-s+1}F^{-1}, \tag{3.19}$$

which leads to the final estimate

$$\|\{\chi,\mathcal{U}\}\|_{\rho-(s-\frac{1}{2}\delta}\le r^{n+3}A\gamma^{-1}\delta_\varphi^{-\eta-1}\delta_I^{-1}\Lambda^{s-1}F\|\mathcal{U}\|_{\rho-(s-1)\delta}. \tag{3.20}$$

Step iii): For the Hamiltonian flow induced by χ one has $\Phi_\chi^{\varepsilon^s}(I',\varphi')\in D_{\rho-(s-\frac{1}{2})\delta}$, with initial data $(I',\varphi')\in D_{\rho-s\delta}$, if

$$\varepsilon^s\left\|\frac{\partial\chi}{\partial\varphi}\right\|_{\rho-(s-\frac{1}{2})\delta}\le\frac{1}{2}\delta_I,\qquad\varepsilon^s\left\|\frac{\partial\chi}{\partial I}\right\|_{\rho-(s-\frac{1}{2})\delta}\le\frac{1}{2}\delta_\varphi, \tag{3.21}$$

i.e., using (3.14) with $\tilde{\delta}=\frac{1}{2}\delta$ and (3.10), if

$$\varepsilon\le(\Lambda r^{-n-3}A^{-1}\gamma\delta_\varphi^{\eta+1}\delta_I F^{-1})^s\Lambda^{-1}. \tag{3.22}$$

This is the required consistency condition, which is clearly satisfied by assumption (3.12), if $\eta=n+1$ and Λ is chosen according to (3.15).

Step iv): Concerning g_s, one has obviously $\|f_{s-1,0}\|_{\rho-s\delta}\le\|f_{s-1}\|_{\rho-s\delta}\le\Lambda^{s-1}F$, and thus

$$\|g_s\|_{\rho-s\delta}\le\|g_{s-1}\|_{\rho-(s-1)\delta}+\left(\frac{\varepsilon}{\Lambda}\right)^{s-1}F. \tag{3.23}$$

By recurrence, recalling $g_0=0$, one has then

$$\|g_s\|_{\rho-s\delta}\le\sum_{l=0}^{s-1}\left(\frac{\varepsilon}{\Lambda}\right)^l F; \tag{3.24}$$

if (3.12) is assumed, then (3.13b) is guaranteed.

Concerning the estimate for f_s, one uses the expressions (3.5) for the remainders appearing in (3.9), and writes:

$$\begin{aligned}\|R_1^{\varepsilon^s}[\chi,\varepsilon g_{s-1}+\varepsilon^s f_{s-1}\|_{\rho-s\delta} &\le \varepsilon^s\|\{\chi,\varepsilon g_{s-1}+\varepsilon^s f_{s-1}\}\|_{\rho-(s-\frac{1}{2})\delta}\\ &\le \varepsilon^{s-1}(2^{n+3}A\gamma^{-1}\delta_\varphi^{-\eta-1}\delta_I^{-1}F)\Lambda^{s-1}\sum_{l=0}^{s-1}\left(\frac{\varepsilon}{\Lambda}\right)^l F\end{aligned} \tag{3.25}$$

and

$$\begin{aligned}\|R_2^{\varepsilon^s}[\chi,\omega\cdot I]\|_{\rho-s\delta} &\le \frac{\varepsilon^{2s}}{2}\|\{\chi\{\chi,\omega\cdot I\}\}\|_{\rho-(s-\frac{1}{2})\delta}\\ &= \frac{\varepsilon^{2s}}{2}\|\{\chi,f_{s-1}-f_{s-1,0}\}\|_{\rho-(s-\frac{1}{2})\delta}\\ &\le \varepsilon^{2s}2^{n+2}A\gamma^{-1}\delta_\varphi^{-\eta-1}\delta_I^{-1}F\Lambda^{s-1}(2\Lambda^{s-1}F)\\ &\le \varepsilon^{s+1}(2^{n+3}A\gamma^{-1}\delta_\varphi^{-\eta-1}\delta_I^{-1}F)\Lambda^{s-1}(\varepsilon/\Lambda)^{s-1}F\,.\end{aligned} \tag{3.26}$$

One has consequently, for $\varepsilon/\Lambda \le \frac{1}{2}$,

$$\|f_s\|_{\rho-s\delta} \le (2^{n+4}A\gamma^{-1}\delta_\varphi^{-\eta-1}\delta_I^{-1}F)\Lambda^{s-1}F\,, \tag{3.27}$$

which gives (3.13a) if Λ is taken as in (3.15).

Step v): To get (3.13c,d), one writes

$$\begin{aligned}\|\alpha_s\|_{\frac{1}{4}\rho} \le \varepsilon^s\left\|\frac{\partial\chi}{\partial\varphi}\right\|_{\frac{1}{4}\rho+\frac{1}{2}\delta} &\le \varepsilon^s\left\|\frac{\partial\chi}{\partial\varphi}\right\|_{\rho-(s-1)\delta-\frac{1}{4}\rho}\\ &\le \varepsilon^s A\gamma^{-1}\left(\frac{\rho_\varphi}{4}\right)^{-\eta-1}\Lambda^{s-1}F\\ &\le \frac{4^{\eta+1}A}{\Lambda\gamma F^{-1}\rho_\varphi^{\eta+1}}\varepsilon^s\Lambda^s \le \frac{1}{4}r^{-n-3}\varepsilon^s\Lambda^s\rho_I\,,\end{aligned} \tag{3.28}$$

and proceeds similarly for β_s.

3.3 Conclusion of the proof. We know from the previous subsection that our perturbative construction can be carried on, for a given ε, up to any order r, such that the inequality (3.12) is satisfied, i.e., recalling (3.11), for

$$r \le r_{\max}(\varepsilon) = \left(\frac{\tilde{\varepsilon}}{2\varepsilon}\right)^{1/m}\,, \tag{3.29}$$

with $\tilde{\varepsilon} = r^{n+3}\Lambda^{-1} = 2^{-5n-17}n^{-2n-3}\gamma\rho_\varphi^{n-2}\rho_I$, and $m = n+3$. Once (3.29) is satisfied, then the final Hamiltonian H_r is well defined in $D_{\rho-r\delta} \subset D_{\frac{1}{2}\rho}$, and has the form

$$H_r(I,\varphi,\varepsilon) = \omega\cdot I + \varepsilon g_r(I,\varepsilon) + \varepsilon^{r+1}f_r(I,\varphi,\varepsilon)\,, \tag{3.30}$$

with $\|g_r\|_{\frac{1}{2}\rho} < 2F$ and $\|f_r\|_{\frac{1}{2}\rho} \le \Lambda^r F = (r^m/\tilde{\varepsilon})^r F$.

Following ref.[5], we now look for the optimal value of r, i.e., that value $r^*(\varepsilon) \le r_{\max}(\varepsilon)$, such that the φ–dependent term of H_r is minimal; this procedure gives quite naturally the exponential estimates claimed in the Nekhoroshev–Gallavotti theorem.

Denote $\mathcal{R}(r) = \varepsilon^{r+1}(r^m/\tilde{\varepsilon})^r$, and think of r as a continuum variable. One immediately recognizes that $\mathcal{R}(r)$ attains its minimum at

$$\tilde{r} = e^{-1}\left(\frac{\tilde{\varepsilon}}{\varepsilon}\right)^{1/m} < r_{\max}\,; \tag{3.31}$$

in order that our optimal value r^* is an integer, we make for it the choice

$$\tilde{r} - 1 < r^* \leq \tilde{r}\,. \tag{3.32}$$

One has then

$$\begin{aligned}\mathcal{R}(r^*) &\leq \varepsilon^{-1}\left(\frac{\tilde{r}^m\varepsilon}{\tilde{\varepsilon}}\right)^{\tilde{r}-1} = (e^{-m})^{e^{-1}(\tilde{\varepsilon}/\varepsilon)^{1/m}-1}\\ &\leq \frac{\varepsilon}{\varepsilon_0}(\varepsilon_0 e^m)e^{-m(\varepsilon_0/\varepsilon)^{1/m}}\\ &< \frac{1}{16}\frac{\varepsilon}{\varepsilon_0}\gamma\rho_I\rho_\varphi e^{-m(\varepsilon_0/\varepsilon)^{1/m}}\,,\end{aligned} \tag{3.33}$$

with $\varepsilon_0 = e^{-m}\tilde{\varepsilon} = C^{-1}\gamma\rho_\varphi^{n+2}\varphi_I F^{-1}$, C satisfying (2.13) (in the last inequality we used $\rho_\varphi \leq 1$, and disregarded a numeric factor less than one).

The above inequality gives us the exponential estimate we had to prove. To complete the proof of the theorem, one must prove the inequalities (2.17). To get (2.17a), one writes

$$\begin{aligned}\|\alpha\|_{\frac{1}{4}\rho} &\leq \sum_{s=1}^{r}\|\alpha_s\|_{\frac{1}{4}\rho} \leq \frac{\rho_I}{4}r^{-n-3}\sum_{s=1}^{r}\left(\frac{\varepsilon}{\Lambda}\right)^s\\ &\leq \frac{1}{2}\rho_I r^{-n-3}\frac{\varepsilon}{\Lambda} = \frac{1}{2}\rho_I\frac{\varepsilon}{\tilde{\varepsilon}}\\ &= \frac{1}{2}e^{-m}\frac{\varepsilon}{\varepsilon_0}\rho_I < \frac{1}{8}\frac{\varepsilon}{\varepsilon_0}\rho_I\,;\end{aligned} \tag{3.34}$$

in a similar way one also gets (2.17b).

4. Nekhoroshev–like results

4.1 This section is devoted to the study of a class of Hamiltonian systems with $n+\nu$ degrees of freedom, representing ν harmonic oscillators weakly coupled to some other Hamiltonian system with n degrees of freedom.

Denoting by $(\pi, \xi) \in \mathbb{R}^{2\nu}$, $\omega \in \mathbb{R}^\nu$ the coordinates and the angular velocity of the oscillators, and by $(p, q) \in B \subset \mathbb{R}^{2n}$ the remaining coordinates, our Hamiltonian has the form

$$H(p,q,\pi,\xi) = \frac{1}{2}\sum_{j=1}^{\nu}(\pi_j^2 + \omega_j^2\xi_j^2) + h(p,q) + f(p,q,\pi,\xi)\,, \tag{4.1}$$

where $h(p, q)$ is any Hamiltonian, and the coupling term f is assumed to vanish for $\xi = 0$. We consider the above Hamiltonian in the limit of large ω, precisely $\omega = \lambda\Omega$, with fixed $\Omega \in \mathbb{R}^\nu$ and $\lambda \to \infty$, at finite total energy; this implies, in particular, $\xi \sim \lambda^{-1}$, so that $f \sim \xi$ is really "small". This class of Hamiltonians has two main applications, which have been already mentioned in Section 1, examples *i.* and *ii.*; here we consider these applications in more detail.

4.2 Example i. Let us use the ordinary spherical coordinates (r, ϑ, φ); the Hamiltonian of an "ideally realized" spherical pendulum of mass m and length l is

$$h(p_\vartheta, p_\varphi, \vartheta, \varphi) = \frac{p_\vartheta^2}{2ml^2} + \frac{p_\varphi^2}{2ml^2 \sin^2 \vartheta} + V_0(\vartheta, \varphi), \tag{4.2}$$

where $V_0(\theta, \varphi)$ is the external potential. Denote $\xi = r - l$, and assume that the constraint is physically realized by a "confining potential" $\lambda^2 W(\vartheta, \varphi, \xi)$, with $W = 0$ for $\xi = 0$ and positive elsewhere. The "physical" Hamiltonian is then

$$H = \frac{\pi^2}{2m} + \frac{p_\vartheta^2}{2m(l+\xi)^2} + \frac{p_\varphi^2}{2m(l+\xi)^2 \sin^2 \vartheta} + V(\vartheta, \varphi, \xi) + \lambda^2 W(\vartheta, \varphi, \xi), \tag{4.3}$$

where π denotes the radial momentum, while $V(\theta, \varphi, \xi)$, with $V(\vartheta, \varphi, 0) = V_0(\vartheta, \varphi)$, is the complete external potential.

Now assume W is homogeneous, i.e. dependent only on ξ, and also, for simplicity, harmonic: $W(\xi, \vartheta, \varphi) = \frac{1}{2} m \Omega^2 \xi^2$. Then, by a Taylor development in ξ, one obtains from (4.3) a Hamiltonian of the form

$$H = \frac{\pi^2}{2m} + m(\lambda\Omega)^2 \frac{\xi^2}{2} + h(p_\vartheta, p_\varphi, \vartheta, \varphi) + f(p_\vartheta, p_\varphi, \vartheta, \varphi, \pi, \xi), \tag{4.4}$$

with $f = 0$ for $\xi = 0$; after a trivial rescaling $\xi \to \sqrt{m}\xi$, $\pi \to \pi/\sqrt{m}$, this Hamiltonian assumes the form (4.1), with $\nu = 1$, $n = 2$, and $p = (p_\vartheta, p_\varphi)$, $q = (\vartheta, \varphi)$. As shown in ref. [15,17], one is naturally led to consider Hamiltonians of the form (4.1), whenever in a dynamical system with $n+\nu$ degrees of freedom one introduces "physically" ν constraints, if one assumes that the confining potential is homogeneous, i.e., independent of the q coordinates.

A very natural question, in connection with this problem of realization of constraints, is the following: let $(p_\lambda(t), q_\lambda(t), \pi_\lambda(t), \xi_\lambda(t))$ be a trajectory with some fixed initial datum $(p^\circ, q^\circ, \pi^\circ, \xi^\circ = 0)$. One would like to know whether, for $\lambda \to \infty$, one has $(p_\lambda(t), q_\lambda(t)) \to (p_\infty(t), q_\infty(t))$, where $(q_\infty(t), p_\infty(t))$ is the trajectory of the ideal Hamiltonian system h corresponding to the initial datum (p°, q°).

As shown in ref.[13-15], the answer (within some assumptions) is positive; in fact, by suitably adapting the existence–uniqueness theorem on the solution of ordinary differential equations, it is not difficult to obtain estimates of the form

$$\text{dist}\,[(p_\lambda(t), q_\lambda(t)), (p_\infty(t), q_\infty(t))] < (\lambda/\lambda_0)^{-1} e^{\kappa t}, \tag{4.5}$$

where λ_0 and κ are suitable positive constants. This inequality shows that, for $\lambda \to \infty$ at fixed t, one has point–wise convergence of orbits: however, the limit is highly non uniform in time, and for a given λ one clearly looses any control on orbits after a very short time–scale, precisely

$$t \sim \ln \lambda/\lambda_0\,. \tag{4.6}$$

As is typical of the existence–uniqueness theorem, the inequality (4.5) is, in general, optimal: clearly, if h has unstable orbits (positive Lyapunov exponents), then any small disturbance due to the transversal vibration is amplified by the dynamics of h, and grows exponentially with t. However, the situation drastically changes if one looks at the integrals of motion of h, in particular at the energy: indeed, as we will see, one can prove that h is almost–constant for a much longer time–scale, precisely that one has

$$|h(p_\lambda(t), q_\lambda(t)) - h(p^\circ, q^\circ)| \sim (\lambda/\lambda_0)^{-1} \tag{4.7}$$

for

$$t \sim e^{\lambda/\lambda_0}\,. \tag{4.8}$$

4.3 Example ii. Consider the collision of a (classical) diatomic molecule with a wall. In a one–dimensional problem one has

$$H = \frac{1}{2}\left(\frac{\pi^2}{\mu} + \mu\omega^2\xi^2\right) + \frac{p^2}{2M} + V(q,\xi)\,, \tag{4.9}$$

where q is the coordinate of the center of mass, ξ is the coordinate of the internal vibrational degree of freedom, M and μ are the total and the reduced mass of the molecule, and V is the collision potential. If one writes $V = V_0(q) + \mathcal{O}(\xi)$, then H assumes the form (4.1) with $\nu = 1$, $n = 1$, and $h = p^2/2M + V_0(q)$; one easily recognizes that, in a three–dimensional problem including rotations, one still obtains H of the form (4.1), with $\nu = 1$ and $n = 5$, while for a collision of N molecules one still gets the above form, with $\nu = N$ and $n = N$ or $n = 5N$, depending on the dimensionality of the problem.

The natural question in this example concerns the efficiency of collisions in producing energy exchanges among the different degrees of freedom. Clearly, the rate of the energy exchange is very relevant for the time–scale necessary to a diatomic gas to reach the thermodynamic equilibrium.

This problem was deeply discussed, at the turn of the century, in connection with the well known difficulties of classical statistical mechanics in providing an explanation for the apparent "freezing" of the high–frequency degrees of freedom. As is well known, on the basis of the principle of equipartition of energy — a cornerstone of classical statistical mechanics — one expects, for diatomic gases, a specific heat $C_v = \frac{7}{2}R$; indeed, each molecule should provide 7 contributions to C_v: 3 for translations, 2 for rotations, and 2 (kinetic + potential) for the vibration (actually, one expects a much higher number of contributions, if one also considers the internal structure of atoms). The experimental value $C_v = \frac{5}{2}R$ indicates that molecules behave as if they were rigid bars, and this is usually considered to be a purely quantum effect, which classical physics is qualitatively unable to explain.

A possible entirely classical explanation was instead proposed by Boltzmann[25] and Jeans[26,27], who conjectured that this, and other phenomena of "freezing" of high frequency degrees of freedom (blackbody spectrum, rotating molecules at low temperatures), could be explained dynamically, by admitting the existence of a very long time–scale ("years", "billions of years", in their own words) for the energy exchanges with the high-frequency degrees of freedom, so that the latter, on the time–scale of ordinary experiments, appear to be frozen.

Jeans, in particular, conjectured and heuristically supported the idea that the energy exchange per collision was exponentially decreasing with the frequency, precisely of the order $e^{-t_0\omega}$, where t_0 is a time of the order of the collision time; consequently, in order to "see" thermodynamically the vibrational degrees of freedom, one should observe the gas for a time–scale $T \sim t_1 e^{t_0\omega}$, t_1 being the averaged time between collisions; using realistic values for t_0, t_1 and ω, time–scales of centuries are easily found.

Hamiltonian systems of the form (4.1) can be used to study the Jeans conjecture; as we will see, exponential laws of the form conjectured by Jeans can be rigorously proven.

4.4 A theorem. We consider here Hamiltonians of the form (4.1) in the simpler case $\nu = 1$ (but any n), i.e. of the form

$$H(p,q,\pi,\xi) = \frac{1}{2}(\pi^2 + \lambda^2\Omega^2\xi^2) + h(p,q) + \xi\hat{f}(p,q,\pi,\xi), \tag{4.10}$$

where the leading ξ-dependence of the coupling term has been put into evidence. Following ref.[17], we show that a natural extension of the techniques introduced in the previous sections allows one to get exponential estimates, although in the weaker form $\exp(\lambda/\lambda_0)^{1/2}$, or $\exp -(t_0\omega)^{\frac{1}{2}}$. One can obtain pure exponentials[28], also working with any $\nu \geq 1$[29], by means of somehow different, more accurate and complicated perturbation techniques; this point will be discussed later.

Let us introduce in Hamiltonian (4.1) the action–angle variables (I,φ) of the oscillator, i.e.

$$\begin{aligned} \pi &= \sqrt{2\lambda\Omega I}\cos\varphi\,, \\ \xi &= \sqrt{2I/\lambda\Omega}\sin\varphi = (\lambda\Omega)^{-1}\sqrt{2\lambda\Omega I}\sin\varphi\,. \end{aligned} \tag{4.11}$$

At finite total energy the quantity $\lambda\Omega I$ is bounded: consequently, our Hamiltonian (for which we do not change the notation) can be written in the form

$$H = (p,q,I,\varphi) = \lambda\Omega I + h(p,q) + \lambda^{-1}f(p,q,I,\varphi,\lambda), \tag{4.12}$$

where $f = (\Omega I)^{-1}\hat{f}(p,q,\sqrt{2\lambda\Omega I}\cos\varphi,\sqrt{2I/\lambda\Omega}\sin\varphi)$ is bounded even for $\lambda \to \infty$, if the total energy is finite.

The above Hamiltonian is naturally defined for $I \in [0, I_0/\lambda]$, $\varphi \in \mathbb{R}$ (f being periodic in φ), and $(p,q) \in B \in \mathbb{R}^{2n}$, where I_0 is a suitable constant with the dimension of an action, and B is the natural domain of h. To proceed as in the previous section, we must assume that H is analytic in a complex neighborhood of the domain of definition; to this purpose, we introduce an "extension vector" $\rho = (\rho_p, \rho_q, \rho_I/\lambda, \rho_\varphi)$ (clearly, to keep H bounded the I-component must be small with λ). As it is not restrictive, we will assume $\rho_p\rho_q = \rho_I\rho_\varphi$, and denote the product by ρ^2.

Unfortunately, the change of variables (4.11) introduces a lack of analyticity at $I = 0$: thus, we must first restrict the real domain of definition of H to $D_0 = B \times [\frac{\rho_I}{\lambda}, \frac{I_0}{\lambda}] \times \mathbb{R}$; then we can consider the polydiscs

$$\begin{aligned} \Delta_\rho(p,q,I,\varphi) = \{&(p',q',I',\varphi') \in \mathbb{C}^{2n+2}; |p'_j - p_j| < \rho_p, \\ &|q'_j - q_j| < \rho_q, j = 1,\ldots,n, |I' - I| < \frac{\rho_I}{\lambda}, |\varphi' - \varphi| < \rho_\varphi\}, \end{aligned} \tag{4.13}$$

and define the extended analyticity domain

$$D_\rho = \bigcup_{(p,q,I,\varphi)\in D_0} \Delta_\rho(p,q,I,\varphi). \tag{4.14}$$

It will also be useful to consider the following domains:

$$\mathcal{D}_\rho(p,q,I) = \bigcup_{\varphi\in\mathbb{R}} \Delta_\rho(p,q,I,\varphi). \tag{4.15}$$

Concerning norms, for any set $D' \subset D$ we will consider, as in Section 3, the supremum norm $\|.\|_{D'}$; for domains D' of the form $D_{\rho'}$, $\rho' \leq \rho$ (the inequality is intended to work separately for each component of ρ), we will also use the short notation $\|.\|_{\rho'}$ in place of $\|.\|_{D_{\rho'}}$. The use of sets of the form $\mathcal{D}_\rho(p,q,I)$, and of the corresponding "local" norms $\|.\|_{\mathcal{D}_\rho(p,q,I)}$, will allow us to give a more detailed statement, which is particularly important for the application of our results to Jeans' problem.

We can now state our

Theorem: *Consider the Hamiltonian*

$$H(p,q,I,\varphi,\lambda) = \lambda\Omega I + h(p,q) + \lambda^{-1} f(p,q,I,\varphi,\lambda)\,, \tag{4.16}$$

and assume:

i. *H is analytic in D_ρ, as defined above, with $\rho = (\rho_p, \rho_q, \rho_I/\lambda, \rho_\varphi)$, $\rho_p\rho_q = \rho_I\rho_\varphi = \rho^2, \rho_\varphi \leq 1$.*

Denote:

$$\begin{aligned} F &= \|f\|_\rho,\ \mathcal{F}_\rho(p,q,I) = \|f\|_{\mathcal{D}_\rho(p,q,I)} \\ E &= \max(F,\ \|k\|_\rho, \Omega\rho^2)\,, \end{aligned} \tag{4.17}$$

and assume:

ii. *One has*

$$\lambda > \lambda_0 = 2^{11} E/(\Omega\rho^2)\,. \tag{4.18}$$

Then:

(a) *There exists a canonical transformation $(p,q,I,\varphi) = \mathcal{C}(p',q',I',\varphi',\lambda)$, $\mathcal{C} : D_{\frac{1}{2}\rho} \to D_\rho$, such that the new Hamiltonian $H' = H \circ \mathcal{C}$ has the form*

$$\begin{aligned} H' = \lambda\Omega I' &+ h(p',q') + \lambda^{-1} g(p',q',I',\lambda) \\ &+ \lambda^{-1} e^{(\lambda/\lambda_0)^{\frac{1}{2}}} f'(p',q',I',\varphi',\lambda)\,, \end{aligned} \tag{4.19}$$

with

$$\|f'\|_{\frac{1}{2}\rho} \leq 8F, \qquad \|g\|_{\frac{1}{2}\rho} \leq 2F\,; \tag{4.20}$$

(b) *The canonical transformation is small with λ^{-1}, precisely one has the (local) estimates*

$$\begin{aligned} |p'-p| &< \frac{1}{8}\frac{\lambda_0}{\lambda}\frac{\mathcal{F}_\rho(p,q,I)}{E}\rho_p \\ |q'-q| &< \frac{1}{8}\frac{\lambda_0}{\lambda}\frac{\mathcal{F}_\rho(p,q,I)}{E}\rho_q \\ |I'-I| &< \frac{1}{8}\frac{\lambda_0}{\lambda}\frac{\mathcal{F}_\rho(p,q,I)}{E}\frac{\rho_I}{\lambda}\rho_\varphi \\ |\varphi'-\varphi| &< \frac{1}{8}\frac{\lambda_0}{\lambda}\frac{\mathcal{F}_\rho(p,q,I)}{E}\rho_\varphi\,. \end{aligned} \tag{4.21}$$

A sketch of the proof is deferred to subsection 6.

4.5 Corollaries. From this theorem one easily deduces two corollaries: precisely, denote by $x(t) = (p(t), q(t), I(t), \varphi(t))$ any orbit with initial datum in D_0, and by

$x'(t)=(p'(t),q'(t),I'(t),\varphi'(t))$ the corresponding orbit $C^{-1}(x(t))$; from (4.21), recalling $\mathcal{F}_\rho(p,q,I)\le F\le E$, $\rho_\varphi\le 1$, one has $x'(0)\in D_{\frac{1}{8}\rho}$. Now, as far as $x'(t)$ remains in the real part of $D_{\frac{1}{2}\rho}$, then one has $|\dot I'_j|\le\|\frac{\partial f'}{\partial\varphi}\|_{\frac{1}{2}\rho}\le\frac{2}{\rho_\varphi}\|f'\|_{\frac{1}{2}\rho}$, and consequently

$$|\lambda\Omega I'_j(t)-\lambda\Omega I'_j(0)|\le\frac{16}{\rho_\varphi}\Omega F e^{-(\lambda/\lambda_0)^{1/2}}|t|\,. \tag{4.22}$$

In the case h admits compact energy surfaces (this is the case of the spherical pendulum, and is typical of constraint problems), then it is meaningful to consider long time–scales, and formulate the following

Corollary I: *One has*

$$|\lambda\Omega I'(t)-\lambda\Omega I'(0)|\le\frac{1}{2}\frac{\lambda_0}{\lambda}F \tag{4.23a}$$

$$|\lambda\Omega I(t)-\lambda\Omega I(0)|\le\frac{3}{4}\frac{\lambda_0}{\lambda}F \tag{4.23b}$$

$$|h(p(t),q(t))-h(p(0),q(0))|\le\frac{\lambda_0}{\lambda}F \tag{4.23c}$$

for $|t|\le\min(\hat T,T)$, *where* $\hat T$ *is the escape time of* $x'(t)$ *from* $D_{\frac{1}{2}\rho}$, *and*

$$T=2^{-5}\Omega^{-1}\frac{\lambda_0}{\lambda}e^{(\lambda/\lambda_0)^{1/2}}\rho_\varphi\,. \tag{4.24}$$

$\hat T$ *is actually greater than* T, *if the intial datum is further restricted according to*

$$\begin{aligned}\rho_I+\frac{3}{4}\frac{\lambda_0}{\lambda}\frac{F}{\Omega}\le\lambda I'(0)\le I_0-\frac{3}{4}\frac{\lambda_0}{\lambda}\frac{F}{\Omega}\\ h(p(0),q(0))\le h_{\min}-\frac{\lambda_0}{\lambda}F\,,\end{aligned} \tag{4.25}$$

where $h_{\min}$ *is the minimum of* h *in the real part of the border of* $D_{\frac{1}{2}\rho}$.

Proof: (4.23a) is obvious; (4.23b) is immediately deduced from (4.21), using $\mathcal{F}_\rho(p,q,I)\le F\le E$; (4.23c) follows from the conservation of energy applied to the Hamiltonian (4.16). The final statement concerning $\hat T$ is also immediate.

Corollary I is symbolically summarized in Figure 1.

The second corollary is instead adapted to Jeans' problem. Assume the potential between the vibrating molecule and the wall decreases exponentially with the distance; then f and in particular the local norms $\mathcal{F}_\rho(p,q,I)$ also decrease exponentially, say as F_0e^{-r/r_0}, where r_0 is the interaction range, and F_0 is a suitable constant; let us assume for simplicity $F_0\ge F$. Let $t=0$ ("before collision") be such that $r/r_0=\lambda/\lambda_0$, and $t=T$ ("after collision") be such that one has again $r/r_0=\lambda/\lambda_0$; one expects that T is of order λ/λ_0, say $T=c\,\Omega^{-1}\rho_\varphi\lambda/\lambda_0$, with c of order unity. One immediately deduces the following

Corollary II: *one has*

$$|\lambda\Omega I'(T)-\lambda\Omega I'(0)|<c\,F_0\,e^{-(r/r_0)^{1/2}} \tag{4.26a}$$

$$|\lambda\Omega I(T)-\lambda\Omega I'(0)|<(c+\frac{1}{4})F_0\,e^{-(r/r_0)^{1/2}} \tag{4.26b}$$

$$|h(p(t),q(t))-h(p(0),q(0))|<(c+\frac{1}{2})F_0\,e^{-(r/r_0)^{1/2}} \tag{4.26c}$$

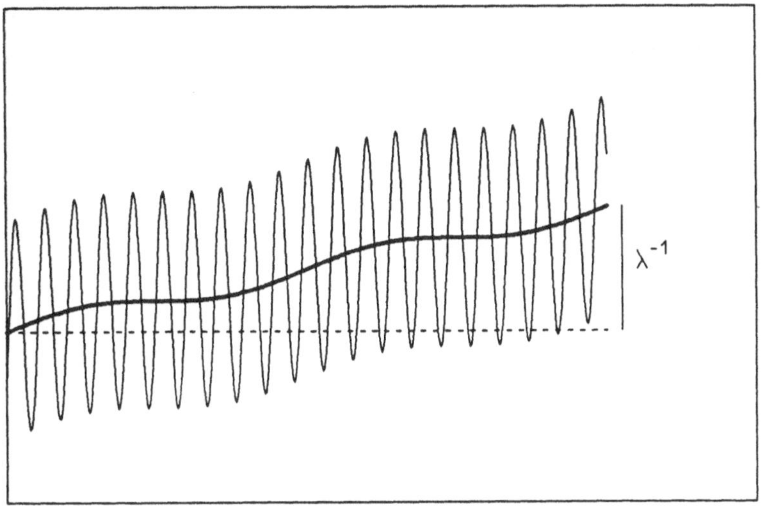

time

Figure 1 – *Illustrating Corollary I. Thick line: $\lambda\Omega I'(t)$ vs. t; thin line: $\lambda\Omega I(t)$ vs. t. The time–scale of the figure is $\sim e^{(\lambda/\lambda_0)^{1/2}}$; on this large time scale the energy of the oscillator is almost constant.*

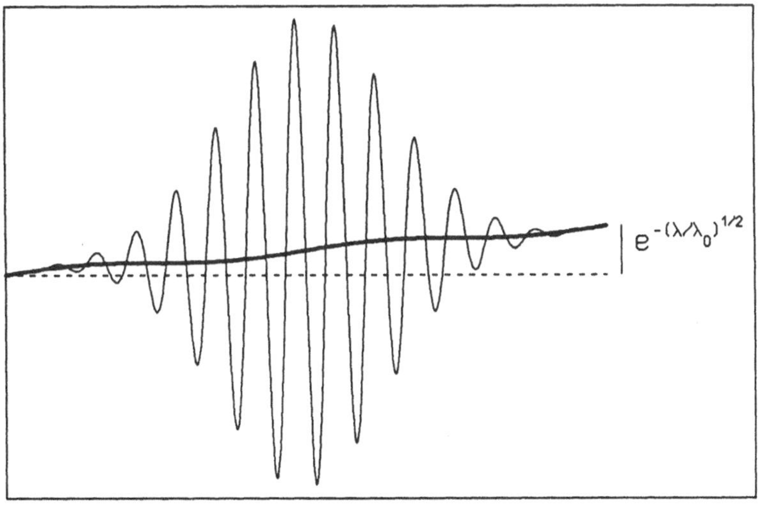

time

Figure 2 – *Illustrating Corollary II. Thick line: $\lambda\Omega I'(t)$ vs. t; thin line: $\lambda\Omega I(t)$ vs. t. The time–scale of the figure is the collision time $\sim \lambda^{-1}$; at the end of the collision the energy exchange with the internal vibrational degree of freedom is exponentially small.*

Proof: one proceeds exactly as in Corollary I, using however the local form of (4.21) to get the exponential estimates (4.26b,c) at $t = T$.

One should stress that (4.26b,c), at variance with (4.26a), hold only at the end of the collision, and not at intermediate times; this fact clearly appears in Figure 2, which symbolically summarizes Corollary II.

4.6 Proof of the theorem (sketch). The proof is similar to the proof reported in section 3; the only delicate point is that the domain of definition of I, and in particular its complex extension, is now small with λ^{-1}: consequently (recall Cauchy inequality) any derivative with respect to I gives an extra factor λ, and the orders in λ get frequently mixed. To make clear this point, let us shortly explain the formal structure of a single step of the canonical transformation $\mathcal{C}$, which, as in section 3, is taken of the form $\mathcal{C} = \mathcal{C}_1 \circ \ldots \circ \mathcal{C}_r$, with $\mathcal{C}_s : D_{\rho-(s-1)\delta} \to D_{\rho-s\delta}$, $\delta = \frac{\rho}{2r}$; $\mathcal{C}_1, \ldots, \mathcal{C}_s$ must be such that the Hamiltonian $H_s = H \circ \mathcal{C}_1 \circ \ldots \circ \mathcal{C}_s$ has the "weak normal form"

$$H_s = \lambda\Omega I + h(p,q) + \lambda^{-1} g_s(p,q,I,\lambda) + \lambda^{-s-1} f_s(p,q,I,\varphi,\lambda) . \tag{4.27}$$

Assume H_{s-1} already has the form (4.27), with $s-1$ in place of s. If $\mathcal{C}_s$ is defined by the Lie method (with $\varepsilon = \lambda^{-1}$), then one can write

$$\begin{aligned} H_s =& \lambda\Omega I + h + \lambda^{-1} g_{s-1} \\ &+ \lambda^{-1}(\{\chi, \lambda\Omega I\} + f_{s-1}) \\ &+ R_2^{\lambda^{-s}}[\chi, \lambda\Omega I] + R_1^{\lambda^{-s}}[\chi, h + \lambda^{-1} g_{s-1} + \lambda^{-s} f_{s-1}] , \end{aligned} \tag{4.28}$$

and in order that the second line is independent of φ, one imposes $\lambda\Omega\frac{\partial\chi}{\partial\varphi} = f_{s-1} - f_{s-1,0}$, i.e.

$$\chi = \lambda^{-1}\Omega \int^{\varphi} (f_{s-1} - f_{s-1,0})\, d\varphi = \mathcal{O}(\lambda^{-1}) \tag{4.29}$$

(no small denominators appear here!). Now, one has $\{\chi, f_{s-1}\}$, $\{\chi, g_{s-1}\} = \mathcal{O}(1)$ (indeed, χ is of order λ^{-1}, but the I–derivative gives a factor λ), and $\{\chi, h\} = \mathcal{O}(\lambda^{-1})$ (now there are no I–derivatives). Then, using expressions (3.5) for the remainders, one immediately checks that the last line gives, at most, terms of order λ^{-s}, as required to get H_s in the form (4.27). The precise estimates of the different terms, and the choice of r as function of λ leading to the exponential estimate (the optimization procedure), are very similar to the corresponding operations of section 3; they are not discussed here, and can be found with all details in ref. [17].

4.7 Improvements. We discuss here two important improvements of the above results, concerning the possibility of obtaining pure exponentials of the form $e^{\pm\lambda/\lambda_0}$ in place of $e^{\pm(\lambda/\lambda_0)^{1/2}}$, as in Jeans' conjecture, and the possibility of working with $\nu > 1$.

Concerning Jean's problem, the most relevant case is $\nu = 2$, i.e., collisions of two molecules; few–molecules collisions can also be taken into account, but certainly, in ordinary conditions, many–molecules collisions need not be considered, if one aims to explain the classical thermodynamics of diatomic gases. Thus, the ν–dependence of the various coefficients, in particular of λ_0, is not very important, while it is very interesting to obtain, for any ν, pure exponentials.

In the case $\nu = 1$, it is not very difficult to obtain pure exponentials: as shown by Fassò[28], one can still follow the perturbation scheme outlined above, and used in ref. [17], but in place of using the Lie method to transform Hamiltonians, one works directly on their derivatives (i.e., one transforms vector fields.) As a rule, in this way any estimate of the form $\exp -(\varepsilon_0/\varepsilon)^{1/m}$, or $\exp -(\lambda/\lambda_0)^{1/m}$, turns into $\exp -(\varepsilon_0/\varepsilon)^{1/(m-1)}$, or respectively $\exp -(\lambda/\lambda_0)^{1/(m-1)}$; this is not so relevant if, as in the Nekhoroshev–Gallavotti theorem, one has $m = n + 3$, but is quite important for the problem at hand, as $m = 2$ turns into $m - 1 = 1$.

Concerning the case $\nu > 1$, let us consider, to be definite, the collision of two identical molecules, i.e., the case of Hamiltonian (4.1), with $\nu = 2$ and $\omega_1 = \omega_2$. Because of this exact resonance relation, the two molecules vibrate in phase, and the transport of energy from one to the other is very easy; consequently, only the overall vibrational energy of the molecules, and not the separate energies of the two oscillators, have a chance to be conserved.

The counterpart in perturbation theory of this physical reasoning is the well known fact that, for each resonance relation admitted by ω, the number of the possibly conserved quantities is decreased by one; in our case, if one has $\nu \geq 2$ identical molecules, then ω admits the $\nu - 1$ resonance relations $\omega_1 = \omega_2 = \ldots = \omega_n$, and it turns out that the only possibly conserved quantity is the total harmonic energy $h_{\mathrm{harm}} = \frac{1}{2}\sum_j(\pi_j^2 + \omega^2 q_j^2)$. This is exactly what is needed if one wants to prove Jeans' conjecture for $\nu > 1$. However, the fact that the internal energies of the molecules are not separately constant, makes now impossible to use, in our perturbation scheme, the action–angle variables. Indeed, single actions cannot be kept away from zero, where the change of variables (4.11) is non–analytic, and the derivatives with respect to the actions diverge.

One is thus forced to work in the original Cartesian coordinates (π, ξ), and the whole perturbation theory becomes painful. Work is still in progress, but I can say that all technical difficulties are now overcome, and moreover, a very careful analysis of the analyticity domains of the different functions entering the proof leads, for any ν, to pure exponentials[29]. A rough statement is the following:

Theorem: *Consider the Hamiltonian*

$$H(p,q,\pi,\xi,\lambda) = h_{\mathrm{harm}}(\pi,\xi\lambda) + h(p,q) + \lambda^{-1} f(p,q,\pi,\xi,\lambda)\,,$$
$$h_{\mathrm{harm}}(\pi,\xi,\lambda) = \frac{1}{2}\sum(\pi_j^2 + \lambda^2\Omega^2\xi_j^2)\,,$$

and denote by D_0 the domain given by $\pi_j^2 + \lambda^2\Omega^2\xi_j^2 < E_0/\lambda^2$, $j = 1, \ldots, \nu$, and $(p,q) \in B \subset \mathbb{R}^n$. Assume:

i. H is analytic in a suitable complex neighborhood of D_0, and there bounded for $\lambda \to \infty$;

ii. f is no more than quadratic in π;

iii. λ is greater than a suitable constant λ_0, depending on all quantities entering the problem.

Then there exists a canonical transformation, small with λ^{-1}, which gives the new Hamiltonian H' the form

$$\begin{aligned} H'(p',q',\pi',\xi') =& h_{\mathrm{harm}}(\pi',\xi',\lambda) + h(p',q') + \lambda^{-1} g(p',q',\pi',\xi',\lambda) \\ &+ \lambda^{-1} e^{-\lambda/\lambda_0} f'(p',q',\pi',\xi',\lambda)\,, \end{aligned} \tag{4.30}$$

where g commutes with h_{harm}, i.e. $\{g, h_{\mathrm{harm}}\} = 0$, and both g and f' are bounded for $\lambda \to \infty$.

Remark: one has then $\frac{d}{dt}h_{\mathrm{harm}} \sim e^{-\lambda/\lambda_0}$, and corollaries analogous to those of section 4.5 are easily drawn.

5. Conclusions

It is not so easy, at this point, to draw conclusions: indeed, the work on Nekhoroshev theorem and Nekhoroshev–like results is in progress and, perhaps, still at the beginning. As remarked in the Introduction, Nekhoroshev–like perturbation theory is open to many physical applications: in particular, in Section 4 we discussed the problem of the realization of constraints, as well as Jeans' conjecture, while in the Introduction we mentioned the study of the adiabatic invariants and the splitting of separatrices in nearly–integrable systems; to complete the list of applications, for which some Nekhoroshev–like results are already available, one should also mention an application to astronomy, precisely a study of the stability of the triangular Lagrangian equilibrium points in the restricted three–body problem[30].

A very important point, in connection with physical applications, is the estimate of the constants entering the exponential laws. For example, for the problem of the "freezing" of the vibrational degrees of freedom in a classical diatomic gas, considered in Section 4, it would be very interesting to study realistic models, and work out for them Jeans–like exponential laws of the form $T \sim t_1 e^{t_0 \omega}$, with optimal estimates for t_0 and t_1. Unfortunately, as it is well known, it is quite hard to get good estimates of constants in classical perturbation theory: indeed, on one side it is difficult to take into account the peculiarities of the particular problem at hand; on the other side, mathematical tools like the Cauchy inequality, which can be used in lack of any detailed information on the functions one needs to estimate, are in most cases terribly pessimistic.

As a remarkable fact, numerical experiments give a rather striking evidence of Jeans' exponential law $T \sim t_1 e^{t_0 \omega}$; in particular, for a very rough one–dimensional model of diatomic gas[31], one finds constants t_0 and t_1 which are compatible with the time–scales of "billions of years" invoked by Jeans. Numerical work is also in progress for the problem of "freezing" of fast rotations; preliminary results show a behavior in good agreement with Jeans' exponential law[32]. One can also mention a numerical work on a one-dimensional model of radiant cavity[33], where no reference is made to Nekhoroshev exponential estimates, but some evidence is produced of a "freezing" of the high–frequency modes of the electromagnetic field, for times rapidly increasing with the frequency of the modes.

All of these numerical studies should be considered with care, and possibly repeated on more realistic models; a great effort should also be made to improve the techniques in perturbation theory. Anyhow, on the basis of the already available results, I am quite confident that the exponential laws conjectured by Jeans at the very beginning of the century, and recently introduced by Nekhoroshev in classical perturbation theory, are an essential deep element of classical mechanics, which could be very relevant for a proper understanding of the relations between classical and quantum physics, and provide in particular a completely classical interpretation of some phenomena, which are usually considered to be purely quantum ones. For a wider discussion on this point, one can look at the lecture by L. Galgani in this volume.

References

[1] N. N. Nekhoroshev: *Usp. Mat. Nauk* **32**, (1977) [*Russ. Math. Surv.* **32**, 1 (1977)].

[2] N. N. Nekhoroshev: *Trudy Sem. Petrows.* No. 5, 5 (1979).

[3] G. Benettin, L. Galgani, A. Giorgilli: *Nuovo Cimento* **89 B**, 89 (1985).

[4] G. Gallavotti: *Quasi-Integrable Mechanical Systems*, in *Critical phenomena, Random Systems, Gauge Theories*, K. Osterwalder and R. Stora editors, *Les Houches*, Session XLIII, 1984 (North–Holland, Amsterdam 1876).

[5] A. Giorgilli and L. Galgani: *Celestial Mechanics* **37**, 95 (1985).

[6] G. Benettin, L. Galgani, A. Giorgilli: *Nuovo Cimento* **89 B**, 103 (1985).

[7] V. I. Arnold: *Méthodes Mathématiques de la Mécanique Classique* (MIR, Moscow 1976).

[8] E. Fermi, J. Pasta and S. Ulam: *Los Alamos Report* No. LA-1940 (1955), later published in *E. Fermi: Collected Papers* (Chicago, 1965), and *Lect. Appl. Math.* **15**, 143 (1974).

[9] M. Hénon and C. Heiles: *Astron. J.* **69**, 73 (1964).

[10] H. Poincaré: *Les Méthodes Nouvelles de la Méchanique Céleste*, Vol. 3 (Gautier–Villars, Paris, 1899).

[11] L. Chierchia and G. Gallavotti: *Nuovo Cimento B* **67**, 277 (1982).

[12] J. Poeschel: *Commun. Pure Appl. Math.* **35**, 653 (1982); *Celestial Mechanics* **28**, 133 (1982).

[13] H. Rubin and P. Hungar: *Commun. Pure Appl. Math.* **10**, 65 (1957).

[14] F. Takens: *Motion under the Influence of a Strong Constraining Force*, in *Global Theory of Dynamical Systems*, Z. Nitecki and C. Robinson editors, *Lect. Notes Math.* No. 819 (Springer, Berlin 1979).

[15] G. Gallavotti: *Meccanica Elementare* (Boringhieri, Torino 1980) [English Edition: *The Elements of Mechanics* (Springer, Berlin 1983)].

[16] Van Kampen: *Phys. Rep.* **124**, 69 (1985).

[17] G. Benettin, L. Galgani and A. Giorgilli: *Realization of Holonomic Constraints and Freezing of High–Frequency Degrees of Freedom in the Light of Classical Perturbation Theory, part I*, to appear in Comm. Math. Phys.

[18] A.I. Neishtadt: *Prikl. Matem. Mekan.* **45**, 80 (1981) [*PMM U.S.S.R.* **45**, 58 (1982)].

[19] A.I. Neishtadt: *Prikl. Matem. Mekan.* **48**, 197 (1984) [*PMM U.S.S.R.* **45**, 133 (1984)].

[20] G. Benettin, D. Casati, L. Galgani, A. Giorgilli and L. Sironi: *Phys. Lett. A* **118**, 325 (1986).

[21] G. Benettin, L. Galgani and A. Giorgilli: *Poincaré's Non-Existence Theorem and Classical Perturbation Theory in Nearly-Integrable Hamiltonian Systems*, in *Advances in Nonlinear Dynamics and Stochastic Processes*, R. Livi and A. Politi editors (World Scientific, Singapore 1985).

[22] J. Moser: *Annali Scuola Norm. Sup. Pisa*, 499 (1966).

[23] H. Ruessmann: *On Optimal Estimates for the Solutions of Linear Partial Differential Equations of first Order with Constant Coefficients on the Torus*, in *Dynamical Systems, Theory and Applications*, J. Moser editor, *Lect. Notes Phys.* No 38 (Springer-Verlag, Berlin 1975).

[24] H. Ruessmann: *Comm. Pure Appl. Math.* **29**, 755 (1976).

[25] L. Boltzmann: *Nature* **51**, 413 (1895).

[26] J.H. Jeans: *Phil. Mag.* **6**, 279 (1903).

[27] J.H. Jeans: *Phil. Mag.* **10**, 91 (1905).

[28] F. Fassò: paper in preparation.

[29] G. Benettin, L. Galgani and A. Giorgilli: *Realization of Holonomic Constraints and Freezing of High–Frequency Degrees of Freedom in the Light of Classical Perturbation Theory, part II*, paper in preparation.

[30] A. Giorgilli, A. Delshams, E. Fontich, L. Galgani and C. Simò: *Effective Stability for a Hamiltonian System near an Elliptic Equilibrium Point, with an Application to the Restricted three Body Problem*, preprint.

[31] G. Benettin, L. Galgani and A. Giorgilli: *Phys. Lett.***A** **120**, 23 (1987).

[32] O. Baldan and G. Benettin: paper in preparation.

[33] G. Benettin and L. Galgani: *J. Stat. Phys.* **27**, 153 (1982).

RELAXATION TIMES AND THE FOUNDATIONS OF CLASSICAL STATISTICAL MECHANICS IN THE LIGHT OF MODERN PERTURBATION THEORY

Luigi Galgani

Dipartimento di Matematica dell'Universitá
Via Saldini 50 - 20133 Milano, Italia

Abstract. We illustrate the point of view of Boltzmann and Jeans, according to which in most cases of interest to statistical mechanics one is in usually in conditions of metaequilibrium rather than in conditions of equilibrium, an show how a mathematical support to such point of view is given by the recent Nekhoroshev theorem of perturbation theory. As an application, we report the deduction of Planck's law in a classical context already given by Nernst, and show moreover how such law should be qualitatively modified at low frequencies to a higher approximation.

1. The problem of relaxation times and the "freezing" of degrees of freedom according to Boltzmann, and Nekhoroshev's theorem

In the present lecture, following the same attitude described first in ref.[1], we discuss some questions of a general character concerning the relevance for statistical mechanics of some recent results of Nekhoroshev's type[2] obtained recently in the framework of classical perturbation theory; a technical illustration of such results can be found in the lecture by Benettin[3], while an application to the significance of an "effective fractal dimension" in conservative dynamical systems, in a spirit very similar to the present one, is given in Giorgilli's lecture[4].

It is well known that by a dynamical system one can describe a certain phenomenon only within some approximation, in particular only within some range of time; for example our planetary system is rather well described by a set of point particles, or rigid bodies, with gravitational interactions only for times such that viscous effects (due e.g. to tides) can be neglected. On the other hand, in investigating the statistical properties of a dynamical system in the framework of statistical mechanics, in order to handle notions which are neatly defined in a mathematical sense it is usually convenient to consider infinite time limits in the spirit of ergodic theory.

It is thus clear that the interpretation of the results of ergodic theory (which involves infinite times) for a particular model (which is meaningful for a finite range of time) requires a careful consideration. For example, one might think of different time scales, with a mathematical model going successively into other ones; one should thus be prepared to consider, so to say, different scales of infinity. This fact was particularly stressed by Boltzmann, the inventor of ergodic theory itself, as one can see from the general discussion he gave in his famous letter to Nature of the year 1895,[5] an excerpt of which is reported here in the Appendix.

In any case, a first essential implication of the existence of different orders of relaxation times is the separation of the internal mechanical energy into a part which is relevant for thermodynamics (*thermodynamic energy*) and another one which is irrelevant for it (*frozen energy*), being non exchangeable within a certain definite observation time; the thermodynamic internal energy is then just a fraction of the mechanical internal energy. This fact is very vividly exhibited in the most typical model of gas theory, namely the beloved Boltzmann model of hard spheres. Indeed, for perfectly smooth spheres it is evident that the distribution of rotational energy is not at all changed by collisions, so that the collisions can only possibly allow for equipartition of translational energy; thus, in such a model one is forced to divide the internal energy into the translational one and the rotational one, the former alone contributing to the termodynamical energy, while the latter is, so to say, frozen, i.e. to all effects as not existing for thermodynamics. In the case instead of rough spheres, rotational energy too would come into play in statistical mechanics, although after a certain relaxation time which should be estimated in any model *. Analogous considerations are in order in the case of polyatomic molecules, with the vibrational degrees of freedom too coming into play after their respective characteristic relaxation times.[6] Judging from the experimental data on the specific heats of gases, one would estimate that the relaxation times should be increasing when one passes from the rotational degrees of freedom to the vibrational ones, and, for the latter ones, they should be increasing with the frequency. Some arguments were given by Jeans in the year 1903 to maintain that they should increase exponentially with the frequency (see the Appendix).

In the same spirit, particularly interesting is the case of the black body, which from the dynamical point of view is just equivalent to a system of harmonic oscillators of all frequencies, coupled by matter; in such a case one should then estimate the relaxation times to equilibrium of the various oscillators as a function of their frequency. In this connection, it was explicitly stated by Planck[7], in discussing his so called "second theory" in the year 1912, that the time required for the oscillators of frequency ν to reach the "first energy level" $h\nu$ at absolute temperature T should be proportional to $\exp(\frac{h\nu}{kT})$, where, as usual h and k are Planck's and Boltzmann's constants.

Such an exponential dependence on the frequency is enormous from a concrete point of view when one looks at the experimental verification of theories. Indeed let us assume that for a certain phenomenon where the frequency ν enters as a parameter we have a relaxation time $T(\nu)$ given by an exponential law

$$T(\nu) = \tau e^{\alpha\nu} ,$$

with two constants τ, α having the dimensions of a time; in an example illustrated below such times are both of the order of of magnitude of 10^{-13} sec. Then τ, which

---

* In connection with the Boltzmann equation the following example is also of interest. For the point particles describing the stars in a galaxy, Chandrasekhar, (see S. Chandrasekhar, *Principles of stellar dynamics*, The University of Chicago Press (1942)) calculated the average collision time to be typically of the order of magnitude of the age of the universe; that's why in such case one uses, even for the translational degrees of freedom, not the complete Boltzmann equation, but rather the corresponding equation with the collision term neglected, namely the Vlasov equation. So, in particular, one does not make use of the standard ensembles for the statistical mechanics of galaxies.

is typically of microscopic size, is just the relaxation time of the soft modes (of zero frequency); on the other hand there will be a frequency $\overline{\nu}$ with a macroscopic relaxation time, say for example one second, i.e. $\exp(\alpha\overline{\nu}) = \tau^{-1}$; but then just doubling the frequency gives a relaxation time $T(2\overline{\nu}) = \tau^{-1}$, namely 10^{13} sec $\simeq 10^5$ years in that example, and analogously $T(3\overline{\nu}) = 10^{26}$ sec, namely a time much larger than the estimated age of the universe. In other words, although one is speaking of relaxation times increasing continuously with the frequency, one is actually confronted with a situation that might appear, and in fact will appear, to a naive experimental investigation as essentially a discontinuous one: frequencies lower than a fraction of $\overline{\nu}$ will appear to relax to equilibrium while frequencies above a fraction of $\overline{\nu}$ will appear to be essentially completely frozen. A careful examination of the experimental data in an appropriate range of low frequencies is necessary in order to reveal the exponential law, and this can be done only if one is well convinced *a priori* of its possibility. From this point of view it is very interesting to look at the reports of the Solvay Conference of the year 1911[8]; Rayleigh, who was in favour of the point of view of Boltzmann, had asked to discuss the experimental data that might confirm or disprove it, and the experimentalists' answer (see the Appendix) shows that they did not take seriously into consideration the possibility of an exponential law for the relaxation times.

However one has to admit that the theoretical support to the point of view of Boltzmann was very poor; although in principle correct, such point of view had no quantitative theory beyond it, and the considerations of Jeans were not really convincing. So, in contrast to the metaequilibrium point of view of Boltzmann there prevailed what might be called the static point of view, according to which in the problems of the specific heats and of the blackbody law one is in presence of a real equilibrium.

We maintain that the situation should be reconsidered today in the light of the modern results of classical perturbation theory for Hamiltonian systems. Such results are of two types. On the one hand there is the famous Kolmogorov (or KAM) theorem on invariant tori (1954), which is in the spirit of ergodic theory, namely refers to manifolds which are invariant for *all* times. Its novelty (foreseen by Poincaré) consists in looking just for invariant surfaces, and not for integrals of motion in the familiar sense which would lead to a family of invariant surfaces giving a continuous foliation (or stratification) of the relevant phase space; however the manifolds are still required to be invariant for all times. On the other hand there is the less famous, but to our purposes more relevant, Nekhoroshev's theorem (1977), which is in a sense complementary to Kolmogorov's theorem. Indeed in Nekhoroshev's theorem one looks for results valid in the whole of an open set of phase space, but only for *finite* times; for example, some variables, as typically the actions in a nearly integrable Hamiltonian systems, are proven to change little (to be frozen) up to a certain time. Now, so stated, such result would be nothing more than the result of ordinary perturbation theory with normalization accomplished up to a finite order, which correspondingly gives results only up to a finite time. The point however is that one has a control on all possible orders of normalization (with analytical estimates for the remainders), and in such a way one can look for the best order of normalization; as a result, stability of the actions is then proven up to times which increase exponentially with some parameter.

Typically, let us consider a nearly integrable Hamiltonian system with Hamiltonian

$$H(I,\varphi) = h(I) + \epsilon f(I,\varphi,\epsilon)$$

where ϵ is a small real parameter, $I = (I_1,\cdots,I_n)$ and $\varphi = (\varphi_1,\cdots,\varphi_n)$ are action angle variables ($I \in \mathcal{B}, \varphi \in \mathbf{T}^n$, $\mathcal{B}$ being an open connected domain of $\mathbf{R}^n$ and $\mathbf{T}^n$ the n-dimensional torus), h and f are smooth, for example analytic. For $\epsilon = 0$ the system is integrable, i.e. one has $I(t) - I(0) = 0$ and $\varphi(t) - \varphi(0) = \omega(I(0))t$, where $\omega = \frac{\partial h}{\partial I}$, namely, as Boltzmann would say, the actions are frozen for all times, while the angles have fast motions. For $\epsilon \neq 0$ one clearly has the *a priori* estimate $\dot{I} = \epsilon \frac{\partial f}{\partial \varphi}$, namely

$$|I(t) - I(0)| \leq \epsilon \, ||\frac{\partial f}{\partial \varphi}||$$

for $|t| \leq 1$ (with a suitable norm $|| \cdot ||$), i.e. the actions are frozen (move little) up to times of order 1. Perturbation theory aims at giving informations for longer times. In order to do this, one looks for a canonical transformation $(I, \varphi) \mapsto (I', \varphi')$ which eliminate the angles up to higher orders, for example by reducing the Hamiltonian to the form

$$H^{(r)}(I', \varphi') = h^{(r)}(I', \epsilon) + \epsilon^r f^{(r)} B(I', \varphi', \epsilon)$$

for a positive integer r, so that one would have

$$|I'(t) - I'(0)| \leq \epsilon \, ||\frac{\partial f^{(r)}}{\partial \varphi}||$$

for $|t| \leq \epsilon^{-r-1}$, i.e. a freezing of the new actions (and essentially also of the old ones) up to times of the order ϵ^{-r-1}. With mild conditions on the unperturbed Hamiltonian h, Nekhoroshev provides uniform estimates in the domain $\mathcal{B}$ for long times if ϵ is sufficiently small: precisely there exist positive constants a, b and ϵ_0, such that one has

$$|I(t) - I(0)| \leq \mathcal{I} \epsilon^b \ , \ \text{for} \quad t \leq T \equiv \tau \frac{1}{\epsilon} \exp\left(\frac{\epsilon_0}{\epsilon}\right)^a \ ,$$

if $\epsilon \leq \epsilon_0$, where $\mathcal{I}$ and τ are dimensional constants. So one has indeed a freezing of the actions up to very long times, which increase essentially in an exponential way with $1/\epsilon$ as $\epsilon \to 0$ (see also ref.[9]).

The exponential dependence of the stability time T on $\frac{1}{\epsilon}$ in Nekhoroshev's theorem leads to a freezing of the actions up to times that in many cases are, in the considered models, essentially indistinguishable from an infinite time; in such a way one might then speak of an *effective* stability. For example in the restricted thrêe boby problem by analogous techniques one can show that, the primaries being Jupiter and the Sun, a point does not escape more than some hundreds kilometers from the Lagrangian triangular equilibrium point up to the estimated age of the Universe[10]. Nearer to the kind of problems discussed here is the problem of the realization of holonomic constraints. Considering for example a physical pendulum as realized by a hard bar, from the equipartition principle on would expect that the oscillatory energy of the pendulum will flow to the vibrational degrees of freedom of the bar, and one would like to estimate the time needed for such a sharing process to occur. The simplest mathematical model is with the bar described by a harmonic oscillator of proper frequency ν , which has the role of $\frac{1}{\epsilon}$ in Nekhoroshev's theorem. In such a case it has been shown that the energy of the oscillator varies little up to a time proportional to $\exp(a\nu)$, namely one gets a result of Nekhoroshev's type, with the constant a being exactly equal to 1[11].

For what concerns the relevance of Nekhoroshev's theorem for statistical mechanics, the common opinion of the few physicists that were aware of it was just that it is *completely irrelevant*, due to the fact that, according to the general *a priori* estimates obtained in the theorem, the power a to which the argument $\frac{1}{\epsilon}$ of the exponential is raised, should be of the form $a \simeq \frac{1}{n^2}$, and in particular should vanish in the limit $n \to \infty$, so that the exponential dependence on $\frac{1}{\epsilon}$ should disappear in the thermodynamic limit. With the words of Zaslavski:[12] *"with increasing n the quantity a becomes very small and the instability time becomes negligible"*.

On the other hand, a rather large group of people had been convinced by numerical computations on several models of the Fermi Pasta Ulam type [13] that some kind of stability should be present in many models of physical interest, even in the thermodynamic limit. The debate whether such stability should be interpreted in the ergodic spirit of Kolmogorov's theorem, or in the spirit of Nekhoroshev's theorem is still open, but eventually there came at least a positive result of Nekhoroshev's type, by numerical computations first, and then by analytical estimates.

The particular model considered is a one–dimensional version of a system of diatomic molecules. Precisely, [14] one considers a system of n equal springs (of frequency ν) on a line, coupled by a repulsive potential acting between the right ends of nearby springs. By numerical solutions of the equations of motion it was found that very rapidly one has equipartition of energy among the translational degrees of freedom of the centers of mass of the springs, while a larger time T_ν is required in order to have equipartition also for the vibrational degrees of freedom; precisely, it was found $T_\nu = A\exp(\alpha\nu)$, so that the quantity a of Nekhoroshev's theorem should be exactly equal to 1. Taking for the masses and the parameters entering the potential typical molecular values, it was found $A, \alpha \simeq 10^{-13}$ sec ; thus, at $\nu = \overline{\nu} \simeq 10^{13}$ sec $^{-1}$ one has $T_{\overline{\nu}} \simeq 1$ sec, while at $\nu = 2\overline{\nu}$ one has $T_{2\overline{\nu}} > 10^5$ years. In such a way, as discussed above in connection with the exponential law, one has, just by doubling a frequency, a dynamic freezing which is clearly indistinguishable from a static one. By a *tour of force* it was then possible very recently to come to an essentially complete analytical proof of such a fact, by just exploiting the particular resonance condition related to the equality of the frequencies of the n oscillators[15].

In conclusion, while by a naive immediate application of Nekhoroshev's theorem one should expect exponential relaxation times with exponents of the order $\nu^{\frac{1}{n^2}}$, where n is the number of degrees of freedom, in reality one meets exponents of the order ν, with a power 1, as was foreseen by Jeans and Planck.

In connection with Nekhoroshev's theorem it is of interest to point out in addition the possible physical significance of the "critical value" ϵ_0 of the perturbative parameter ϵ, below the theorem can be applied. To understand this, let us recall that, in performing any of the canonical transformations described above in order to get normalization up to order r, one has to require a bound $\epsilon < \epsilon_r$, with $\epsilon_{r+1} \leq \epsilon_r$; this is due to the fact that, in general, in a canonical transformation defined in the classical way using mixed variables one has to make use of the implicit function theorem; from another (essentially equivalent) point of view, the canonical tranformations, even at the first step, will necessarily be series in ϵ, which should be required to converge. So one has the following picture. Above ϵ_0 one cannot even perform a single step of perturbation theory, and one has just no freezing, apart from the trivial *a priori* one, which is effective just up to times of order 1; below ϵ_0, instead, the actions are frozen up to times which become larger and larger the more ϵ is close to 0. In this sense, ϵ_0 can be considered as a first estimate of some stochasticity threshold, above which the fluctuations of the actions are essentially free, while below ϵ_0 the actions become more and more frozen as if there were integrals of motion. More precisely, there are in general integrals of motion in some generalized sense, i.e. for very special initial conditions (on KAM tori), and the possible "Arnold diffusion" in the complement is very slow, being just estimated by Nekhoroshev theorem, and thus requiring exponential times.

In particular, let us come now to the case of weakly coupled harmonic oscillators. This is just the simplest of the cases of applicability of Nekhoroshev theorem, as was stressed first by Gallavotti, [16] because the so called geometric part of the theorem does not enter at all. In this case the unperturbed Hamiltonian $h(I)$ is linear in the actions, $h(I) = \sum_j \omega_j I_j$, $\omega = (\omega_1, \cdots, \omega_n) = \text{const}$, where $I_j = \frac{p_j^2+q_j^2}{2}$, if q_j is the displacement of the j-th oscillator from the equilibrium position and p_j its conjugate momentum. The Hamiltonian has in general the form $H = \sum_j \omega_j \frac{p_j^2+q_j^2}{2} + H^{(3)} + H^{(4)} + \cdots$, where $H^{(k)}$ is a polynomial of degree k in $q_1, \cdots, q_n$; namely, one is in fact considering a Taylor expansion about an (elliptic) equilibrium point, and in a sense there is no parameter ϵ, the role of ϵ being taken by the 'distance' $|I|$ from the origin. So, now the analogue of diminishing ϵ is to restrict I to balls of decreasing radius about the origin. This can be seen formally by introducing the trivial rescaling (blowing up) $\epsilon p' = p, \epsilon q' = q$, which allows to consider the actions I' still defined as before in a fixed domain $\mathcal{B}$ (for example the ball of radius 1) with an Hamiltonian (neglecting accents) $H(I,\varphi) = \sum_j \omega_j I_j + \epsilon H^{(3)}(I,\varphi) + \epsilon^2 H^{(4)}(I,\varphi) + \cdots$.

So one has for oscillators, instead of the critical parameter ϵ_0, a critical action A_0 estimating the onset of stochasticity: for $|I(0)| \leq A_0$ one can apply perturbation theory which guarantees a freezing of the actions for longer and longer times the more $|I(0)|$ tends to zero, while for $|I(0)| > A_0$ one can expect large fluctuations. Recalling that the energies E_j of the oscillators are given in a first approximation by $E_j = \omega_j I_j$, one can also say that there are critical energies $E_j^{(c)} = A_0 \omega_j$, such that one has frozen motions for $E_j < E_j^{(c)} (j = 1, \cdots, n)$, and possibly chaotic motions for $E_j > E_j^{(c)} (j = 1, \cdots, n)$. Moreover, let us consider an initial condition in which one excites many oscillators all essentially of the same angular frequency ω with energy $E_j(0) < E_j^{(c)}$; then, in the energy exchanges, each of them will be frozen up to a time of the order $\exp(\frac{A_0}{I_j(0)}) = \exp(\frac{A_0\omega}{E_j(0)})$, which increases exponentially with ω. And this seems strongly to support the opinion expressed by Boltzmann, concerning the freezing of the high frequency oscillators.

2. Nernst's deduction of Planck law, and possible modifications due to Arnold diffusion

The stochasticity threshold $A_0\omega$ for oscillators of angular frequency ω can be conceived as the analog of the quantum zero–point energy $\hbar\omega \equiv h\nu$ ($\hbar = \frac{h}{2\pi}$). [17] This was firts stated almost explicitly by Planck himself (see the Appendix) in the paper of the year 1912 where he discussed his so called "second theory" and also introduced the concept of zero–point energy, and more explicitly by Nernst,[18] in the paper where he gave his classically minded deduction of Planck's law. From this point of view, the quantity h (critical action for stochasticity) should be in principle computable in a definite model, but no clear progress has been made on this at present. Let us then assume that in a system of weakly coupled oscillators the nonlinearity produces a stochasticity action $A_0 = \hbar$ as described above, so that the oscillators of frequency ν have chaotic motions for $E > h\nu$ and ordered (frozen) motions for $E < h\nu$, as Nernst did.

Nernst's deduction of Planck's law is the following one. Let the oscillators, of frequency ν, be distributed according to the Maxwell–Boltzmann law at absolute temperature T; so, the probability density for the energy E of each oscillator $p_{\beta,\nu}(E) = p_\beta(E)$ is independent of the frequency ν and is given by

$$p_\beta(E) = \beta e^{-\beta E} \ , \quad 0 \leq E < \infty$$

where $\beta = \frac{1}{kT}$. Clearly one has normalization $\int_0^{+\infty} p_\beta(E)dE = 1$ and moreover equipartition $\int_0^{+\infty} E\, p_\beta(E)dE = \frac{1}{\beta}$, i.e. an average energy independent of frequency. But now, due to the dynamical properties of motions as described by the existence of an energy threshold $E^{(c)}(\nu) = h\nu$, one has, in the spirit of Boltzmann's ideas (think of the case of the rotational energy for perfectly smooth hard spheres), to separate the mechanical energy into two parts: the energy which can be exchanged, pertaining to oscillators having energy $E > h\nu$, and the energy which is frozen, pertaining to oscillators with energy $E < h\nu$. For a given frequency ν one will have a fraction $n_1(\beta,\nu)$ of oscillators above threshold, and a fraction $n_0(\beta,\nu)$ of oscillators below threshold, with

$$n_1 = \int_{h\nu}^{\infty} p_\beta(E)dE \;\; , \;\; n_0 = \int_0^{h\nu} p_\beta(E)dE \; ,$$

and $n_0 + n_1 = 1$; moreover, the average energy of any of the oscillators above threshold will be $E_1(\beta,\nu)$, while the average energy of any of the oscillators below threshold will be $E_0(\beta,\nu)$, where

$$E_1 = \frac{1}{n_1}\int_{h\nu}^{\infty} E\, p_\beta(E)dE \;\; , \;\; E_0 = \frac{1}{n_0}\int_0^{h\nu} E\, p_\beta(E)dE \; ,$$

and one has clearly $n_0E_0 + n_1E_1 = \frac{1}{\beta} \equiv kT$, namely equipartition of the mechanical energy. By elementary integrations one finds

$$n_1 = e^{-\beta h\nu}, \;\; n_0 = 1 - n_1, \;\; E_1 = kT + h\nu \; , \;\; E_0 = kT - \frac{h\nu}{e^{\beta h\nu} - 1} \; .$$

The most elementary model for the thermodynamic energy u, or the energy available for thermal exchanges in Boltzmann's sense, which is just a fraction of the mechanical energy, is then clearly

$$u(\beta,\nu) = n_1(E_1 - E_0) \; ;$$

indeed one thinks that only the n_1 oscillators above threshold can loose energy, and in fact each of them can just loose the quantity $E_1 - E_0$, because it would be frozen in falling to the lower energy level E_0. Thus, with the expressions reported above, one gets immediately

$$u(\beta,\nu) = \frac{h\nu}{e^{\beta h\nu} - 1} \; ,$$

namely exactly Planck law (and without zero-point energy, of course). In the original paper of Nernst, one finds for u the expression $u = kT - E_0$, which, using $kT = n_0E_0 + n_1E_1$ and $n_0 + n_1 = 1$, is immediately checked to coincide with $n_1(E_1 - E_0)$.

So, the conception of the zero-point energy $h\nu$ as a stochasticity threshold immediately leads to Planck law for the energy u exchangeable in the sense of Boltzmann, due to the fact that only the n_1 oscillators above threshold (on the average on the level E_1) can exchange energy, while the n_0 oscillators below threshold (on the average on the level E_0) are frozen. Now, in such a reasoning, the oscillators below threshold were considered to be frozen for all times, while, in the spirit of Nekhoroshev's theorem, as also predicted by Planck, they should be frozen only up to a time of the order $\exp\left(\frac{h\nu}{kT}\right)$, and this is large for the large frequencies but not for the low ones. Thus, if one fixes an observation time t, it is clear that to a better approximation one can conceive that there exists a frequency $\hat{\nu} = \hat{\nu}(t)$, such that the large frequencies ($\nu > \hat{\nu}$) are frozen, while the low frequencies ($\nu < \hat{\nu}$) are not at all frozen up to that time. In such a way, for the thermodynamic energy u one should have a plateau $u = kT$ for $\nu < \hat{\nu}$, followed by an exponential tail, for example just a planckian $\frac{h\nu}{e^{\beta' h\nu} - 1}$ for $\nu > \hat{\nu}$;

the effective inverse temperature $\beta' = \beta'(\hat{\nu},\beta)$, which is the temperature apparent from the side of the high frequencies, is then determined by the matching condition $\frac{h\nu}{e^{\beta' h\hat{\nu}}-1} = \frac{1}{\beta}$, which gives $\beta' h\hat{\nu} = \log(1+\beta h\hat{\nu})$.

Now, as time goes on, the plateau front at $\hat{\nu}(t)$ also advances towards the high frequencies; however, due to the fact that the Arnold diffusion time increases exponentially with the frequency, such an advancement should be negligible on a linear time scale at any given time, giving the impression of a stationary distribution. Moreover, this phenomenon of apparent stationarity should be enhanced if one looks at the distribution from the side of the high frequencies and if the energy u is plotted versus the relevant quantity, i.e. the dimensionless parameter $x' = \beta' h\nu$; indeed the inflection point occurs at $x' = \beta' h\hat{\nu} = \log(1+\beta h\hat{\nu})$, which depends only logarythmically on $\hat{\nu}$.

So far the reasoning was of "canonical" type, with a fixed inverse temperature β, because we assumed a Maxwell–Boltzmann distribution with parameter β. Analogous considerations can also be made for an isolated system, with a fixed energy. In such a case, the height of the plateau should diminish as $\hat{\nu}$ increases, due to energy conservation. In fact, the numerical computations performed up to now refer to the latter case. Particularly impressive are the computations on the Bocchieri-Loinger model for a blackbody cavity [19]: see especially fig.6, with a very nitid plateau followed by an exponential tail, and fig.7, where the exponential time for the invasion of the high frequencies is very beautifully exhibited. In this connection, a very illuminating contribution came from Parisi et al. [20], who transported to Hamiltonian systems some techniques previously used by Frisch and Morf in connection with the problem of turbulence in fluids.

Let us now come to the problem of the experimental verification of the blackbody radiation law. In this connection, recall first that the energy density observed is not u itself, but rather $\nu^2 u$, because of the factor $\nu^2 d\nu$ which is proportional to the number of oscillators between ν and $\nu + d\nu$ in a cavity; so the observed energy tends to zero for $\nu \to 0$, and the corrections alluded to above should not be easy to observe. In any case, it is a fact that the observations on the spectrum of the blackbody radiation law are incredibly poor: essentially, there are no (no, means no one) observations after the year 1921 [21], and such observations fit Planck law only within 3 (three) per cent [22]; moreover, they all refer to $\frac{h\nu}{kT} > 0.2$; for a history of the problem see [23]. It is curious to notice that for the cosmic background radiation too one has data only for $\frac{h\nu}{kT} > 0.2$. Data for lower values of $\frac{h\nu}{kT}$ are available for the sun (which very well might not be a good blackbody), and they are just a kind of exponential tail at high frequencies with a very nitid plateau at low frequencies, the inflection point being around $\beta' h\nu \simeq 1$. So, independently of the theoretical considerations reported here, it seems appropriate to point out that the incredibly poor status of the experimental data on the blackbody radiation law, which historically led to the discovery of quantum mechanics, is a scandal for the credit of general physics.

3. Conclusions

So, in the present lecture it has been illustrated how the modern results of classical perturbation theory, along the lines of Nekhoroshev's theorem, lead to the conception of the existence of exponentially long relaxation times. This suggests a reconsideration of the formalization of some notions, such as typically the notions of stability and of ergodicity, which usually are stated with reference to infinite times. A "softened" formalization consists in considering just finite, but very long, and in a sense qualitatively long, times. Some applications to the three body problem and to the realization of holonomic constraints have been mentioned; for another application to the existence of "effective fractal dimensions" in conservative dynamical systems, see the lecture of Giorgilli at this school. Here the emphasis was put instead on the application to statistical mechanics; the point of view of Boltzmann was illustrated, according to which the lack of equipartition for the high frequencies (ultraviolet cutoff) is due to a situation of metaequilibrium, instead of the usually assumed situation of equilibrium. In particular it was shown how from this point of view, in the case of the blackbody, Planck's law appears just as a first approximation: neglecting all stability times leads to equipartition, while considering the oscillators below threshold to be frozen for all times leads to Planck's law, and finally a more careful consideration of the involved times would lead to equipartition for the low frequencies followed by an exponential or a planckian tail at higher frequencies.

As a last comment, the considerations reported here suggest also an appropriate softening of the notion of structural stability in the general theory of dynamical systems.

APPENDIX: Some excerpts from the literature of the turn of the century concerning equipartition of energy

L.Boltzmann, *On certain questions of the theory of gases,* Nature 51, 413 (1895).

> The generalized coordinates of the ether, on which these vibrations depend, have not the same *vis viva* as the coordinates which determine the position of a molecule, because the entire ether has not had time to come into thermal equilibrium with the gas molecules, and has in no respect attained the state which it would have if it were enclosed for an infinitely long time in the same vessel with the molecules of the gas.
>
> But how can the molecules of a gas behave as rigid bodies? Are they not composed of smaller atoms? Probably they are; but the *vis viva* of their internal vibrations is transformed into progressive and rotatory motion so slowly, that when a gas is brought to a lower temperature the molecules may retain foe days, or even for years, the higher *vis viva* of their internal vibrations corresponding to the original temperature. This transference of energy, in fact, takes place so slowly that it cannot be perceived amid the fluctuations of temperature of the surrounding bodies. The possibility of the transference of energy being so gradual cannot be denied, if we also attribute to the ether so little friction that the Earth is not sensibly retarded by moving through it for many hundreds of years."

L.Boltzmann, *Lectures on gas theory,* translated by S.G.Brush, University of Cal. Press (1966).

See the whole section 45: *Comparison with experiments.*

J.H.Jeans, *A comparison between two theories of radiation,* Nature 72, 293 (1905).

He compares Planck's law with equipartition, which is obtained from the former by letting Planck's constant go to zero, and adds:

"Of course, I am aware that Planck's law is in agreement with experiment if h is given a value different from zero, while my own law, obtained by putting $h = 0$, cannot possibly agree with experiment. This does not alter my belief that the value $h = 0$ is the only value which it is possible to take, my view being that *the supposition that the energy of the ether is in equilibrium with that of matter is utterly erroneous in the case of ether excitations of short wave-length under experimental conditions.*"

By the way, Jeans thus expects equipartition for the low frequencies, which is also the modification proposed in this lecture. This is even more explicitly stated in the next quotation.

J.H.Jeans, *On the partition of energy between matter and aether,* Phil.Mag. 10, 91 (1905).

"We may say that the transfer of energy between the (3N) material degrees of freedom and s (low frequency) degrees of aether freedom (of n degrees of freedom) is comparatively rapid, while that to the remaining $n - s$ degrees is very slow. *For an enormous time these $n - s$ degrees of freedom will not receive their due share of the energy, while the energy will rapidly equalise itself between the remaineng $3N + s$ degrees of freedom.* During this time, the ratio of the energy of the aether to that of the material system is $\frac{s}{3N}$, and this will generally be very small."

J.H.Jeans, *On the vibrations set up in molecules by collisions,* Phil.Mag. 6, 279 (1903).

"A steel ball dropped on to a rigid steel plate will rebound perhaps half a dozen times before its energy is appreciably lessened; this is because of the great elasticity of the steel. If the kinetic energy of gases is true, a system of molecules must rebound from one another and from rigid walls many billions of times before the total energy is appreciably lessened. The aim of the present paper is to show that, in so far as the data available enable us to judge, molecules will possess sufficient elasticity for this to occur."

He considers a model of molecules with the internal vibrations described by harmonic oscillators (isochronous oscillators) of frequency p, and finds with some considerations, for the vibrational energy acquired by a molecule after a collision, a quantity proportional to $\exp(-2ap)$, with a certain constant a ..., and adds:

"The appropriate unit of time in this case is of course a. If p in these units has a value 200 we see that the "elasticity" of the molecules has introduced a factor e^{-400} ... This means that if the molecules were all moving with average velocity the number of collisions required to dissipate a given fraction of energy would be increased by the "elasticity" in a ratio of about e^{400}:1. In other words, the "elasticity" could easily make the difference between dissipation of energy in a fraction of a second and dissipation in billions of years."

M.Planck, *On the foundation of the blackbody law,* Ann.d.Phys. 37, 642 (1912).

In this paper Planck for the first time conceived of zero-point energy: as one knows, the energy levels of a harmonic oscillator of frequency ν (or angular frequency or pulsation $\omega = 2\pi\nu$) are considered to be $E_n = (n+\frac{1}{2})h\nu = (n+\frac{1}{2})\hbar\omega$. So, according to standard quantum mechanics, using the quantum Gibbs ensemble at zero absolute temperature (or zero-point, from the german *nullpunkt*) all oscillators should have the zero-point energy $\frac{1}{2}h\nu$. Let us not discuss here whether the zero-point energy according to quantum mechanics should be $\frac{1}{2}h\nu$ or $h\nu$ (or $\alpha h\nu$ with any positive $\alpha \leq 1$ as Enz and Thellung [24] would say), and whether such energy is measurable or not, which would lead us into a rather complicated subject.

What is quite astonishing in any case, is that Planck conceived of $h\nu$ as a stochasticity threshold more or less in the sense described in this lecture, with $\hbar$ in place of A_0. Indeed, he assumes that an oscillator of frequency ν absorbs energy in a kind of diffusive process, requiring a time $\exp \frac{h\nu}{kT}$ in order to reach the level $h\nu$. On reaching such level, it could emit the complete energy $h\nu$ or proceed by absorbing energy in a continuous way as before, according to some probabilistic law. In his words:

> "This will not be regarded as implying that there is no causality for emission; but the processes which cause the emission will be assumed to be of such a concealed nature that for the present their laws cannot be obtained by any but statistical methods. Such an assumption is not at all foreign to physics; it is e.g. made in the atomistic theory of chemical reactions and in the disintegration theory of radioactive substances."

So, one should not be allowed to use the standard methods of statistical mechanics if $E < h\nu$, i.e., as we said above, zero-point energy is somehow an energy threshold for the onset of stochasticity. Moreover, the diffusion time for reaching such energy is exponentially increasing with frequency, as essentially comes out of Nekhoroshev theorem. Planck then proceeds, deducing his law in a way that we don't intend to follow here; in addition, he quite stangely ascribes zero-point energy only to material oscillators and not to the oscillators of the electromagnetic field.

W.Nernst, *On an attempt to understand quantum mechanics through continuous changes of energy*, Verh.d.Deutsch.Phys.Ges. 4, 83 (1916).

More interesting from the present point of view are the ideas introduced four years later by Walter Nernst, the inventor of the third principle of thermodynamics. First of all, he reconfirms Planck's interpretation of $h\nu$ as a stochasticity threshold in an even more direct way; he says in fact explicitly that the motions of the oscillators are ordered (*geordnete*) below $h\nu$ and disordered (*ungeordnete*) above $h\nu$. His reason for such an idea is quite interesting in itself. According to the third principle, in lowering temperature one eventually reaches a situation where the internal energy coincides with the free energy. Now, free energy is just that part of internal energy which according to thermodynamics can be used to do macroscopic work, and so it has, according to Nernst, an ordered character. That's why, with an intuition coming from thermodynamics, he comes to the striking proposal that the microscopic motions too have an ordered character below $h\nu$, by introducing the conception of a stochasticity threshold as we are accustomed to conceive now, after the impressive evidence coming from the numerical works started by Hénon and Heiles [25]. He also explicitly ascribes the threshold to the oscillators of the electromagnetic field too, and furthermore conceives of h as a constant deduced from other elementary quantities. Namely, thinking of the equations of motion of the electromagnetic field interacting with an electron of charge e, which also are the equations for a system of weakly coupled harmonic oscillators, assume that one can find a critical action such as the action A_0 described above. Then by dimensional considerations one clearly has $A_0 = \frac{e^2}{c}\mathcal{A}$, where c is the velocity of light and $\mathcal{A}$ a dimensionless quantity; if one finds $\mathcal{A} = 137$, then one has $A_0 = \hbar$.

P.Langevin and M.de Broglie eds., *La théorie du rayonnement et les quanta*, (Paris, 1912).

This is the conference where the quantum theory was officially accepted. From the point of view discussed here, it is of interest to consider the letter of Rayleigh (page 49) where he says:

> "D'une autre coté, Boltzmann et Jeans considèrent qu'il s'agit seulement là d'une question de temps et que les vibrations nécessaires pour l'equilibre statistique complet pourraient ne s'etablir qu'après des milliers d'années. Les calculs de Jeans semblent montrer qu'une telle opinion n'a rien d'arbitraire. Je voudrais savoir si elle est contredite par des faits expérimentaux précis. Autant que je puis savoir, les expériences ordinaires de laboratoire n'apportent rien de décisif à ce sujet."

There followed an interesting discussion, reported at the pages 51–52. Nernst says that in the measurements of specific heats there are two methods, namely the explosion method which requires some milliseconds, and the Regnault method which rerquires minutes or months; and in both cases

> " on n'a jamais observé de telles valeurs constamment croissantes dans la mesure des chaleurs spécifiques".

Furthermore he adds:

> "La considération suivante est encore plus frappante. D'après la thermodynamique, la température de fusion comme la tension de vapeur serait considérablement modifiée si la chaleur spécifique et, par suite, le contenu d'énergie se modifiait avec le temps; mais on n'a jamais remarqué une différence de température de fusion entre les minéraux naturels et les composés synthétiques. Il faut supposer que l'état d'équilibre demandé par la loi de l'equipartition n'aurait pas encore lieu après 400 millions d'années, tandis qu'une autre partie de l'énergie se met immédiatement en équilibre; c'est bien peu probable."

But then he quite strangely adds:

> "On pourrait soutenir l'opinion qu'une partie de l'energie est absorbée dans des temps si longs que cela deviendrait sans intéret pour l'expérimentateur. Pour l'autre partie qui serait la seule intéressante, la théorie classique de l'equipartition ne suffirait pas; il faudrait admettre une autre théorie comme, par exemple, celle des quanta."

Very interesting is also the comment of Kamerlingh Onnes at page 32.

References

[1] G.Benettin, L.Galgani and A.Giorgilli, Nature 311, 444-445 (1984).

[2] N.N.Nekhoroshev, Russian Math.Surv. 32,1-65 (1977); Trudy Sem. Petrovs. n.5, 5-50 (1979), in russian; see also G.Benettin, L.Galgani and A.Giorgilli, Celestial Mechanics, 37, 1-25 (1985).

[3] G.Benettin, this volume.

[4] A.Giorgilli, this volume.

[5] L.Boltzmann, Nature 51,413-415 (1895).

[6] L.Boltzmann, *Lectures on gas theory*, translated by S.G.Brush, University of Cal. Press (1966); see especially section 45:*Comparison with experiments*.

[7] M.Planck, Ann.d.Phys. 37, 642-656 (1912); *The theory of heat radiation*, Dover (1959).

[8] P.Langevin and M.de Broglie eds., *La théorie du rayonnement et les quanta*, (Paris, 1912).

[9] A.Giorgilli and L.Galgani, Cel. Mech. 37, 95-112 (1985)

[10] A.Giorgilli, A.Delshams, E.Fontich, L.Galgani and C.Simó, *Effective stability for a Hamiltonian system near an elliptic equilibrium point, with an application to the restricted three body problem*, preprint.

[11] G.Benettin, L.Galgani and A.Giorgilli, *Realization of holonomic constraints and freezing of high frequency degrees of freedom in the light of classical perturbation theory, Part I*, Comm.Math.Phys., in print.

[12] H.M.Zaslavski, *Stochasticity of dynamical systems*, Nauka, Moscow (1984), in russian: at the end of page 249; see also B.V.Chirikov, Phys.Rep. 52, 263 (1979).

[13] P.Bocchieri, A.Scotti, B.Bearzi and A.Loinger, Phys.Rev.A, 2, 2013 (1970); for a review of later works see G.Benettin, in *Molecular-dynamics simulation of statistical-mechanical systems*, 97th Varenna Course (1986).

[14] G.Benettin, L.Galgani and A.Giorgilli, Phys.Lett.A, 120, 23–27 (1987).

[15] G.Benettin, L.Galgani and A.Giorgilli, *Realization of holonomic constraints and freezing of high frequency degrees of freedom in the light of classical perturbation theory, Part II*, preprint.

[16] G.Gallavotti, in *Lecture notes at the 1983 Erice summer school*, A.Wightman and G.Velo eds., Plenum Press (1985); *Quasi integrable mechanical systems*, in 1984 Les Houches summer school, session 43, K.Osterwalder and R.Stora eds., Elsevier (1986); see also Comm.Math.Phys. 87, 365 (1982).

[17] C.Cercignani, L.Galgani and A.Scotti, Phys.Lett. A38, 403 (1972); L.Galgani and A.Scotti, Phys.Rev.Lett. 28, 1173 (1972); L.Galgani and A.Scotti, Rivista Nuovo Cim. 2, 189 (1972).

[18] W.Nernst, Verh.d.Deutsch.Phys.Ges. 4, 83 (1916); see also L.Galgani, Nuovo Cim. B62, 306 (1981); L.Galgani, Lett.Nuovo Cim. 31, 65-72 (1981); L.Galgani and G.Benettin, Lett.Nuovo Cim. 35, 93-96 (1982).

[19] G.Benettin and L.Galgani, J.Stat.Phys. 27, 153 (1982).

[20] F.Fucito et al., J. de Physique 43, 707 (1982); R.Livi et al., Phys.Rev. A28, 3544 (1983).

[21] H.Rubens and G.Michel, Phys.Zeitschr. 22, 569 (1921).

[22] L.Crovini and L.Galgani, Lett.Nuovo Cim. 39, 210 (1984).

[23] L.Galgani, Annales Fond.L.de Broglie 8, 19 (1983).

[24] C.P.Enz and A.Thellung, Helv.Phys.Acta 33, 839 (1962); C.P. Enz, *Is zero-point energy real ?*, contribution to a book dedicated to Jauch.

[25] M.Hénon and C.Heiles, Astron.J. 69, 73 (1964); M.Hénon, in Les Houches summer school (1982).

RELEVANCE OF EXPONENTIALLY LARGE TIME SCALES IN PRACTICAL APPLICATIONS: EFFECTIVE FRACTAL DIMENSIONS IN CONSERVATIVE DYNAMICAL SYSTEMS

Antonio Giorgilli

Dipartimento di Fisica dell'Università
Via Celoria 16, 20133 – Milano (Italia)

Abstract. The problem of evaluating the fractal dimension of a chaotic orbit in a conservative dynamical system is revisited in the light of exponentially large time scales rigorously introduced by recent results of classical perturbation theory. The possible relevance for the problem of comparing theoretical previsions with experimental results in statistical models is pointed out.

1. Introduction

Starting from the celebrated KAM theorem on the persistence of invariant tori in quasi integrable Hamiltonian systems,[1] modern perturbation theory has produced very beautiful and significant results. Among them, a central role is played by Nekhoroshev's theorem on exponential bounds on Arnold's diffusion.[2] The result is essentially the following: consider a nearly integrable Hamiltonian system, namely an integrable one with the addition of a small perturbation of size ε; then the change with time of the action variables I relative to the integrable system is bounded by a certain power of ε up to a time $T \simeq T_0 \exp\left(\varepsilon^*/\varepsilon\right)^a$, with suitable constants T_0, ε^* and a. Thus, one renounces to look for results valid for all times, as in KAM theory on invariant tori, but finds results which are valid for an open set of initial conditions.

Now, these results could hardly be used directly in practical applications, since the general theoretical estimates on the size of the allowed perturbation are ridiculously small, and in particular decrease to zero when the number of the degrees of freedom goes to infinity. However, at least in particular cases, non completely unrealistic results can be achieved. For example, considering the Lagrangian equilibrium point L_4 of the restricted three body problem in the circular spatial case, and taking as a model problem the Sun–Jupiter case, one can prove that an orbit starting in a neighbourhood of radius R_0 of the point L_4 cannot leave a neighbourhood of radius $2R_0$ in a time of the order of the estimated age of the universe.[3] In such case R_0 can be taken of the order of a few hundreds kilometers (the estimate given in the reference above is about 1 to 10 kilometers, but can be improved at least by a factor 100). A case relevant to statistical mechanics is under investigation. Thus, it seems quite reasonable to hope that a suitable theory adapted to phenomena which require exponentially large time scales can be developed.

An example of the relevance of such kind of results is found in the computation of fractal dimensions[4] for conservative dynamical system. Although several works have been devoted in recent years to the computation of fractal dimensions in dissipative systems, the case of conservative systems is less popular: in the case of two dimensional mappings the dimension of the chaotic orbits, which are the analogs of the strange attractors in dissipative systems, was just assumed to be two,[5] while in the case of a Hamiltonian system of three degrees of freedom a non integral dimension was observed.[6]

In fact, there is a diffuse opinion that the Hausdorff dimension of chaotic orbits in conservative systems should be an integral number.[7] This is supported, at least in the case of a two dimensional mapping, by a theorem of Newhouse:[8] considering an area preserving diffeomorphism of a two dimensional compact connected orientable manifold, Newhouse proves that the closure of the set of transverse homoclinic points of an hyperbolic fixed point has "generically" Hausdorff dimension two.

If such a diffuse opinion is correct, then the non integral dimension observed in refs. [6] quoted above should be considered as a wrong experimental result, possibly due to the fact that the numerical integration has been carried over a too short time interval. I will try to show instead that, in essence, results of such a type should be considered as correct, giving a dimension which is to most purposes the effective one, just in virtue of stability properties of Nekhoroshev's type.

The present lecture aims at investigating in more detail the limitations to which a numerical experiment on the measure of the fractal dimension of a chaotic orbit may be subjected, and the reasons of the discrepancy, if any, between theoretical expectations and experimental results, in the light of the Nekhoroshev like results. For completeness, the definition of fractal dimension and the box counting algorithm to evaluate it are also recalled. The lecture is based on the papers [9], [10] and [11].

2. The model problem

We consider the μ-dependent family of mappings[12] of the torus $\mathbf{T}^2$ into itself

$$\begin{cases} x' = x + y \\ y' = y - \mu \sin(2\pi x') \end{cases} \pmod 1 , \tag{1}$$

where μ is a real parameter. For $\mu = 0$ the mapping is integrable, with invariant curves $y = \text{const}$. For $\mu > 0$ one has two trivial fixed points, namely the origin $(0,0)$, which is elliptic, and the point $\left(\frac{1}{2}, 0\right)$, which is hyperbolic.

Taking an initial point close to the unstable manifold and computing a number of iterates one has an orbit in the chaotic region surrounding that unstable manifold. For example, the figures 1a and 1b represent such an orbit for $\mu = 0.04$ and $\mu = 0.12$ respectively. It is immediately seen that for $\mu = 0.04$ such region looks as a one dimensional curve, while for $\mu = 0.12$ the points are clearly scattered in a two dimensional region.

However, the two figures look more similar if one makes an enlargement of a suitable small region close to the unstable fixed point of fig. 1a. This is done in fig. 2, where one clearly sees that the enlargement reveals a very complicated structure, and that the points still appear to be scattered in a two dimensional region.

Things look even more complicated if one dynamically sees the figure on the screen of a graphic device. The successive points of an orbit can stay for a long time in a region, and give at first sight the impression that they are uniformly filling a curve (or a very narrow region), making for example several rounds around one of the chains of islands which appear in the figures. But, superimposed to such fast motion, there is also a slow diffusion, which tends to increase the thickness of such curves; moreover, one observes sudden changes of the trajectory, which cause the successive

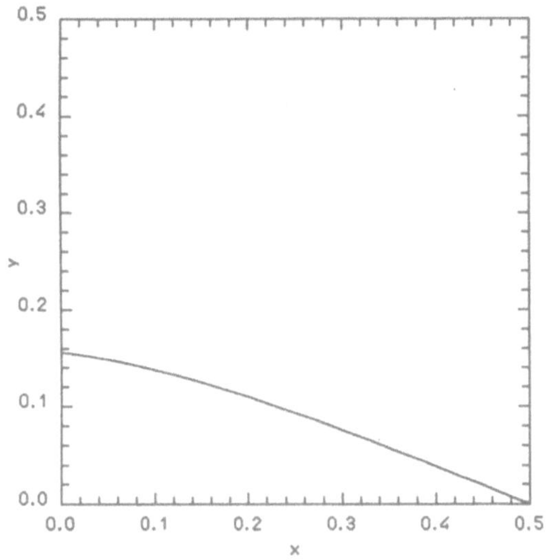

Fig. 1a. The orbit of the mapping for $\mu = 0.04$ starting from the initial point (0.500001, 0.000001) up to 50000 iterations, at a macroscopic scale. Only the lower left quarter of the torus is represented.

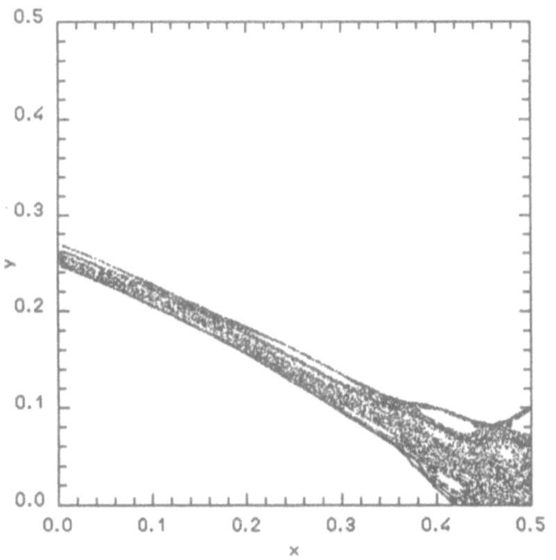

Fig. 1b. Same as fig. 1a, for $\mu = 0.12$.

points to invade new regions by drawing new curves, and to continue the diffusion process described above, so that the orbit fills larger and larger areas.

Such diffusive behaviour should ultimately stop, because in a two dimensional model, as is well known, KAM tori exist, at least for small values of the perturbation parameter, which isolate different chaotic regions. The situation would be well different, and worse, in higher dimensional models, because in the latter case the KAM tori do not split the phase space into separated regions, and a chaotic orbit can successively invade all the available phase space.

These rather simple remarks suggest that the fractal dimension of a chaotic orbit can well be an integral number, but also that finding such number could require smaller and smaller observation scales, and consequently larger and larger number of iterations, which seems to be a hopeless task. This is, roughly speaking, the result

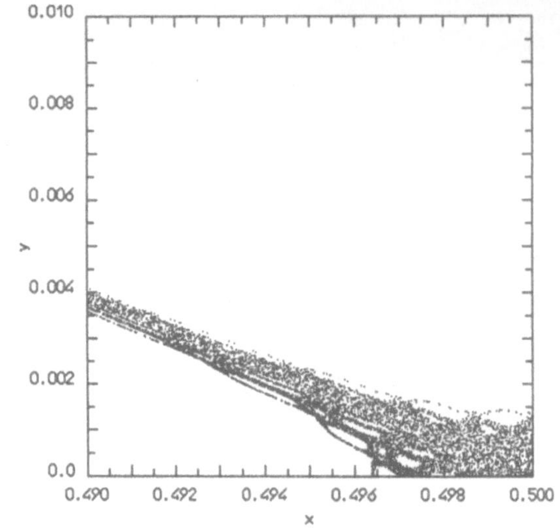

Fig. 2. Enlargement of a small region of fig. 1a close to the fixed point, to be compared with fig. 1b. The number of iterations has been increased to 130000, in order to obtain that the number of points inside the represented region be roughly the same as in fig. 1b.

I'm going to discuss, and the additional nontrivial information will be given that the observation scale decreases exponentially with the inverse of the parameter μ.

3. The box counting algorithm

Let me recall the definition of fractal dimension, usually called the capacity.[13] Given an open set $\mathcal{A} \subset \mathbf{R}^p$, and the "strange" subset S of $\mathcal{A}$ whose dimension must be computed, let $N(\varepsilon)$ be the minimum number of hypercubes of side ε (ε–hypercubes) needed in order to cover the set S. Then the fractal dimension is defined as

$$d = \lim_{\varepsilon \to 0} \frac{\log N(\varepsilon)}{\log(1/\varepsilon)} \ . \tag{2}$$

In practical applications, $\mathcal{A}$ is typically a hypercube of side 1, and S an orbit of a discrete dynamical system exhibiting a strange chaotic behaviour. In fact, only a finite subset S_n of S can be actually computed, and taking $N(\varepsilon)$ as the minimum number of hypercubes is impractical. So, one uses the following box counting algorithm.

1. Subdivide $\mathcal{A}$ into $(1/\varepsilon)^p$ ε–hypercubes.
2. Build S_n as the set of n successive iterates of a discrete dynamical system.
3. Compute $N_n(\varepsilon)$ as the number of ε–hypercubes visited by S_n (i.e., containing at least a point of S_n).
4. Try to evaluate the double limit

$$\begin{aligned} d(\varepsilon) &= \lim_{n \to \infty} \frac{\log N_n(\varepsilon)}{\log(1/\varepsilon)} \ , \\ d &= \lim_{\varepsilon \to 0} d(\varepsilon) \end{aligned} \tag{3}$$

In working out such a program one encounters several difficulties due to the need of memory and computing time. Indeed, n should be chosen large enough, so that the computed value of $N_n(\varepsilon)$ is quite well saturated and can be taken as a reliable approximation of the true value, $N_\infty(\varepsilon)$ say; on the other hand ε should be chosen small enough, so that the second limit in (3) can be reliably evaluated; moreover,

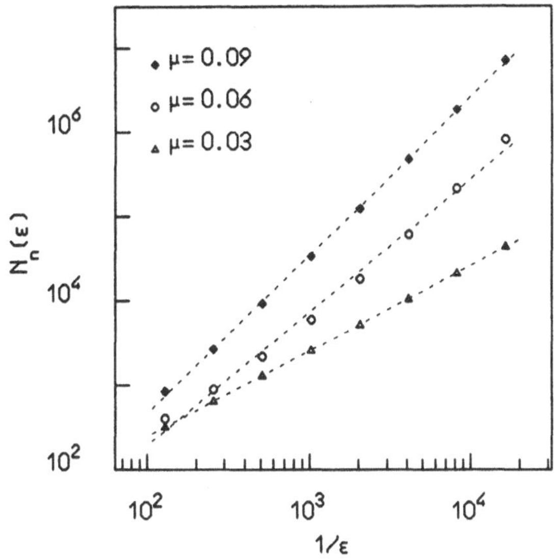

Fig. 3. Curves $\log N_\varepsilon$ vs. $\log(1/\varepsilon)$ for three values of the parameter μ.

n and ε must obviously be related by taking n as a suitably increasing function of ε. Of course, taking larger and larger n's increases the computation time, while taking smaller and smaller ε's increases the amount of memory needed to store all the informations about the visited ε–hypercubes. Such difficulties can be overcome by essentially two methods.

i. Using efficient algorithms to manage memory, by taking into account the fact that the total number of ε–hypercubes is $(1/\varepsilon)^p$, p being the dimension of the embedding space, while the number of actually visited ε–hypercubes is proportional to $(1/\varepsilon)^d$, i.e., much less than the one above if d is smaller than p.
ii. Trying to reveal the information hidden in low values of n.

These methods were studied in refs. [14] and [9], to which the reader is referred for more details.

In practice, a computational algorithm is the following.

1. Choose a finite number of values of ε, say $\varepsilon_0, \ldots, \varepsilon_r$; a convenient choice is to take the largest side ε_0 as a fraction of the side of the whole cube, and $\varepsilon_j = \varepsilon_{j-1}/2$ for $1 \leq j \leq r$.
2. Compute $N_n(\varepsilon_j)$ for $0 \leq j \leq r$ by choosing n in such a way that the value $N_n(\varepsilon_j)$ looks well saturated.
3. Plot $\log N_n(\varepsilon)$ vs. $\log(1/\varepsilon)$, as in fig. 3.
4. Evaluate the fractal dimension d as the slope of the straight line interpolating the points of the graph above.

Such procedure can be improved by the methods used in refs. [14] or [9], but we do not worry about these technical details. In fact, all the subsequent computations have been made by the procedure illustrated in ref. [9].

4. Numerical results

Let me first report the results of a naive computation of the fractal dimension for μ in the interval (0.04, 0.12). By "naive", I mean here that the computation is performed by using the procedure described in sect. 3, which has proven to be useful in the case of dissipative systems, without taking into account the theoretical expectations discussed

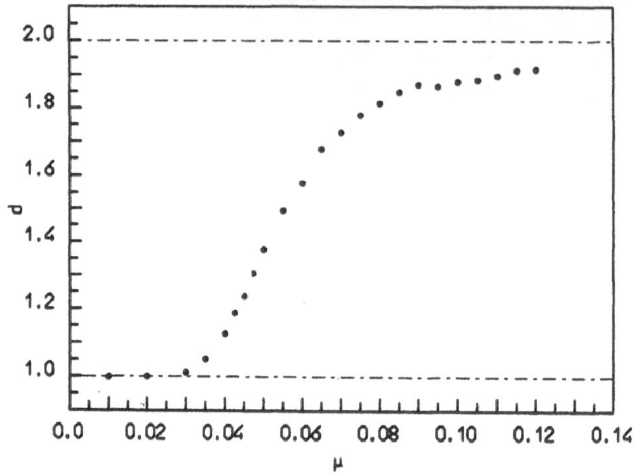

Fig. 4. Estimated dimension d as a function of the perturbative parameter μ.

in sect. 1. All the computations have been done for the orbit starting from the initial point (0.500001, 0.000001) up to a number of iterations $n = 10^8$, and with boxes of side $\varepsilon = 2^{-7}, \ldots, 2^{-14}$. For each value of μ the value $d(\mu)$ of the fractal dimension has been evaluated, and the results are summarized in fig. 4, where the estimated dimension d is plotted versus μ. As is seen, the estimated dimension is close to 1 up to $\mu \simeq 0.03$, and then smoothly rises towards 2.

The numerical experiment described above seems to give satisfactory results, which could be hardly improved because of the considerations made in sect. 3 about the computational algorithm. Nevertheless, these results are inconsistent with the Newhouse theorem, which leads to expect a dimension two for all $\mu > 0$ (notice that the fractal dimension computed here is greater than or equal to the Hausdorff dimension). So, it seems quite natural to conclude that the numerical experiment discussed here is simply wrong.

However, let's look more carefully at the numerical method used, in particular at fig. 3. For the smallest and the largest value of μ (0.03 and 0.09) the points seem to be well aligned on the interpolating straight line, while for the intermediate value $\mu = 0.06$ they clearly show a concavity. So, the limit slope of the curve for $\varepsilon \to 0$, which is the relevant one by the definition, can be expected to be larger than the slope of the interpolating line. This means, on one hand, that the smallest value of ε used in the computation above is not small enough, but, on the other hand, also that using smaller and smaller values of ε will not produce substantial improvements.

A better approach is to try to reveal the information hidden in the available experimental data. To this end, look for the slope s of the curves of fig. 3 as a function of ε, by computing the slope of the segment joining the points corresponding to two nearby values of ε, and plot s vs. $\log 1/\varepsilon$. This is done in fig. 5, and makes evident that $s(\varepsilon)$ smoothly increases to the final value 2 as $\varepsilon \to 0$, but curves corresponding to smaller values of μ start to raise only for smaller values of ε.

Such qualitative remark suggests the following description: for fixed values of μ and ε, the naive computation gives a definite value of the fractal dimension, but "naive" should now be interpreted as "the best we can do", in the sense that the computed value really depends on the observation scale ε, and that we cannot do better, unless we change the scale.

These remarks can be made quantitative by the following procedure: fix s, with $1 < s < 2$, and, for any $\mu > 0$, look for the value of ε such that the estimated dimension is s; then plot $\log 1/\varepsilon$ vs. $\log 1/\mu$. This can be done by interpolation on the data of fig. 5, and the plot is reported in fig. 6 for several values of s. The points

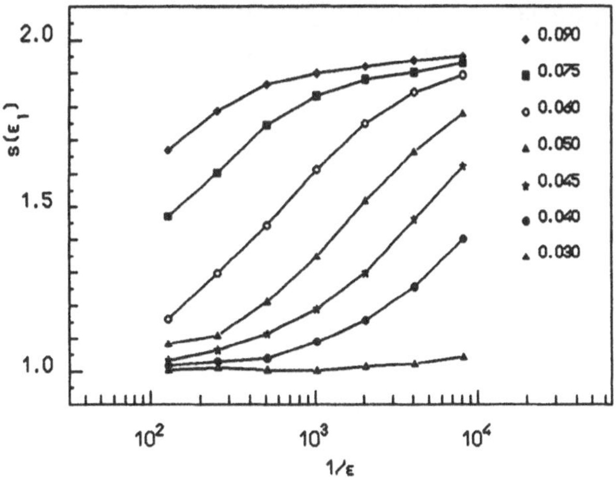

Fig. 5. Slopes of the curves $\log N_\varepsilon$ vs. $\log(1/\varepsilon)$ obtained from fig. 3, for several values of the parameter μ.

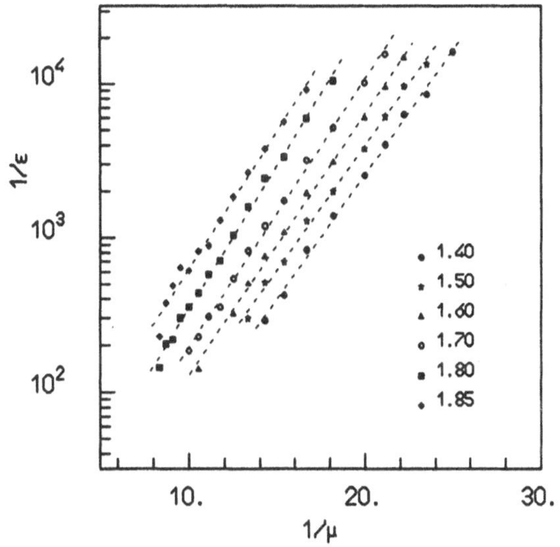

Fig. 6. The values of $\log(1/\varepsilon)$ required to get the slope s, as a function of the perturbative parameter μ, for different values of s.

corresponding to a given value of s look quite well aligned on a straight line, and we can interpolate them by the function

$$\frac{1}{\varepsilon} = A(s)e^{\mu_0/\mu} \, . \tag{4}$$

Here, μ_0 is a constant which can be evaluated, by interpolation, to be $\mu_0 = 0.395 \pm 0.060$, and the function $A(s)$ can also be empirically determined. More details, which are unessential for what follows, can be found in ref. [9].

So, the remark made above that the observation scale ε must be a decreasing function for $\mu \to 0$ is made more precise by the statement that ε must decrease exponentially fast with $1/\mu$. As a consequence, if one works with a minimal ε, there will exist a finite value $\overline{\mu}$ below which the apparent or effective dimension will be 1, while intermediate values between 1 and 2 will be observed for slightly larger values of μ. Such remark might in particular explain the non integral dimension found in refs. [6].

5. The Neishtadt's theorem

The experimental results of sect. 4 can be interpreted in the framework of classical perturbation theory by estimating the size of the splitting of the separatrices. This is given, on a dynamical model similar to the mapping considered in our paper, by a theorem due to Neishtadt,[11] which states that the splitting of the separatrices is exponentially small with the inverse of the perturbative parameter.

Consider a non autonomous one–dimensional Hamiltonian

$$H(I,\varphi,t) = H_0(I) + \mu H_1(I,\varphi,t) \ , \tag{5}$$

where I,φ are conjugate canonical variables, μ a small parameter, and H_1 is 2π periodic in φ and t and analytic in all the variables. For $\mu = 0$ the system is integrable, and the motion is periodic with frequency $\omega(I) = \frac{\partial H_0}{\partial I}$. Consider now I^* such that $\omega(I^*)$ assumes a rational (or resonant) value, $\frac{m}{n}$ say, with relatively prime integers m and n.

Following Neishtadt, we introduce local coordinates in the neighbourhood of the resonance via the time–dependent canonical transformation $p = (I - I^*)/\sqrt{\mu}$, $q = \varphi - \frac{m}{n}t$, so that the Hamiltonian in the new variables p, q, t reads

$$E = \sqrt{\mu}\left[\frac{a}{2}p^2 + V_1(q,t)\right] + \mu V_2(p,q,t,\mu) \ , \tag{6}$$

where $V_1(q,t) = H_1(I^*,q,t)$, and E is 2π periodic in q and $2\pi n$ periodic in t. This is the system to be compared with the mapping (1). In making the comparison, one should notice that the actual perturbation parameter in the Hamiltonian (6) is $\sqrt{\mu}$, and not μ as in the mapping. By neglecting $V_2(p,q,t,\mu)$, and introducing the averaged potential with respect to time $U(q) = \langle V_1(q,\cdot)\rangle$ instead of $V_1(q,t)$, the Hamiltonian $\frac{a}{2}p^2 + U(q)$ looks as a pendulum–like one, being $\frac{2\pi}{n}$ periodic in q. Assuming that $U(q)$ has a unique nondegenerate maximum, such maximum is an unstable equilibrium point, and unstable equilibria are connected by separatrices. The intersection of the separatrices with the plane $t = 0$ gives the unstable manifold W_u and the stable one W_s, which are invariant under the mapping at time $2\pi n$.

Considering now the complete Hamiltonian (6) for small μ, the unstable equilibria persist, but the invariant manifolds W_u and W_s generally split, giving rise to homoclinic intersections, and so to a chaotic region.

Neishtadt's theorem then states that the manifolds W_u and W_s lie in a strip of size $K_2\exp(-K_1/\sqrt{\mu})$ which is the union of the neighbourhoods of two analytic curves that are 2π periodic and uniquely projected on the q axis; here, K_1 and K_2 are positive constants.

The proof of the theorem is a beautiful combination of classical perturbation techniques leading to Nekhoroshev–like results with the modern methods of KAM theory. First, the Hamiltonian (6) is transformed into

$$F = \sqrt{\mu}\left[\frac{a}{2}P^2 + U(Q)\right] + \mu R(P,Q,\mu) + \alpha(P,Q,t,\mu) \ , \tag{7}$$

via a near to identity canonical mapping $p,q \longrightarrow P,Q$. If $|p| < K_3$, then one has $|R| = O(1)$ and $|\alpha| = O(\exp(-K_4/\sqrt{\mu}))$, with constants K_3 and K_4; moreover, the canonical mapping is $2\pi n$ periodic in t and 2π periodic in q, and the function R is $\frac{2\pi}{n}$ periodic in Q.

The system obtained from (7) by neglecting $\alpha(P,Q,t,\mu)$ is still integrable: its separatrices S_1 and S_2 do not split, and divide the phase plane into separated regions, which in turn are filled up by invariant tori of the mapping at time $2\pi n$.

The next step is to apply KAM theory to the whole system (7), and to prove that in each of these regions there exist invariant tori which are preserved under the perturbation $\alpha(P, Q, t, \mu)$, and lie at a distance $O(\exp(-K_1/\sqrt{\mu}))$ from the union of the curves S_1 and S_2. These curves bound an invariant region containing the unstable fixed points and their stable and unstable manifolds, and is the union of neighbourhoods of size $K_2 \exp(-K_1/\sqrt{\mu})$ of the curves S_1 and S_2, as stated by the theorem.

6. Conclusions

Let's now come back to the problem of the inconsistency between the experimental result of sect. 4 and the theoretical expectations discussed in the introduction.

If one looks at the problem in the light of Neishtadt's result, then one immediately realizes that the experimental result cannot be improved, even if one tries to evaluate the dimension by different methods or definitions. Indeed, the evaluation of the dimension always involves the measure of a distance, and Neishtadt's results states that such distance is exponentially small. Thus, the exponential dependence of the observation scale on the inverse of the parameter makes completely hopeless any attempt to improve the experimental apparatus.

On the other hand, one cannot simply conclude that an experiment cannot be performed: in fact, the experiment has been done, and has given coherent results which are the only valid ones over a finite observation scale and time. The possible inconsistency with the theoretical expectations should be interpreted not as an indication that the experiment is wrong, but rather as a strong suggestion that the theory is insufficient, in the sense that the theoretical approach gives results which are valid for the whole orbit, i.e. for infinitely many points observed with an observation scale which really goes to zero, but gives essentially no previsions for very large but finite realizations of the orbit, and for very small but finite observation scales.

The analogy with the problem of the equipartition of energy in classical statistical mechanics is striking. Experimental facts as the values of the specific heats of polyatomic gases or the form of the blackbody spectrum are clearly inconsistent with the theoretical expectations coming from the equipartition principle. This is often considered as an internal inconsistency of classical physics, which seems to be unable to explain why the high frequency degrees of freedom do not take part in the energy sharing. But recent numerical investigations[15] and theoretical results[16] strongly support the conjecture, proposed at the turn of the century by Boltzmann[17] and Jeans,[18] that such energy sharing can only occur over exceedingly large time scales ("days or years" according to Boltzmann, "billions of years" in the very words of Jeans), and, although being the true one for an actually infinite time, it can nevertheless be unobservable over a finite but large time scale, for example the age of the universe.

Thus, the problem of comparing theoretical versus experimental results in statistical models should be carefully reconsidered. Precisely, one should make a full distinction between theoretical results which are valid over a really infinite time, and the apparent or effective ones, which are the only valid results in any application to a physical problem. In this connection, see also the contribution of Galgani to this school.

References

[1] A.N. Kolmogorov, Dokl. Akad. Nauk SSSR, **98**, 527 (1954).
V.I. Arnold, Usp. Mat. Nauk, **18**, 13 (1963); Russ. Math. Surv., **18**,9 (1963);

Usp. Math. Nauk **18** N.6, 91 (1963); Russ. Math. Surv. **18** N.6, 85 (1963).
J. Moser, *Proceedings of the International Conference on Functional Analysis and Related Topics*, p. 60 (Tokio, 1969).

[2] N. N. Nekhoroshev, Usp. Mat. Nauk. **32**, (1977) [Russ. Math. Surv. **32**, 1 (1977)]; Trudy Sem. Petrows. No. 5, 5 (1979);
G. Benettin, L. Galgani and A. Giorgilli, Cel. Mech. **37**, 1–25 (1985).
G. Benettin and G. Gallavotti, J. Stat. Phys. **44**, 293–338 (1986);

[3] A. Giorgilli, A. Delshams, E. Fontich, L. Galgani and C. Simò, *Effective stability for a Hamiltonian system near an elliptic equilibrium point, with an application to the restricted three body problem*, preprint.

[4] B. B. Mandelbrot, *The fractal geometry of nature*, Freeman, San Francisco (1982).

[5] D. K. Umberger and J. D. Farmer, Phys. Rev. Lett. **55** N. 7, 661–664 (1985);
C. Grebogi, S. W. McDonald, E. Ott and J. A. Yorke, Phys. Lett. 110A, 1, 1–4 (1985).

[6] M. Pettini and A. Vulpiani, Phys. Lett. A, **106**, 5,6, 207–211 (1984);
A. Malagoli, G. Paladin and A. Vulpiani, Phys. Rev. A, **34**, 1550–1555 (1986).

[7] E. Ott, Rev. Mod. Phys. **53**, 4, 655–671 (1981).

[8] S. E. Newhouse, Soc. Math. de France, Astérisque 51, 223–334 (1978).

[9] A. Giorgilli, D. Casati, L. Sironi and L. Galgani, Phys. Lett. A, **115**, 5, 202–206 (1986).

[10] G. Benettin, D. Casati, L. Galgani, A. Giorgilli and L. Sironi, Phys. Lett. A, **118**, 325–330 (1986).

[11] A.I. Neishtadt, PMM U.S.S.R, **48**, N. 2, 133–139 (1984).

[12] G. Benettin, C. Cercignani, L. Galgani and A. Giorgilli, Lett. Nuovo Cim., **28** N. 1, 1–4 (1980); **29** N. 6, 163–166 (1980).

[13] L. Pontrjagin and L. Schnirelmann, Ann. Math. **33**, 156 (1932);
A.N. Kolmogorov and V.M. Tihomirov, Usp. Mat. Nauk **14**, 3 (1959) (english trans.: Am. Math. Soc. Transl., series 2, **17**, 277 (1961)).

[14] P. Grassberger, Phys. Lett. A, **97** N. 6, 224-226 (1983).

[15] G. Benettin, L. Galgani and A. Giorgilli, Phys. Lett. A, **120**, 23–27 (1987).

[16] G. Benettin, L. Galgani and A. Giorgilli, *Realization of holonomic constraints and freezing of high frequency degrees of freedom in the light of classical perturbation theory*, Comm. Math. Phys., to appear.

[17] L. Boltzmann, Nature **51**, 413 (1885).

[18] H. Jeans, Phil. Mag. **6**, 279 (1903).
H. Jeans, Phil. Mag. **10**, 91 (1905).

NUMERICAL RESULTS FROM TRUNCATED NAVIER-STOKES EQUATIONS

V. Franceschini

Dipartimento di Matematica
Università di Modena
Via Campi 213/13
41100 Modena
Italy

Franceschini described numerical methods for determining orbits, return maps, structure of attractors, etc. ecc. In one case by slightly varying appropriate parameters he found, very close together, a torus, a pseudo-periodic orbit of period 29 and a strange attractor. The numerical procedures were more-or-less standard, involving Newton's method and iteration; the results were exciting.

A SIMPLE AND COMPACT PRESENTATION OF BIRKHOFF SERIES

Michel Vittot*

Centre de Physique Théorique+
CNRS- Luminy, Case 907
F-13288 Marseille Cedex 09 (France)

Abstract: A new, and more algebraic method for the construction of the perturbation series of the classical mechanics is given. The use of Poisson bracket operators yields a compact expression, which actually is a formal summation of the recurrence formulas usually obtained for the normal form of a quasi-integrable hamiltonian.

Talk given at the school: "Non Linear Evolution and Chaotic Phenomena"-Noto, June 87
*Supported by the Deutsche Forschunnggemeinschaft (DFG).
+Laboratoire propre, Centre National de la Recherche Scientifique.

A. *INTRODUCTION*

Let H be an integrable hamiltonian system with Λ degrees of freedom, ie its phase space $R^\Lambda \times T^\Lambda$ admits an action-angle coordinate system (A,θ) such that $H = H(A)$ is independent of the angles θ. We also consider a perturbation of H, given by an hamiltonian $E = E(A,\theta)$ and a small parameter ϵ; and we want to study the dynamics of $\tilde{H} \equiv H + \epsilon E$. Instead of solving the Hamilton equations directly, Jacobi and also Hamilton noticed that it is less intricate to look for a (canonical) change of variables f such that in the "new" coordinates $(A',\theta') \equiv f(A,\theta)$, $\tilde{H}$ becomes integrable: that is we want f such that the composition $\tilde{H} \circ f$ is independent of θ. This independence can be expressed, for instance, by saying that $\tilde{H} \circ f$ is equal to its average over the angles since it must remain constant when θ varies in T^Λ:

$$(1) \qquad (\tilde{H} \circ f)(A,\theta) = (\tilde{H} \circ f)(A) = \int_T \frac{d^\Lambda\theta}{(2\pi)^\Lambda} (\tilde{H} \quad f)(A,\theta)$$

We designate by $M(\tilde{H} \circ f)$ the last member, thus introducing the "Mean" operator M. This is the linear operator: "average over the angles".

Hence the equation for the unknown f (Hamilton-Jacobi equation) may be written as: $\tilde{H} \circ f = M(\tilde{H} \circ f)$.
In order to simplify, we set $K = 1-M$, and we introduce a linear operator T instead of f, by writing for any hamiltonian H: $T(H) = H \circ f$.
Since f is a canonical transformation, T is an invertible operator and satisfies for any hamiltonians H and E: $T(H.E) = T(H).T(E)$ (for the pointwise product) and $TL_H T^{-1} = L_{TH}$ where $L_H = (\partial H/\partial A)(\partial/\partial \theta) - (\partial H/\partial \theta)(\partial/\partial A) = \{.;H\}$ is the "Poisson bracket operator with H".

So the problem is the following: given $\tilde{H}$, find T such that: $KT\tilde{H} = 0$ $(K = 1-M)$.

B. HAMILTON-JACOBI EQUATION FOR T

From now, we assume that T may be expanded as a formal power series in ϵ:

$$(1) \qquad T = T_\epsilon = \sum_{n=o}^{\infty} \frac{\epsilon^n}{n!} T^n$$

such that for any "small" ϵ:

$$(2) \qquad KT_\epsilon(H + \epsilon H) = 0 \quad \text{(with H, E given as above)} .$$

For instance we get, when $\epsilon=0$: $T_o = T^o = 1$ (since H is integrable: $KH = 0$). In order to compute T_ϵ the first step is to explicit the Hamilton-Jacobi equation.

Proposition 1

If there exists $T = T_\epsilon$ verifying (1) & (2)
Then necessarily the function $\epsilon \to T_\epsilon$ is a solution of the following integral equation:

$$(3) \qquad T_\epsilon(H + \epsilon E) = H + M \left[\int_o^\epsilon ds T_s \right] E$$

From (3) it is clear that the transformed hamiltonian $T_\epsilon(H + \epsilon E)$ is integrable, since the right hand side is the sum of H and the mean of some hamiltonian. This one is obtained in applying the linear operator in the bracket to the perturbation E. The bracket itself is nothing else than the sum of the canonical transformations T_s when the parameter s varies from 0 to ϵ. Let us note that this sum is approximately equal to $\epsilon 1$,ie proportional to the identity operator, since $T_o = 1$. By definition, the "normal form" of the perturbed hamiltonian $H + \epsilon E$ is the left hand side of (3), ie is equal to this hamiltonian expressed in its adapted coordinates. Hence we deduce from (3) that the normal form of $H + \epsilon E$ is approximately $H + \epsilon ME$, that is the perturbation E has to be replaced by its average over the angles. This remark is at the origin of the "averaging method" of Arnold [Arn2]. Before proving (3), we notice that it is an integral form of the Hamilton-Jacobi equation, and gives a compact, closed expression (and not a recurrence formula) of the normal form of any quasi-integrable hamiltonian.

Proof of proposition 1:

To simplify we set: $H_\epsilon \equiv T_\epsilon(H + \epsilon E)$. By assumption (2):

(4) $$H_\epsilon = MH_\epsilon, \quad \text{ie: } H_\epsilon' = MH_\epsilon'$$

where the prime is the formal derivation with respect to ϵ, of the formal power series H_ϵ.

Let us compute this derivation:

(5) $$H_\epsilon' = T_\epsilon'(H + \epsilon E) + T_\epsilon \quad \text{where } T_\epsilon' \equiv \sum_{n=o}^{\infty} \frac{\epsilon^n}{n!} T^{n+1} .$$

We will now prove that:

(6) $$MT_\epsilon'(H + \epsilon E) = 0$$

which complete the proof of proposition 1, since (5) plugged into the rhs of (4) yields:

(7) $$H_\epsilon' = M\left[T_\epsilon'(H + \epsilon E) + T_\epsilon E \right] = MT_\epsilon E$$

so that, by integration: $H_\epsilon - H_o = \int_o^\epsilon dsMT_sE \quad (H_o \equiv H)$.

The first step in the proof of (6), is the following computation of T_ϵ'

Proposition 2

If ϵ is small enough, then there exists a "generating function" $G_\epsilon = G_\epsilon(A,\theta)$ such that:

(8) $$T_\epsilon' = L_{G_\epsilon} T_\epsilon$$

Proof: When ϵ is small enough, there exists a real function $F_\epsilon = F_\epsilon(A,\theta)$ (sometimes also called "generating function") such that the canonical transformation T_ϵ is the flow generated by this "hamiltonian" F_ϵ, for a unit time:

(9) $\quad T_\epsilon = \exp(L_{F_\epsilon}) = \sum_{n=o}^{\infty} \frac{1}{n!} (L_{F_\epsilon})^n .$

There remains to compute the derivative with respect to ϵ of this exponential. We temporarily write L instead of L_{F_ϵ} :

$$[\exp L]' = \sum_{n=1}^{\infty} \frac{1}{n!} \sum_{k=1}^{n} L^{k-1} L' L^{n-k} .$$

Let us define L^* and L_* as the left and right multiplication by L, ie for any X: $L^*(X) = LX$ and $L_*(X) = XL$. Since the commutator with L is: $ad(L) = L^* - L_*$, the binomial formula yields:

$(L^*)^{k-1} = \sum_{p=o}^{k-1} \binom{k-1}{p} [ad(L)]^p (L_*)^{k-1-p}$. So plugging it into the last equation:

$$[\exp L]' = \sum_{n=1}^{\infty} \sum_{k=1}^{n} \sum_{p=o}^{k-1} \frac{1}{n!} \binom{k-1}{p} [[ad(L)]^p (L')] L^{k-1-p} L^{n-k}$$

$$= \sum_{n=1}^{\infty} \sum_{p=0}^{n-1} \frac{1}{n!} \binom{n}{p+1} [[ad(L)]^p (L')] L^{n-1-p}$$

(we used: $\sum_{k=p+1}^{n} \binom{k-1}{p} = \binom{n}{p+1}$)

$$= \sum_{p=o}^{\infty} \frac{1}{(p+1)!} [[ad(L)]^p (L')] \sum_{m=o}^{\infty} \frac{1}{m!} L^m = \left[\frac{\exp[ad(L)]-1}{ad(L)} (L')\right] [\exp L]$$

$$= L_{G_\epsilon} [\exp L] \quad \text{where:} \quad G_\epsilon = \frac{\exp(L_{F_\epsilon}) - 1}{L_{F_\epsilon}} (F'_\epsilon) \qquad \therefore$$

To complete the proof of (6), ie of proposition 1, there remains to show that: $M(L_{G_\epsilon} T_\epsilon)(H + \epsilon E) = 0$. But this rewrites as $ML_{G_\epsilon} H_\epsilon$, or even $ML_{G_\epsilon} MH_\epsilon$, (cf (4)).

And it is obvious that for any hamiltonian G, the following operator vanishes:

(10) $\quad ML_G M = 0 .$

Indeed when applied to a test function E, this rewrites as:

(11) $\quad ML_G ME = \int_{T^\Lambda} \frac{d^\Lambda \theta}{(2\pi)^\Lambda} [(\partial G/\partial A)(\partial ME/\partial \theta) - (\partial G/\partial \theta)(\partial ME/\partial A)]$

$$= 0 - (\partial ME/\partial A) \int_{T^\Lambda} \frac{d^\Lambda\theta}{(2\pi)^\Lambda} (\partial G/\partial\theta) = 0$$ since G is periodic in θ ∴

C. RECURSIVE RELATIONS FOR THE COEFFICIENTS T^n OF T_ϵ

Let us write proposition 1 under the following equivalent form:

Proposition 3

If there exists $T = T_\epsilon$ verifying B-(1) & B-(2)

Then necessarily the function $\epsilon \to T_\epsilon$ is a solution of the following integral equation:

(1) $\quad L_H G_\epsilon = K W_\epsilon E$

(2) $\quad$ where $\quad W_\epsilon \equiv T_\epsilon + L_{G_\epsilon} M \int_0^\epsilon ds T_s$

We introduced the linear operator W_ϵ only for the sake of simplicity. The unknown in (1) is T_ϵ (or its generating function G_ϵ).

Proof: According to B-(7), B-(8) & B-(3): $MT_\epsilon E = H'_\epsilon = T'_\epsilon(H + \epsilon E) + T_\epsilon E$

$$= (L_{G_\epsilon} T_\epsilon)(H + \epsilon E) + T_\epsilon E = L_{G_\epsilon} \left[H + M \int_0^\epsilon ds T_s E \right] + T_\epsilon E$$

Hence: $\quad - L_{G_\epsilon} H = (1-M) T_\epsilon E + L_{G_\epsilon} M \int_0^\epsilon ds T_s E$

$$= K T_\epsilon E + K \left[L_{G_\epsilon} M \int_0^\epsilon ds T_s E \right] + M \left[L_{G_\epsilon} M \int_0^\epsilon ds T_s E \right]$$

where we replaced 1 by K+M. Finally the last term vanishes (cf B-(10)), and the antisymmetry of the Poisson bracket reads: $- L_{G_\epsilon} H = L_H G_\epsilon$ ∴

We claim that (1) is easier to solve than B-(3). First of all we define the frequency: $\omega \equiv \omega(A) \equiv H'(A) \equiv \partial H/\partial A$ so that: $L_H G_\epsilon = (\partial H/\partial A)(\partial G_\epsilon/\partial\theta) \equiv \omega(A)(\partial G_\epsilon/\partial\theta)$. Then we denote the Fourier decomposition of any function $E = E(A,\theta)$ by: $E(A,\theta) = \sum_{\nu \in Z^\Lambda} E(A,\nu) e^{i\nu\theta}$

so that equation (1) reads in Fourier components:

(3) $\quad i(\omega(A)\nu) G_\epsilon(A,\nu) = (KW_\epsilon E)(A,\nu) \quad \begin{cases} \equiv (W_\epsilon E)(A,\nu) & \text{if } \nu \neq 0 \\ \equiv 0 & \text{if } \nu = 0 \end{cases}$

We see that the presence in (3) of the operator K is a condition of consistency, since the lhs vanishes when $\nu = 0$. And the formal solution of (1) is given by:

$$(4) \qquad G_\epsilon(A,\theta) = -i \sum_{\nu \in Z^\Lambda - \{0\}} \frac{(W_\epsilon E)(A,\nu)}{\omega(A)\nu} e^{i\nu\theta} \qquad \text{or:} \quad G_\epsilon = L_H^{-1} K W_\epsilon E.$$

For the moment, the meaning of the inverse L_H^{-1} of the operator L_H is purely formal. In section *E* below we give it a precise meaning in a particular case.

Let us now define the coefficients of the formal series ϵ (cf B-(1)):

$$(5) \qquad G_\epsilon = \sum_{n=o}^{\infty} \frac{\epsilon^n}{n!} G^n \qquad W_\epsilon = \sum_{n=o}^{\infty} \frac{\epsilon^n}{n!} W^n \qquad L_{G_\epsilon} = \sum_{n=o}^{\infty} \frac{\epsilon^n}{n!} L^n$$

ie $L^n \equiv L_{G^n}$. These series may be considered as "Birkhoff series". From B-(8) we get:

$$(6) \qquad T^{N+1} = \sum_{n=0}^{N} \binom{N}{n} L^n T^{N-n} \qquad (\text{and } T^o = 1) \; .$$

The definition of W_ϵ gives immediately:

$$W^{N+1} = T^{N+1} + \sum_{n=0}^{N} \binom{N+1}{n} L^n M T^{N-n} \qquad \text{ie:}$$

$$(7) \qquad W^{N+1} = \sum_{n=0}^{N} \binom{N}{n} L^n \left[1 + \frac{N+1}{N+1-n} M\right] T^{N-n} \quad (\text{and } W^0 = 1; \text{ we used (6)}).$$

Finally the solution (4) above yields

$$(8) \qquad G^N = L_H^{-1} K W^N E \; .$$

So the algorithm for the computation of these coefficients is the following:

$$G^0 = L_H^{-1} KE \Rightarrow L^0 = L_{Go} \Rightarrow T^1 = L^0 T^0 = L^0 \text{ and } W^1 = L^0(1+M)T^0 = L^0(1+M)$$

$$\Rightarrow G^1 = L_H^{-1} K W^1 E \quad (= L_H^{-1} K\{(1+M)E, G^0\} = L_H^{-1} K\{(1+M)E, L_H^{-1} KE\})$$

$$\Rightarrow L^1 = L_{G1} \Rightarrow T^2 = L^0 T^1 + L^1 T^0 = (L^0)^2 + L^1 \text{ and } W^2 = L^0(1+M)L^0 + L^1(1+2M) \; .$$

And so on: we deduce G^2, then L^2, T^3 and W^3...The recurrence relations (6)(7)(8) are sufficient for the whole computation. The advantage of this algebraic method over the usual ones is that here the recurrence formulas are only sums over a single integer n, and not a multi-integer. The use of "Poisson bracket operators" turns out to be a partial summation of the usual recurrence relations. Indeed these ones are sums of terms like: $(\partial^a h_1/\partial A^b \partial\theta^{a-b})(\partial^c h_2/\partial A^d \partial\theta^{c-d})$ (...), where h_1, h_2 are some hamiltonians E, G^0, G^1... and a,b,c,d are integers. And any operator $(L_{Gn})^K$ is a sum of $2^k k!$ such terms.

D. GENERALIZATION TO THE RESONANT CASE

From now we replace the linear projector M introduced in A-(1), by the following one. For any test function $E = E(A,\theta)$ defined on $R^\Lambda \times T^\Lambda$ we set:

(1) $$(ME)(A,\theta) = \sum_{\nu\in M} E(A,\nu)e^{i\nu\theta} \quad \text{where } M \text{ is a subgroup of } Z^\Lambda .$$

For instance when $M = \{0\}$, this operator is the same as in previous sections. Property B-(10) is again true, but with a slight restriction:

(2) $$\text{if } MG = 0 \text{ then } ML_G M = 0 .$$

The proof is trivial, and uses only the group property of M (or: $ML_G M = ML_{MG}M$). So if we restrict ourselves to look only for solution G_ϵ verifying $MG_\epsilon = 0$, then the same calculations as above yield the formal solution: $G_\epsilon = L_H^{-1} KW_\epsilon E$. And we verify that this solution satisfies $MG_\epsilon = 0$, since it can also be written: $G_\epsilon = KL_H^{-1} KW_\epsilon E$. This generalization is physically relevant when the frequencies ω introduced in Section C are "resonant" ie $\omega(A)\nu$ vanishes for some values of A and ν Indeed in this case it is impossible to eliminate by a canonical transformation these resonant harmonics in the perturbation. The normal form obtained by the above theory is (cf B-(3)):

(3) $$T_\epsilon(H + \epsilon E) = H + M\left[\int_0^\epsilon ds T_s\right]E$$

that is, in the "renormalized" perturbation there remains only resonant harmonics. For instance if the integrable part is an harmonic oscillator $H = \omega A$ (where ω is a fixed vector of R^Λ) then we have to use: $M = \{\omega\}^\perp \cap Z^\Lambda$ ie the set of ν orthogonal to ω. Let us notice that the above expression (3) is well suited for the formal computation of prime integrals of $\tilde{H} = H + \epsilon E$. Indeed for any hamiltonian $P = P(A)$ such that $P'(A) \in M^{\perp}$ (ie its derivative is orthogonal to any vector of M) the transformed hamiltonian $T_{\epsilon}^{-1} P$ is a prime integral of $\tilde{H}$:

$$\{T_{\epsilon}^{-1}P,\tilde{H}\} = T_{\epsilon}^{-1}\{P,T_{\epsilon}\tilde{H}\} = T_{\epsilon}^{-1}\{P,H + M\int_{0}^{\epsilon} dsT_{s}E\} = 0$$ since P is independent of θ. There are as many such functions P as the dimension of $M^{\perp}$.

E. ESTIMATES ON THE NORM OF T^N-NEKHOROSHEV THEOREM

In order to study the convergence of the series B-(1) & C-(5) we need to introduce some topology. So we designate by $H_{r,\rho}$ the Banach algebra of hamitonians $E = E(A,\theta)$ defined on $D \times T^{\Lambda}$ (D a union of open balls of R^{Λ} with a radius r), real analytic in A, and that can be extended to a complex neighbourhood $T(\rho)$ of T^{Λ}: $|\mathrm{im}\theta_x| < \rho$ for any x. The norm we choose is the following:

$$||E||_{r,\rho} = \sup_{A\in D} \sum_{\nu\in Z^{\Lambda}} \sup_{\theta\in T(\rho)} |E(A,\nu)e^{i\nu\theta}|.$$ See [VitBel] for a detailed study of this algebra. One of its advantage is that the Poisson bracket operator L_G is a bounded derivation between $H_{r,\rho}$ and $H_{r',\rho'}$ for any $r' < r$ and $\rho' < \rho$. From now we study the case where $H = A\omega$ is an harmonic oscillator. We choose a subgroup M of Z^{Λ} and we assume the existence of $C > 0$, $\tau > \Lambda$ such that

(1) for any multi-integers $\nu \in M$: $|\omega\nu| \geq C/|\nu|^{\tau-1}$ with

$$|\nu| = \sum_{x=1}^{\Lambda} |\nu_x| .$$

For the norm on operators induced by the norm above, we find from C-(6)(7)(8) that:

(2) $$||\epsilon^N T^N/N!|| \leq (\text{const.}\epsilon)^N N^{\tau N}(\text{LogN})^N .$$

This estimate yields the well known divergence of the "Birkhoff series" and has also been recently established in [BeGaGi2, GDFGS, BenGal]. The details of the proof of (2) will be published elsewhere.

Let us now consider an approximation $\tilde{G}_{\epsilon}$ of G_{ϵ} which is convergent. We choose an integer m, and we set:

$\tilde{G}_\epsilon = \sum_{n=o}^{m} \frac{\epsilon^n}{n!} G^n$. Then by formulas C-(6)(7)(8) we construct $\tilde{T}^n$ (and $\tilde{W}^n$).

When $n \le m$, we have: $\tilde{T}^n = T^n$. And a calculation analogous to that proving (2) gives for any $N \ge m$:

(3) $$||\epsilon^N \tilde{T}^N / N!|| \le (\text{const.}\,\epsilon)^N m^{\tau N} (\text{Log } m)^N .$$

The canonical transformation $\tilde{T}_\epsilon$ generated $\tilde{G}_\epsilon$ is given by:

$\tilde{T}_\epsilon = \sum_{n=o}^{\infty} \frac{\epsilon^n}{n!} \tilde{T}^n$ and is a convergent series (cf(3)). Let us write the transformed hamiltonian $\tilde{T}_\epsilon(H + \epsilon E)$ as:

$$\tilde{T}(H+\epsilon E) = H + \sum_{n=1}^{m} \frac{\epsilon^n}{n!} \tilde{T}^n H + \sum_{n=m+1}^{\infty} \frac{\epsilon^n}{n!} \tilde{T}^n H + \sum_{n=1}^{m} \frac{\epsilon^n}{n!} n\tilde{T}^{n-1} E + \sum_{n=m+1}^{\infty} \frac{\epsilon^n}{n!} n\tilde{T}^{n-1} E.$$

We designate by P the sum of the second and fourth term, and by R the sum of the third and fifth term. By definition:

$P = \sum_{n=1}^{m} \frac{\epsilon^n}{n!}(T^n H + nT^{n-1}E)$ which may be written as MP since

$P = \sum_{n=1}^{m} \frac{\epsilon^n}{n!}(MT^{n-1}E)$ (identify the coefficient of ϵ^n in the 2 members of B-(3)). Of course we cannot do the same for R since $\tilde{T}^n \ne T^n$ when $n > m$. But we can use estimate (3) and sum the geometric series. So we get the following:

Proposition 4

We consider a perturbed harmonic oscillator $\tilde{H} = A\omega + E$ where ω satisfies condition (1). If

(4) $$\epsilon \le \text{const}/m^\tau \text{ Log } m$$

then there exists a convergent canonical transformation $\tilde{T}_\epsilon$ such that:

(5) $$\tilde{T}_\epsilon(H + \epsilon E) = H + MP + R$$

where $P - MP$ is some hamiltonian of order ϵ, and R is a "small" hamiltonian:

(6) $$||R|| \le (\text{const.}\,\epsilon m^\tau \text{ Log } m)^{m+1} .$$

It is natural to look for the value of m which minimizes R. It is approximately given by:

(7) $m_0 \simeq$ const. $\epsilon^{-1/\tau}$ and the corresponding remainder is bounded by:

(8) $||R_0|| \le$ const. $\exp-(1/\epsilon|\text{Log } \epsilon|)^{1/\tau} \equiv R_c$.

Hence the perturbation E will significantly modify the dynamics only after a time of the order $1/R_c$, which is bigger than any power of $1/\epsilon$. This is a particular case of the Nekhoroshev theorem which bounds the "Arnold diffusion", ie the separation between a perturbed trajectory and the non-perturbed one. For more details see [Arn1, Nek, BeGaGi1, BeGaGi2, BenGal, Gal, GDFGS].

Actually it is possible to get better results in what concerns the dependence of R_c in the number Λ of degrees of freedom, when Λ is big. In [BeGaGiVi] it is found that:

(9) $||R_0|| \le$ const. $\exp-(|\text{Log } \epsilon|^2/b \text{ Log } \Lambda)$

when $\epsilon \le$ const. Λ^{-c} (for some positive constants b and c, independent of Λ). This threshold is bigger than that obtained by the method (4)-(8) described above: $\epsilon \le$ const. $c^{-\Lambda}$ (for the same c). Actually the upper bound (9) for $||R_0||$ is better than (8) when $\epsilon \ge$ const.$e^{-a\Lambda \text{ Log } \Lambda}$, ie when $\Lambda \ge$ const.$|\text{Log } \epsilon|/ \text{ Log}|\text{Log } \epsilon|$. Otherwise (8) is better, although (9) is still of order ϵ^{∞}, ie smaller than any power of $1/\epsilon$. A non-resonant condition different that (2), has been applied in [BeGaGiVi] in order to decrease the dependence of the estimates in the number Λ of degrees of freedom. This method was introduced in [Vit, VitBel] in order to extend the Kolmogorov-Arnold-Moser theorem to an infinite number of freedoms, in some particular cases.

Acknowledgment: The author would like to express his gratitude to R. Seiler, V. Enβ, M. Loss and R. Schrader for their useful advices and their kind hospitality in Berlin while this paper was prepared.

F. REFERENCES

[Arn1] V.I. ARNOLD, Instability of dynamical systems with many degrees of freedom, (Russian), Dokl. Akad. Nauk SSSR, 156, (1964), 9.

[Arn2] V.I. ARNOLD, Méthodes mathématiques de la mécanique classique, Ed. Mir, Moscou (1976).

[BeGaGi1] G. BENETTIN, L. GALGANI, A. GIORGILLI, Boltzmann's ultraviolet cut-off and Nekhoroshev's theorem on Arnold's diffusion, Nature, 311, (1984), 444-446.

[BeGaGi2] G. BENETTIN, L. GALGANI, A. GIORGILLI, Rigorous estimates for the series expansions of hamiltonian perturbation theory, Preprint Università de Milano, 1985.

[BeGaGiVi] G. BENETTIN, L. GALGANI, A. GIORGILLI, M. VITTOT, Perturbation theory and large numbers of degrees of freedom, Preprint Università di Milano, 1985.

[BenGal] G. BENETTIN, G. GALLAVOTTI, Stability of motions near resonances in quasi-integrable hamiltonian systems, Preprint 1985.

[GDFGS] A. GIORGILLI, A. DELSHAMS, E. FONTICH, L. GALGANI, C. SIMÓ, Effective stability for a hamiltonian system near an elliptic equilibrium point, with an application to the restricted three body problem. Preprint Università di Milano, 1987.

[Gal] G. GALLAVOTTI, Quasi-integrable mechanical systems, "Critical Phenomena, Random Systems, Gauges Theories", Proceedings of "Les Houches" summer school 1984, Ed. K. Osterwalder & R. Stora, North Holland 1986.

[Nek] N. N. NEKHOROSHEV, Exponential estimates of the time of stabilty for nearly integrable hamiltonians, Russ. Math. Surveys, 32, (1977), 1-63.

[Vit] M. VITTOT, Théorie classique des perturbations et grands nombres de degrés de liberté, Thèse, (Université de Provence, Marseille), Preprint CPT 85/P. 1838.

[VitBel] M. VITTOT, J. BELLISSARD, Invariant tori for an infinite lattice of coupled classical rotators, Preprint CPT 85/ P. 1796 (Marseille).

TWO LECTURES ON CHAOTIC DYNAMICS IN THE SOLAR SYSTEM

Jack Wisdom

Dept. of Earth, Atmospheric and Planetary Sciences
Massachusetts Institute of Technology
Cambridge, Massachusetts 02139 U.S.A.

The material presented in these lectures has to a large extent been published elsewhere. In particular, reviews of my work on chaotic behavior in the solar system are published in Wisdom (1987) *Proc. Roy. Soc. Lond.* **A413**, 109 and in somewhat greater detail in Wisdom (1987) *Icarus* **72**, in press. Only an abstract of the topics will be given here, with references to the original literature. References to associated literature may be found in the cited references.

The solar system is generally perceived as evolving with clockwork regularity, yet there are several physical situations in the solar system where chaotic solutions of Newton's equations play an important role. There are physical examples of both chaotic rotation and chaotic orbital evolution.

I. Chaotic Rotations

Saturn's satellite Hyperion is currently tumbling chaotically, its rotation rate and spin axis orientation undergo significant chaotic variations on a timescale of only a couple of orbit periods (Wisdom, Peale, Mignard 1984 *Icarus* **58**, 137). Voyager 2 images of Saturn's satellite Hyperion showed it to be unexpectedly out-of-round. Hyperion is nearly twice as long as it is across. The orbit of Hyperion also has a rather large eccentricity near 0.1. Taken together these two facts lead to quite interesting classical dynamics. The equations governing the motion of a satellite are simply Euler's equations for the motion of a rigid body, with the addition of gravity gradient torques from the host planet. The gravity gradient torques are time dependent due to the non-uniform Keplerian motion in the elliptical orbit.

Over very long times, tidal friction tends to drive the spin of a satellite near synchronous rotation (where the same face of the satellite is always pointed toward the planet), and at the same time drives the spin axis perpendicular to the orbit plane. It is thus natural to consider a model problem where the spin axis is taken to be exactly normal to the orbit plane. It is also natural to take in our model problem the orbit to be fixed since the timescale for significant changes in the orbit are much longer than the rotational dynamical timescale which is of order the orbital period. The dynamical problem which remains has only a single degree of freedom which is the orientation of the body about its spin axis, but there is still explicit time dependence through the Keplerian motion of the body in its orbit. Surfaces of section for this model problem reveal a large chaotic zone in the rotational phase space. The origin of this large chaotic zone can be understood qualitatively in terms of the resonance overlap criterion.

A more complete model includes all three degrees of freedom for the motion of the rigid body. In particular, the dynamical stability of the special orientation of the spin axis is questioned. It turns out that rotation in the chaotic zone with the spin axis perpendicular to the orbit is attitude unstable, i.e., the spin axis falls away from the orbit normal under the slightest disturbance. The resulting motion is chaotic tumbling, as demonstrated by the computation of the Lyapunov exponents. Dynamical states which are stable for other satellites are unstable for Hyperion, the most important of these is synchronous rotation which is also attitude unstable to chaotic tumbling.

Tidal friction acting over the age of the solar system is responsible for bringing Hyperion to this chaotic region of phase space. Astronomical observations are consistent with the conclusion that Hyperion is currently tumbling chaotically.

The chaotic tumbling of Hyperion is not an isolated curiosity. Many other satellites in the solar system have had chaotic rotations in the past (Wisdom 1987 *Astron. J.* **94**, in press). It is not possible to tidally evolve into synchronous rotation without passing through a chaotic zone. For irregularly shaped satellites this chaotic zone is large even if the orbital eccentricity is relatively small. In every case studied this chaotic zone is attitude unstable and chaotic tumbling ensues. This episode of chaotic tumbling probably lasts on the order of the tidal despinning timescale. For example, the Martian satellites Phobos and Deimos tumbled chaotically before they were captured into synchronous rotation for a time interval on the order of 10 million years and 100 million years, respectively. This episode of chaotic tumbling could have had a significant effect on the orbital histories of these satellites.

II. Chaotic Orbits

The asteroids provide a near ideal dynamical laboratory for the study of the long-term evolution of Newton's equations. There are several thousand asteroids with well determined orbits. There are qualitative features in the distribution of the orbital elements which have been known for some time yet defied dynamical explanation. The most important of these qualitative features are the Kirkwood gaps in the distribution of semimajor axes near low order commensurabilities between the mean orbital motion of the asteroid and that of Jupiter.

By deriving a symplectic mapping of the phase space onto itself which approximates the dynamical evolution near these resonances it was possible to study the motion over much longer times than was previously possible with ordinary numerical integration (Wisdom 1982 *Astron. J.* **87**, 577, Wisdom 1983 *Icarus* **56**, 51). The phase space near one of these resonances where the orbital period of the asteroid is one-third the orbital period of Jupiter was found to have a large chaotic zone. Trajectories in this chaotic zone displayed surprising variations in orbital eccentricity when viewed over timescales of order several million years. Orbits could spend as much as a million years with near circular orbits and then over a period of only a few tens of thousands of years could suddenly become significantly eccentric. These large eccentricity increases explain the origin of this Kirkwood gap because the large eccentricity increases give the asteroids planet crossing orbits which eventually lead to removal by collision or close encounter with the planet. The predicted boundary of the Kirkwood gap is in close agreement with the observed population of asteroids. These eccentricity increases have subsequently been verified by conventional numerical integration as well as by a novel perturbation theory (Wisdom 1985 *Icarus* **63**, 272).

The meteorites provide important clues to the conditions present at the time of formation of the solar system. Though it has long been believed that most meteorites originate in the asteroid belt, a dynamical transport mechanism consistent with the properties of the meteorites had not been found. Chaotic trajectories at the 3:1 resonance have sufficiently large eccentricity increases that the orbits cross the orbit of Earth (Wisdom 1985 *Nature* **315**, 731). Thus this chaotic zone provides a dynamical route for the tranport of asteroidal debris from the asteroid belt to Earth.

Numerical exploration of other resonances has shown a similar correspondence between chaotic zones and depletions of the asteroid population, though the agreement is not as striking in all cases as it is for the 3:1 gap. More work has to be done.

Mean motion commensurabilities between the Uranian satellites are accompanied by significant chaotic zones (Tittemore and Wisdom 1987 submitted to *Icarus*). It is not yet known whether chaotic orbital evolution during passage through these resonances is related to the exotic surface features on the Uranian satellites which were revealed by the Voyager images. The anomalously high inclination of Miranda arises naturally during passage through one of these resonances.

We have investigated the long-term stability of the solar system through a 200 million year integration of the outer planets (Sun, Jupiter, Saturn, Uranus, Neptune, and Pluto) with a special purpose computer built by Gerald J. Sussman (MIT) and coworkers called the Digital Orrery (Applegate, Douglas, Gursel, Sussman, Wisdom 1986 *Astron. J* **92**, 176). The long-term evolution of Pluto is suspiciously complicated, but objective criteria (i.e., Lyapunov exponents) have not yet indicated that the motion is chaotic. A billion year integration of the outer planets is currently being carried out with the Digital Orrery.

QUANTUM CHAOLOGY OF ENERGY LEVELS

NOTES BASED ON LECTURES BY MICHAEL BERRY

Jonathan Keating and Raul Mondragon

H.H.Wills Physics Laboratory, Royal Fort, Tyndall Avenue, Bristol BS8 1TL, U.K.

The subtle and complex nature of classical Hamiltonian mechanics is now well recognised: long-time predictability and the topologies of the orbits of a system are known to depend critically on the form of the Hamiltonian and the phase space may support regions of regular and chaotic motion interwoven on all scales.[1] A natural question is: how does this classical complexity manifest itself in the corresponding quantum system? Sometimes this question is put in the form: what is Quantum Chaos?

One approach to this question is to study the dynamics of quantum systems which are classically chaotic. Experiments on one such system, the ionization of hydrogen in strong microwave fields[2], have been supported by theoretical work on models such as the quantum kicked rotator, in which a particle on a ring is kicked periodically with an impulse which depends on where it is.[3] This, and other computational work on quantised maps[4], has led to the growing understanding that quantum systems are not chaotic in the way that classical systems are; such systems may be said to exhibit the general phenomenon of 'quantum suppression of classical chaos'[5]. This phenomenon has yet to receive a full analytic treatment, but may be understood by considering quantal phase space to be smoothed out over areas the size of Planck's constant h, so that the quantum mechanics is insensitive to classical phase space structure on scales smaller than h. Although there is no long-time chaotic quantum evolution there are elements of the quantal description which are sensitive to the details of the classical orbits and which emerge in the semiclassical limit (i.e. as $h \rightarrow 0$), hence Quantum Chaology may be defined as[6]

> The study of semiclassical - but nonclassical - behaviour characteristic of systems whose classical motion exhibits chaos.

In these lectures we shall consider the quantum chaology

of spectra i.e. of the eigenenergies $\{E_i\}$ of time independent systems which are either classically integrable (i.e. in a system with N freedoms there are N independent constants of the motion) or classically chaotic (i.e. energy is the only constant of the motion and N>1).[7] Ideally one would like an explicit asymptotic formula giving $\{E_i\}$ with an error that decreases as $h \to 0$ faster than the mean level spacing, however such a formula exists only for integrable systems where all the classical orbits are regular and are confined to N-dimensional tori in the 2N-dimensional phase space. In this case one can make a global transformation to action-angle variables $[(p,q) \to (I,\theta)]$[7] where H=H(I) and then the eigenenergies are

$$E_{\{n_i\}} = H\left(\left\{ I_i = \left(n_i + \frac{\alpha_i}{4}\hbar\right)\right\}\right) \qquad (1)$$

where $\{n_i\}=n_1 \ldots n_N$ are the quantum numbers, $\{I_i\}$ are the actions

$$I_i \equiv \oint_{\gamma_i} \mathbf{p} \cdot d\mathbf{q} \qquad (2)$$

$\{\gamma_i\} = \gamma_1 \ldots\ldots \gamma_N$ are the N topologically irreducible cycles of the torus and the α_i are constants (Maslov indices)[8]. These eigenenergies are the correct topological generalisations of the WKB solution of the Schrodinger equation and (1) is known as the EBK quantisation formula. Such a semiclassical formula has so far not been found for systems which are not integrable and so one must consider statistical properties of the distribution of energy levels.

We may define the density of states by

$$d(E) \equiv \sum_n \delta(E-E_n) = Tr(E-\hat{H}) \qquad (3)$$

the simplest characteristic of the spectrum is the average of (3) which is given semiclassically by the 'Weyl formula'[9]

$$\langle d(E) \rangle = \frac{1}{h^N} \frac{d}{dE} \Omega(E) \quad , h \to 0 \qquad (4)$$

where $\Omega(E)$ is the volume of that part of classical phase space whose points have energies less than E

$$\Omega(E) = \iint d^N q \, d^N p \, \Theta(E - H(q,p)) \qquad (5)$$

One of the most frequently used model systems is the family of billiards which take the form of a particle moving freely in a finite domain D, with boundary ∂D. The shape of the boundary determines the nature of the classical dynamics. In this case the spectral staircase

$$N(E) \equiv \sum_n \Theta(E-E_n) = \int_{-\infty}^{E} dE' \, d(E') \qquad (6)$$

has the average

$$\langle N(E)\rangle = \frac{A}{4\pi}K^2 \quad , \quad K = \frac{\sqrt{2mE}}{\hbar} \tag{7}$$

A=area of D
One may include further corrections due to the boundary to give[9]

$$\langle N(E)\rangle = \frac{A}{4\pi}K^2 - \frac{L}{4\pi}K + \frac{1}{6} \tag{8}$$

L= length of ∂D
which provides a remarkably good fit to the actual staircase and may be used, when computing eigenenergies, to check that no states have been missed.

Of course <N(E)> is unaffected by the underlying dynamics as it depends only upon geometrical properties of D, hence it cannot distinguish between integrable and chaotic systems. To do this one must consider the fine scale fluctuations which are described by statistics that involve correlations between nearby levels, that is on scales $\hbar^N$. In fact it is best to initially magnify the spectrum $\{E_i\}$ by <d(E)> to obtain $\{X_i\}$ which then has a mean density of 1. Then the most important statistics used are[10]:

(a) Spacing distribution, P(S), defined as the probability that there exists a spacing

$$S = X_{j+1} - X_j$$

(b) Rigidity, which is defined as the best mean square fit to a straight line of the staircase over a range in X of length L, averaged over the whole spectrum

$$\Delta(L) \equiv \left\langle \min_{A,B} \frac{1}{L}\int_{-L/2}^{L/2} dx \left[N(\tau+x) - Ax - B\right]^2 \right\rangle \tag{9}$$

(c) Number Variance,

$$\Sigma(L) \equiv \langle (n(L)-L)^2 \rangle$$

where n(L) is the number of elements of $\{X_i\}$ in a length L, averaged over the whole spectrum.

What these statistics exhibit is a remarkable universality such that the statistics of the levels of a given system fall into one of a small number of universality classes determined by the chaology of the classical orbits of that system. These universality classes are just those found in the study of the statistics of eigenvalues of large random matrices.

Random matrix theory[11] developed as a model of the eigenenergies of nuclei, where the Hamiltonians involved are so complex that one could not, at present, hope to find the exact eigenenergies. The idea is that this theory might work not only for systems with complex Hamiltonians but also for systems with simple Hamiltonians and complex underlying dynamics. Formulae may be obtained for the statistics of the eigenvalues of NxN hermitian matrices (in the limit $N \to \infty$) if the matrix elements are assumed randomly distributed with respect to a given ensemble. If the elements are real the most general ensemble is that which is invariant under orthogonal transformation: the Gaussian Orthogonal Ensemble (GOE). If the elements are complex then the most general ensemble is that which is invariant under unitary transformations: the Gaussian Unitary Ensemble (GUE). One then finds the following universal classification of spectra:

(a) if the system is classically integrable the spectral statistics are those of a Poisson Distribution of levels[12] i.e. the levels act like uncorrelated random numbers. These statistics are characterised by "level clustering" (i.e. $P(S) \to 1$ as $S \to 0$) and one finds

$$P(s) = \exp(-s) \quad \text{and} \quad \Delta(L) = L/15 \tag{11}$$

(b) if the system is classically chaotic and there exists an antiunitary operator

$$A = UK \tag{12}$$

(U is a unitary operator and K is the complex conjugation operator) which commutes with the Hamiltonian. Then the level statistics will be GOE[13]:

$$P(s) \approx \frac{\pi}{2} s \exp(-\pi s^2/4)$$

and

$$\begin{aligned} \Delta(L) &\to L/15 \quad (L \ll 1) \\ &\to \frac{\ln L}{\pi^2} - 0.00695 \quad (L \gg 1) \end{aligned} \tag{13}$$

One of the most commonly encountered antiunitary symmetries is time-reversality i.e. the time reverse of an orbit of the classical system is also an orbit[14].

(c) If the system is classically chaotic and there is no antiunitary operator which commutes with the Hamiltonain then the level statistics will be GUE:

$$P(s) \approx \frac{32}{\pi^2} s^2 \exp(-4s^2/\pi)$$

$$\begin{aligned} \Delta(L) &\to L/15 \quad (L \ll 1) \\ &\to \frac{\ln L}{2\pi^2} + 0.05902 \end{aligned} \tag{14}$$

Levels with these statistics have been studied in chaotic billiards with a single line of magnetic flux passing through D (out of the plane), such systems are known as Aharonov-Bohm billiards[15]; the effect of the flux is to break time reversal symmetry, but since it is a single line only a measure zero of the classical orbits actually pass through the magnetic field, hence the chaology of the classical orbits is unaltered. In such systems one can view the effect of antiunitary symmetries by turning off the field or by introduing geometrical symmetries, causing the statistics to switch to GOE[16]. These statistics are also observed in the 'neutrino billiard' i.e. a billiard domain containing a massless fermion whose quantum mechanical description is via the Dirac equation[17].

(d) There does exist a further universality class of the statistics of random matrices: those associated with quaternion real Hermitian matrices whose elements are distributed according to an ensemble which is invariant under symplectic transformations (GSE statistics). However these have as yet been little studied.

Having discussed the universal nature of semiclassical quantum spectra it must now be revealed that there are ways in which universality may be compromised. Firstly we have discussed systems which are completely regular or completely irregular, whereas in fact the generic Hamiltonian system has regular and irregular regions in its phase space[1]. Attempts have been made to understand the statistics of these systems as a mixture of the pure statistics with an appropriate weighting which depends upon the given system (i.e. is non--universal)[18]. Secondly, even for purely regular or purely chaotic systems universality breaks down over energy ranges exceeding a quantity of order $\hbar$[12], i.e. a range which semiclassically shrinks to zero but which contains infinitely many levels. There now exists the beginnings of a general theory to explain this non-universal character which has so far proved successful in modelling statistics.

We must now look at how these properties of levels can be modelled by theory. Any semiclassical theory must be based on the classical dynamics and a beautiful description has been developed[19] which relates the quantum density of states to a sum over the classical closed orbits:

$$d(E) = -\frac{1}{\pi}\lim_{\varepsilon\to 0} \mathrm{Im}\,\mathrm{Tr}\left(\frac{1}{E+i\varepsilon-\hat{H}}\right)$$

$$\underset{\hbar\to 0}{\approx} \langle d(E)\rangle + \sum_j A_j(E)\exp\left[\frac{i}{\hbar}S_j(E)\right] \qquad (15)$$

where j labels a classical closed orbit, $S_j(E)$ is the classical action around the orbit

$$S_j(E) = \oint_j \underline{p}\cdot d\underline{q} \qquad (16)$$

and $A_j(E)$ is the amplitude of the contribution of the jth orbit and is related to its stability. Here j labels all the closed orbits including repetitions; the paths which shrink to zero as $\underline{r}\to\underline{r}'$ in the trace operation contribute to $\langle d(E)\rangle$ and can be shown to give the Weyl Formula. There are various problems associated with (15) in that to obtain d(E) at a given energy all closed orbits must be used (i.e. a given energy level is built up from the contributions of all closed orbits)[20] and secondly (15) is usually divergent and really only makes sense for complex energies. However (15), along with semiclassical sum rules for the amplitudes A_j[21] can be used to build a theory of spectral statistics.

We shall conclude by exhibiting a model which illustrates clearly all of the features discussed, but for which, unfortunately, the underlying classical system is not known![22]

Consider the function

$$\zeta(z) \equiv \sum_{n=1}^{\infty} \frac{1}{n^z} = \prod_p (1-p^{-z})^{-1} \qquad Re(z) > 1 \tag{17}$$

this is the Riemann Zeta Function and can be continued to the whole plane (except for a pole at z=1). It is the Riemann Hypothesis (RH)[23] that the zeros of this function lie along the line Re(z)=1/2

i.e. $$\zeta(1/2 + i E_j) = 0 \tag{18}$$

is true only for $Im(E_j)=0$. It is an old idea going back to Hilbert and Polya that one could prove the hypothesis if one could find a Hamiltonian whose eigenvalues were the $\{E_j\}$ of (18). We now have 1.5×10^9 zeros[24] which all satisfy the RH and so one may ask: if there is such a Hamiltonian what do the statistics of its eigenvalues (the Riemann zeros) tell us about its classical limit?

One finds that the statistics of the $\{E_j\}$ are very accurately modelled by the GUE statistics over the range where one would expect universality[25], suggesting that the classical dynamics are chaotic and that the system has no antiunitary symmetry. Furthermore one has that for this (unknown) system

$$d(E) - \langle d(E) \rangle = \frac{1}{\pi} \operatorname{Im} \frac{d}{dE} \log \zeta(1/2 + iE) \tag{19}$$

which may be expanded in the (non convergent) series

$$d(E) = \langle d(E) \rangle - \frac{1}{2\pi} \sum_{m=-\infty}^{\prime\,\infty} \sum_{p=2}^{\infty} \ln p \, \exp\left[-\frac{|m| \ln p}{2}\right] \exp\left[iEm \ln p\right] \tag{20}$$

giving the closed orbits of the (unknown) system labelled by the prime numbers and repetition number, the actions around the closed orbits

$$S_{p,m} = mE \ln p \tag{21}$$

and information about their stability from the amplitude term. Using (20) one can obtain asymptotic formulae for the statistics which model, with amazing accuracy, the universal and non--universal statistics obtained by computation[26].

In summary, we now have good evidence of the effect of the chaology of the classical orbits on the statistics of the eigenvalues and we have the basic means of description of the spectral fluctuations in the closed-orbit summation formula. However there remains much more to be done to understand the subtle interplay between the classical orbits which allows this formula to sum up to a series of delta-functions with the correct statistical properties.

REFERENCES

1 A.J.Lichtenberg and M.A.Liberman, 'regular and Stochastic Motion", Springer: New York (1983)
M.V.Berry, Regular and Irregular Motion in "Topics in Nonlinear Mechanics", S.Jorna, Ed.,Am.Inst.Phys.Conf. Proc. 46:16-120 (1978)
H.G.Schuster, "Deterministic Chaos", Physik-Verlag GMBH: Weinheim (1984)

2 J.E.Bayfield, L.D.Gardner and P.M.Koch, Phys.Rev.Lett. 39: 76-79 (1977)

3 G.Casati, B.V.Chirikov, and D.L.Shepelyansky, Phys.-Rev.Lett. 53 2525-2528 (1984)
G.Casati, B.V.Chirikov, D.L.Shepelyansky and I.Guarneri Phys.Rev.Lett. 57 823-826 (1986)
G.Casati, B.V.Chirikov, J.Ford and F.M.Izraelev in "Stochastic Behaviour in Classical and Quantum Hamiltonian Systems," G.Casati, J.Ford, ed., Springer Lecture Notes in Physics 93: 334-352 (1979)

4 M.V.Berry, N.L.Balazs, M.Tabor and A.Voros, Ann.Phys.-(N.Y) 122:26-63 (1979)
H.J.Korsch and M.V.Berry 3D: 627-636 (1981)
M.V.Berry, Physica 10D: 369-378 (1984)
B.V.Chirikov,F.M.Izraelev and D.L.Shepelyansky, Sov.Sci.-Revs. 2C: 209-267 (1981)
D.L.Shepelyansky, Physica 8D: 208-222 (1983)
Shmuel Fishman, D.R.Grempel and R.E.Prange,Phys.Rev.Lett. 49: 509-512 (1982)

5 B.V.Chirikov, F.M.Izraelev and D.L.Shepelyansky, Sov.-Sci.Revs. 2C 209-267 (1981)
Shmuel Fishman, D.R.Grempel and R.E.Prange Phys.Rev.Lett. 49 509-512 (1982)
D.R.Grempel, Shmuel Fishman and R.E.Prange Phys.Rev.A 29 1639-1647 (1984)

6 M.V.Berry "Quantum Chaology" Proc.Royal Soc. 1987, in press.

7 V.I.Arnold, "Mathematical Methods of Classical Mechanics," Springer, New York (1978)

8 V.P.Maslov and M.V.Fedoriuk, "Semiclassiccal Approximation in Quantum Mechanics," D.Reidel: Dordrecht, (1981)
M.V.Berry, Semiclassical Mechanics of Regular and Irregular Motion in "Chaotic Behavior of Deterministic Systems," Les Houches Lectures XXXVI, G.Iooss, R.H.G.Helleman and R.Stora, eds., North-Holland, Amsterdam, pp 171-271 (1983)
I.C.Percival, Adv.Chem.Phys. 36:1-61 (1977)

9 H.P.Baltes and E.R.Hilf, Spectra of finite systems (B.I.Wissenschaftsverlag: Mannheim)

10 C.E.Porter,Statistical Theories of Spectra: Fluctuations (Academic Press: New York).

11 F.J.Dyson, and M.L.Mehta, J.Math.Phys. 4 701-712 (1963)

12 M.V.Berry and M.Tabor Proc.Roy.Soc.Lond. A356 375-394
M.V.Berry Proc.Roy.Soc.Lond. A400 229-251 (1985)

13 E.Wigner "Group Theory and its Application to the Theory of Atomic Spectra" Academic: New York (1959)

14 M.Robnik J.Phys.A: Math.Gen. 17 1049-74 (1984)

15 M.V.Berry and M.Robnik J.Phys.A 19 649-668 (1986a)

16 M.Robnik and M.V.Berry J.Phys.A. 19 669-682 (1986)

17 M.V.Berry and R.J.Mondragon Proc.Roy.Soc., in press. (1987)
18 G.Casati, B.V.Chirikov and I.Guarneri Phys.Rev.Lett. 54 1350-1353 (1985)
Berry, M.V. and M.Robnik J.Phys.A. 17 2413-2421 (1984)
19 M.C.Gutzwiller J.Math.Phys. 12 343-358 (1971)
M.C.Gutzwiller in "Path Integrals and their Applications in Quantum Statistical and Solid-State Physics", eds. G.J.Papadopoulos and J.T.Devreese) Plenum,N.H.163-200 (1978)
R.Balian and C.Bloch Ann.Phys.(N.Y) 69 76-160 (1972)
20 J.P.Keating and M.V.Berry. In preparation (1987)
21 J.H.Hannay and A.M.Ozorio de Almeida,A.M. J.Phys.A. 17 3429-3440 ( )
22 M.V.Berry in "Quantum Chaos and Statistical Nuclear Physics" (eds.T.H.Seligman and H.Nishioka) Springer lecture notes in physics No. 263, pp 1-17 (1986b)
23 H.M.Edwards, Riemann's Zeta Function Academic Press: New York and London (1974)
24 J. Van de Lune, H.J.J.te Riele and D.T.Winter Math.of Comp. 46 No.174 667-681 (1986)
25 A.M.Odlyzko, Math.of Comp. Vol. 48 No.177, 273-308 (1987)
26 M.V.Berry and A.M.Odlyzko, Classical Formula for the Variance of the Riemann Zeros, in preparation.

QUANTUM MECHANICS AND CHAOS

Giulio Casati

Dipartimento di Fisica, Università di Milano
Via Celoria 16, 20133 Milano - Italy

1. INTRODUCTION

Our understanding of the qualitative features of classical systems has greatly improved after the discovery of the so called deterministic chaotic motion. This type of motion is characterized by exponential instability of almost all orbits with respect to initial conditions. In turns this instability leads to loss of memory of initial conditions, decay of correlations and approach to statistical equilibrium.

The question whether chaotic behaviour survives, in some way, in quantum mechanics is of primary importance both for the foundation of quantum statistical mechanics as well as for several applications. Certainly, at first sight, this question seems to receive a negative answer if one consider, bounded, conservative, finite-particles systems. Indeed these systems possess a discrete spectrum no matter whether the corresponding classical system is chaotic or not. This implies almost-periodicity in time of the wavefunction $\psi(t)$ and therefore absence of "chaotic" or irreversible behaviour.

On the other hand, on the basis of Ehrenfest's theorem, one expects that, in the semiclassical region, a narrow packet will follow the classical trajectory and one is led to suspect that quantum systems have different qualitative properties depending on whether the corresponding classical systems are integrable or chaotic.

As a matter of fact, analytical computations, extensive numerical evidence and experimental results on heavy nuclei, atoms and molecules show (even if a complete rigorous proof is still lacking) that the energy eigenvalues distributions of classically chaotic systems fall in different universality classes depending on whether the dynamics possesses the time-reversal invariance property or not. A remarkable fact is that these distributions have statistical properties which ar independent on the dynamics of the particular system considered and on the type of interaction (nuclear, electromagnetic, ecc.). Indeed these statistical properties are the same as those of the eigenvalues of random matrices (Gaussian Orthogonal Ensemble, Gaussian Unitary Ensemble).

We will not discuss here this interesting subject and we refer the interested reader to [1] in which a present day state of the art can be found. Instead we will concentrate here on systems subject to periodic time dependent, external perturbations. This choice is motivated by the two following reasons:

i) for these systems the solution of Schroedinger equation may be written in the form

$$\psi(t) = P(t)\, e^{iGt}\, \psi(o) \tag{1}$$

where P is periodic and G is self-adjoint. Since the spectrum of G, or the quasi-energy spectrum, may be continuous, there is here the possibility, in principle, of some chaotic motion.

ii) These systems describe physically interesting situations such as atoms and molecules under electromagnetic fields and therefore the results may be compared with real laboratory experiments.

2. THE KICKED ROTATOR

Let us consider the classical Hamiltonian

$$H = p^2/2 + \omega^2 \cos\theta \sum_j \delta(t-jT),. \tag{2}$$

where p is the rotator momentum, θ is the angular coordinate, T the kick period and ω the perturbation strength.
Due to the presence of the δ-function, the classical equations of motion can be integrated and reduced to the mapping.

$$P_{n+1} = P_n + K\sin\theta_n,$$
$$\theta_{n+1} = \theta_n + P_{n+1}, \qquad (3)$$

where $K = \omega^2 T$, n is time measured in number of kicks and P is the dimensionless angular momentum $P_n = p_n T$.

Mapping (3) is the well-known "standard map", extensively discussed in the literature and frequently used, at a tutorial level, to illustrate the great complexity of motion for simple dynamical systems. For K=0 this mapping is integrable and all orbits lie on smooth curves. For $0<K<K_c$, with $K_c \sim 1$ most orbits continue to lie on smooth curves and, in particular, the kinetic energy remains bounded with a variation $(\Delta P) \sim K$. On the other hand, when K exceeds the critical value K_c, most of the invariant curves disappear and the mapping orbits become chaotic: the system performs a random walk-like motion in momentum space leading to a diffusive growth of the kinetic energy

$$\overline{P^2} \simeq (K^2/2)t, \qquad (4)$$

and to the angular momentum distribution of the Gaussian type

$$f(P,t) = (K^2 \pi t)^{-1/2} \exp(-P^2/K^2 t). \qquad (5)$$

where t is integer time measured in number of kicks periods.

Indeed in this regime of motion the exponential instability of orbits with respect to initial conditions, leads to a rapid randomization of the phase variable and hence to the diffusive motion described by eqs. (4) and (5).
Without entering in further details, we stress the fact that the classical motion of system (2) exhibits completely different qualitative features depending on whether parameter K is less or larger than $K_c \sim 1$.

Turning to the quantum description, we consider the quantum Hamiltonian [2]

$$H = -\hbar^2 \partial^2/\partial\theta^2 + \omega^2 \cos\theta \sum_j \delta(t - jT). \qquad (6)$$

By letting

$$\psi(\theta,t) = \sum_n c_n e^{in\theta}, \quad (7)$$

we may write the solution of the Schroedinger equation as a mapping $\psi \to \Psi$. Then for the wave function after one step, which includes a free rotation of time T and a kick, we find

$$\Psi(\theta) = S\,\psi(\theta) = \exp(-ik\cos\theta)\Sigma_n c_n \exp[i(n\theta-2\pi n^2\tau)] \quad (8)$$

where

$$k = \omega^2/\hbar\,, \quad \tau = \hbar T/4\pi. \quad (9)$$

From eq. (8) we see that it suffices to consider values of τ within the interval [0,1]. Notice also that $K=4\pi k\tau$ and that the classical limit is reached by letting $k\to\infty$ $\tau\to 0$ keeping the classical parameter K fixed.
We may now perform numerical iterations of the quantum mapping (8), and inquire about its properties for values of the classical parameter K below or above $K = K_c$. Starting from a given set of $\{c_n(0)\}$ we computed the probability distribution

$$\rho(n) = |c_n|^2$$

and the average energy

$$\langle E\rangle = (1/2)\,\Sigma_n n^2 \rho(n).$$

In order to investigate the extent to which the numerically computed quantum distribution $\rho(n)$ mimics the classical stochastic distribution we write the quantum version of eq. (5) as

$$f(n,t) = (k^2\pi t)^{-1/2} \exp(-n^2/k^2 t) \quad (10)$$

where t is integer time measured in number of kicks periods.

In analogy with the classical results, it was expected [2] that for $K>K_c$, the distribution (10) would be obtained. However, this was not the case. Instead, for typical values of the period τ, the quantum system exhibited a strong stability character. In particular it was found that the quantum motion can mimic the classical diffusive energy growth yielding

$$\langle E\rangle = \sum_m (m^2/2)\,|c_m|^2 \simeq (k^2/4)t = (D/2)t$$

only up to a *break time* t_b. Empirically it appears that $t_b \to \infty$ as $\hbar \to 0$ (or equivalently $k \to \infty$), for constant $K=4\pi k\tau$. For times greater than the break time, the quantum energy appears to enter a steady-state oscillatory regime. In short, the kicked quantum rotator apparently introduces a limitation to classical diffusion and randomness.

We recall that the character of the time evolution for ψ is known once the quasi energy (abbreviated q.e., hereafter) spectrum is known. It can be shown [3] that, the q.e. spectrum for our model can be pure point only when the rotator energy remains bounded for all times, in which case the initial wave packet remains localized in momentum space. Thus the very existence of a break time after which energy growth stops completely would ensure that the quasi energy spectrum is pure point and that the full quantum motion is almost periodic.

Chirikov, Izrailev, and Shepelyansky [4] have advanced an heuristic argument which yields a quantitative estimate for this quantum limitation of diffusion due to the break-time phenomenon. In essence, this argument is the following: assume that the q.e. spectrum is pure point since otherwise there will be no finite break time. By the quantum theory of measurement, this pure-point character of the spectrum can become apparent only after a certain time t^*. Prior to this time the system will behave as if its spectrum were continuous and hence the limitation on diffusion will occur no later than t^*. Chirikov et al. therefore identify t^* with the break time t_b in order of magnitude.

To estimate t^*, recall that the total number of q.e. levels is infinite and that all these levels are located within a bounded interval. In addition, recall that each particular state ψ of the system can be obtained as a superposition of quasi energy eigenstates; while, generally speaking, the superposition involves an infinite number of q.e. states, we may assume that for an initially localized packet, the finite time evolution of ψ is dominated by a finite number N_ψ of them. This effective number N_ψ is also the number of eigenfrequencies actually occuring in the time evolution of ψ. Therefore the spectrum of the motion corresponding to state ψ will consist of N frequencies with average spacings about N_ψ^{-1}. It is now clear that $t^* \approx N_\psi$ and we must estimate N_ψ. First, we assume that N_ψ roughly coincides with the number of unperturbed eigenstates significantly involved in the motion. Next, we identify N_ψ with $\Delta P(t^*)$, that is, with the spread in momentum achieved at time t^*. Experimentally we have $\Delta P(t^*) \sim (k\, t^*)^{1/2}$ (for $k>1$, $K>1$). Since $N_\psi \sim t^* \sim \Delta P(t^*)$, we finally get that, in order of magnitude, $t^* \sim k^2$. In other words, both the break time and the localization length $l = N_\psi$ are of the order of the classical diffusion coefficient $D=k^2/2$

Localization of quantum wave packets in cases where classical mechanics would predict a diffusive behaviour is a well-known phenomenon in solid-state physics also. Indeed, the (time - independent) Hamiltonian of a particle in a random potential on a line can be proved to have a pure-point spectrum with eigenfunctions exponentially localized in space; this fact is known as Anderson localization. Along this line, Grempel, Fishman, Prange [5] gave an important contribution in understanding the quantum kicked rotator problem. They were able to show that the quantum kicked rotator is related to Anderson's problem of motion of a quantum particle in a one-dimensional lattice in a static potential. In this second problem, if the potential is random, quantum interference effects suppress diffusion and lead to a localization of the wave packet. By establishing a mapping between the two systems they conclude that a similar exponential localization should take place in the quantum rotator also.
More precisely the kicked rotator problem (6) can be mapped into a tight-binding model

$$T_m u_m + \Sigma W_r u_{m+r} = E u_m \qquad (11)$$

where the hopping coefficient W_r is the Fourier transform of $W(\theta)=tg\left((k/2)\cos\theta\right)$, $E = - W_0$ and where the diagonal potential $T_m = tg(\lambda-\tau m^2/2)$ with λ the quasi-energy. The angular momentum in the quantum problem corresponds to the lattices sites m in the solid state problem. In the tight binding problem, eq. (11) is an eigenvalue equation for the energy E while the quasi-energy λ plays the role of a parameter in the potential.

Whether the similarity with the kicked rotator problem is close enough that Anderson's result can be invoked in this case is a point which turns out to depend on the "degree of randomness" of the number sequence $T_m=tg(\lambda-\tau m^2/2)$. If T_m is periodic in m (τ rational) then the corresponding eigenstates are Bloch extended states. If T_m is a random sequence then the spectrum is discrete and the states are exponentially localized. In our case, for irrational τ, T_m is not strictly random but can be considered as pseudorandom in the sense that, to some extent, it mimics a random sequence.

As a matter of fact, for typical irrational values of τ, the picture of Anderson localization in momentum space fits very well the rotator problem also. In Fig. 1, for two different close values of τ (expressed in continued fraction), we show the shape of the squared modulus of the wave function in momentum space, after the break time t* and for an initial δ-like state.

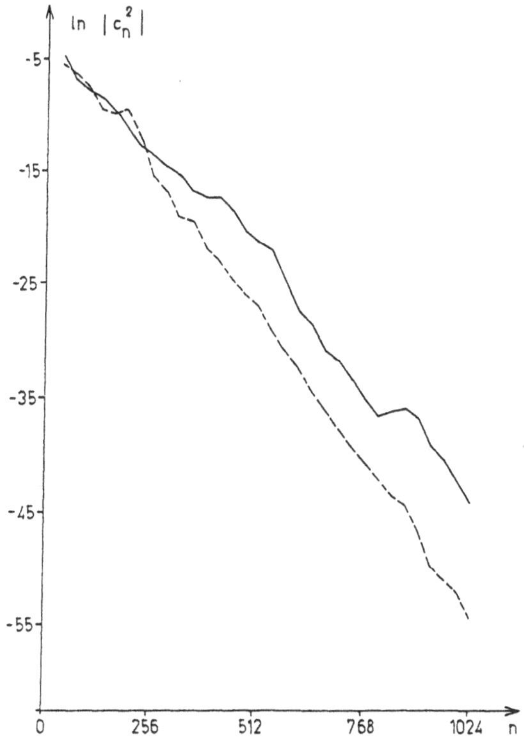

Fig. 1- Distribution function $|c_n|^2$ after 50000 iterations of the quantum mapping for the case k=10 and τ=[0,25,1,1,1,....] (dashed line), τ=[0,25,100,200,400,800,.....] (solid line). Notice the fairly good exponential localization in momentum space over several orders of magnitude.

The exponential behaviour is apparent here; also, the localization length is in excellent agreement with the estimate given above by Chirikov et al. [4]. This is surprising because Anderson's result is concerned with the asymptotics for n->∞ of eigenfunctions of q.e.; here, we find very good exponential decay even for small n and for a nonstationary state. Evidently some mechanism must be working here that requires further investigation.

The above described behaviour holds for a typical irrational τ. For rational values of the kick period τ, the rotator energy resonantly increases asymptotically as t^2. This is a strictly quantum phenomenon and does not occur in the corresponding classical system. This quantum peculiarity arises because the unperturbed quantum motion, as opposed to the classical, has the same period independent of initial state. In this rational, or resonant case, we have specific information on the q.e. spectrum. In particular, it has been shown [6] that, at resonance, the q.e. spectrum has an absolutely continuous component. Moreover, it has been shown that the growth of rotator energy with time t is proportional to t^2. But perhaps the most remarkable feature of resonance is that its quadratic growth of energy is the fastest possible [3].

In conclusion, for rational τ, the q.e. spectrum of the kicked rotator is continuous while for irrational τ the spectrum appears to be pure point. This situation is remindfull of the familiar process of destruction of invariant KAM tori in classical mechanics: rational tori are destroyed by the pertubation together with irrational-very close to rational-tori; morever as the perturbation strength increases, more and more tori are gradually destroyed. One may therefore wonder whether a similar picture takes place for the quantum motion. In our case we know that in the absence of perturbation (ω=0) the q.e. spectrum is pure point; moreover, for any small perturbation and for rational τ the q.e. spectrum is continuous. What happens for irrational-very close to rational - τ? The analysis in ref. [6] shows that for τ=p/q the q.e. spectrum is made, typically, by q bands and that in the limit p $\to\infty$, q $\to\infty$ the width Δs of each band appears to be exponentially small ($\Delta s \sim (k/q)^q$). This would imply that the total width of all the q bands tends to zero as q $\to\infty$. Therefore for irrational τ the q.e. spectrum can be discrete or continuous with zero measure (singular continuous).

A recent analysis [7] has shown that, no matter how small is the perturbation, there is a non empty set of irrationals τ for which the spectrum is continuous. This set is formed by those irrationals which are very well approximated by rationals. We do not know how large is this set and in particular the dependence of its size from the perturbation strength. It may happen that the size of the set of irrationals τ yielding continuous spectrum increases with the perturbation strength or, on the contrary, it is possible that for a typical irrational τ, the spectrum remains pure point no matter how large is the perturbation. To our knowledge this is still an open question.

3. THE HYDROGEN ATOM IN A MICROWAVE FIELD

In the previous section, by analysing a simple model, we have shown that quantum mechanics suppresses the classical diffusive energy growth and that only a finite number of unperturbed levels are significantly excited during the whole course of quantum evolution. Moreover it was shown that the localization length l of the quantum distribution is equal to the diffusion coefficient D of the classical motion. This quantum limitation of chaos may be considered as a dynamical version of the Anderson localization in one-dimensional disordered solids [5].

Two main questions are now open:

i) Is the quantum suppression of classical chaos only a property of the peculiar model studied, namely the kicked rotator, or is it a general occurence in quantum mechanics?

ii) Is it possible to observe this quantum suppression in real laboratory experiments?

Even though the quantum suppression of chaos was mainly investigated on the very particular rotator model, nonetheless the nature of the arguments supporting it, and especially the localization picture, indicate that quantum mechanics should indeed have an inhibitory effect on classical chaos even for generic quantum systems subject to time-periodic perturbations. This is in itself a remarkable discovery, bringing into light once more the deep fundamental difference between quantum and classical mechanics.

However, there are sound reasons to believe that this quantum suppression of chaos must suffer significant exceptions. As a matter of fact, the existence of a kind of quantum motion retaining some features of classical chaotic diffusion is at the present time the only possible explanation for available experimental results on the ionization of highly excited H-atoms (principal quantum number $n \sim 60$) in microwave fields of frequency $\omega/2\pi \sim 10$GHz with peak intensity $\epsilon \sim 10$ V/cm, in conditions where ionization would require the absorption of ~ 100 photons [8,9].

In order to answer the above questions we consider the ionization mechanism for the hydrogen atom when a linearly polarized monochromatic electric field induces transitions from initial states having principal quantum number $n_0 \gg 1$. For simplicity, we restrict ourselves to the study of very elongated quantum states having parabolic quantum numbers $n_1=n_0$, $n_2 = 0$, and magnetic quantum number $m=0$. Since to a good approximation these wave functions have nonzero values only along the direction of the applied field, we are at liberty here to treat the hydrogen atoms as if it were one dimensional, having the Hamiltonian:

$$H= p^2/2 - 1/x + \epsilon\, x \cos \omega t; \quad x>0,\ \omega=2\pi/T \tag{12a}$$

which, in action-angle variables (n,θ) reads,

$$H= -1/2n^2 + \epsilon\, x\,(n,\theta) \cos \omega t,\ x>0 \tag{12b}$$

where ϵ and ω are the microwave electric field strength and frequency, respectively, in atomic units. Here, $x\,(n,\theta)$ is the x coordinate of the electron, expressed as a function of action angle variables of the unperturbed atom. The validity of this one-dimensional approximation is due to the small value of matrix elements for transition having $\Delta n_2 \neq 0$.
As a consequence, the atom remains one dimensional during the relevant interaction times [10,11]. This important fact has also been verified in laboratory experiments which produce such states and excite them by microwave fields [12]. Even more important is the fact that, as recently shown [13], Hamitonian (12) describes quite acurately, as far as the excitation process is concerned, also the two-dimensional model.

Let us consider first the classical behaviour (14). In order to get a more transparent illustration of the latter, it is convenient to resort to a simplified description, which, however, retains all the essential features of the problem. To this end we integrate the classical equations of motion given by (12b) over one unperturbed orbital period of the electron. By substituting the unperturbed motion in the field-dependent terms and keeping only the resonant term $s \sim \omega n^3$ one can easily obtain [15] the change $\Delta n \approx (2\pi n^3 \epsilon\ 0.411/\omega^{2/3}) \sin \phi$ over one orbital period, where $\phi = \omega t - s\theta$. By introducing the quantity $N=E/\omega = -1/(2n^2\omega)$ one obtains the variations $\Delta N=\Delta n/(n^3\omega)$, $\Delta\phi= 2\pi\ n^3\omega= 2\pi\omega(-2\omega N)^{-3/2}$ over one orbital period. This leads to the following "Kepler map"

$$\begin{aligned} \bar{N} &= N + k \sin \phi \\ \bar{\phi} &= \phi + 2\pi\ \omega\ (-\ 2\omega\bar{N})^{-3/2} \end{aligned} \tag{13}$$

The classical map (13) is area-preserving. If we linearize this map around the initial value $N_0 = -\ 1/(2n_0{}^2\omega)$ we obtain the well-known standard map or kicked rotator map

$$\begin{aligned} \bar{N} &= N + k \sin \phi \\ \phi &= \phi + T\bar{N} \end{aligned} \tag{14}$$

where $k = 0.822\pi\epsilon/\omega^{5/3}$, $T = 6\pi\omega^2 n_0^5$, and $K = kT \sim 49\,\epsilon\,\omega^{1/3}\,n_0^5$.

According to the analysis of section 2, mapping (14) becomes chaotic for $K>1$ namely for $\epsilon > 1/(49\omega^{1/3}\,n_0^5)$. By introducing the rescaled parameters $\epsilon_0 = \epsilon n_0^4$, $\omega_0 = \omega n_0^3$, the critical field value for transition to chaos reads.

$$\epsilon_c = 1/(49\omega_0^{1/3}). \tag{15}$$

Therefore for $\epsilon_0 > \epsilon_c$, strong excitation and ionization takes place in the hydrogen atom. As explained in ref. [16], the approximation involved in deriving mapping (13) is good for $\omega_0>1$. It is possible to generalize mapping (13) to frequencies $\omega_0 < 1$ [16]. We will not enter in these details here. For the purpose of the present paper it is sufficient to remark that expression (15) is valid for $\omega_0 > 1$ and that a correction to this formula can be obtained which is valid down to $\omega_0 \approx 0.5$. For still smaller frequencies ω_0, we are not able to compute the critical classical chaos border; however this border will approach the critical value for ionization in static field (dotted line in Fig. (3)).

As explained in section 2 and according to analytical estimates [17,18] and to numerical data [17] the quantum excitation process reaches a steady-state distribution which can be satisfactory described by the formula

$$f\,(N) \approx (1/2l)\,(1+2|N|/l)\,\exp\,(-\,2N/l) \tag{16}$$

where l is the localization length which, in semiclassical conditions, is [17,18]

$$l \approx D \approx k^2/2 = 3.33\,\epsilon^2/\omega^{10/3} \tag{17}$$

Notice that exponential localization takes place in N, namely in the number of absorbed photons; therefore only in the vicinity of the initially excited level n_0 there may be appropriate exponential localization on unperturbed levels also.

The above theoretical predictions have been numerically checked [15,16] for different values of ω_0, ϵ_0, n_0. For each case we numerically determined the photonic localization length l and plotted it in Fig. 2 versus the field intensity ϵ. As is clearly seen from the figure, the numerical data are in good agreement with the theoretical expression (17) for a wide range of ϵ.

The phenomenon of photonic localization described above shows that actual suppression of strong excitation and ionization in the hydrogen atom takes place if the localization length is small. Instead if the localization length l is larger than the number of photons required for ionization $N_I = E/\omega = 1/(2n_0^2\omega)$ then probability will flow into the continuous part of the spectrum and strong excitation will take place as in the classical case [11,15,16,19]. Therefore the condition $l = N_I$ leads to the critical threshold value

$$\epsilon_q \approx \omega_0^{7/6} (6.6\, n_0)^{-1/2} \tag{18}$$

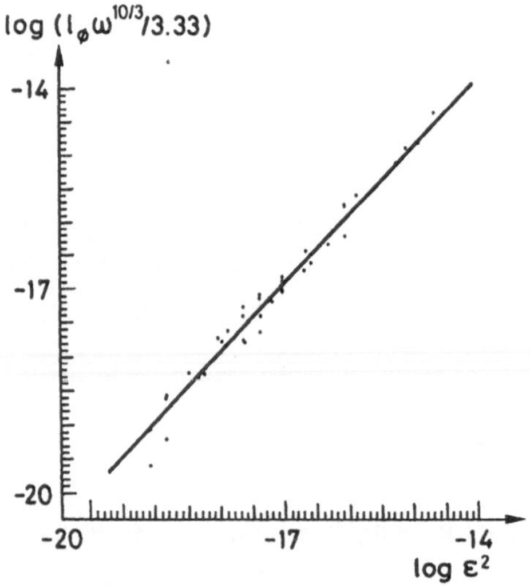

Fig. 2 The rescaled photonic localization length as a function of the field strength in logarithmic scale. The dots are obtained from quantum numerical integration with different parameter values while the straight line is drawn according to the theoretical expression (9).

This is the critical delocalization border described in refs. [11,19].

In order to have strong ionization in the hydrogen atom one needs a field intensity larger than both the threshold value ϵ_c for transition to classical chaos and the threshold value ϵ_q for transition to quantum delocalization.

The two critical borders, classical and quantum, are reported in Fig. 3 in the ϵ_0, ω_0 plane for a fixed frequency $\omega/2\pi \approx 9.9$ GHz. The dots are experimental results [20] and give the threshold field value ϵ_0 for 10%

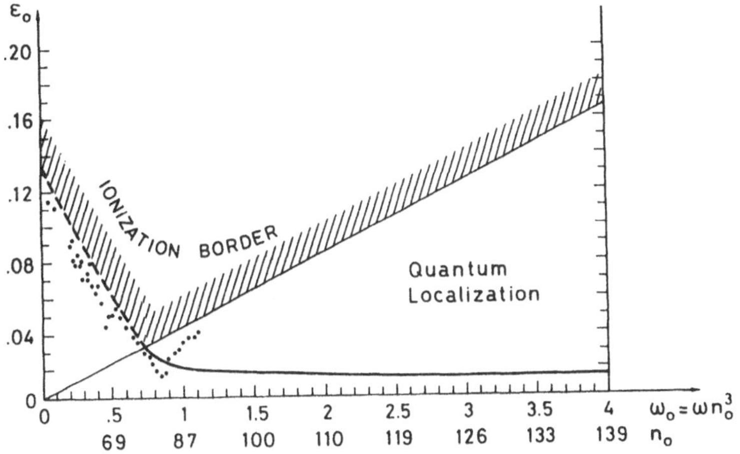

Fig. 3 - Ionization threshold as a function of the initially excited unperturbed state n_0 at fixed frequency $\omega \approx 9.9$GHz. The dots are experimental data obtained in ref. [20] for the threshold field value for 10% ionization. The thin curve is the classical border for transition to chaos given by expression (15). As explained in the text the dotted line is just a smooth continuation of the curve from $\omega_0 \approx 0.5$ up to the value 0.13 for ionization in static field. The gross line is the quantum delocalization border (18) which for fixed ω=9.9 GHz writes $\epsilon_q \approx (\omega^{1/6}/\sqrt{6.6})\ \omega_0$= 0.0417 ω_0.

ionization. The ϵ_0 values for 10% ionization are typically very close to the critical border for ionization and therefore our theory predicts that experimental points must be close to the ionization border evidentiated by the dashed area in Fig. 3.

Notice that for $\omega_0 < 1$ the ionization border is determined by the classical chaos border. This is the reason why good agreement has been found between experimental data and the result of numerical computations on the classical model. The most interesting region lies in the frequency range $\omega_0 > 1$ where according to our theoretical predictions one should observe the effect of quantum suppression of classical chaos and of diffusive excitation. This can be done either by increasing the initially excited state n_0, which appears to be difficult, or by increasing the microwave frequency ω. Experiments with larger ω are now under preparation [21].

It is necessary to mention that the experimental data reported in Fig. 3 were obtained on full three-dimensional atoms, which due to the conservation of magnetic quantum number are actually two-dimensional, while our theoretical predictions refer to a one-dimensional model. However, as we have recently shown [13,16], the energy excitation process in the two-dimensional atom can be well described by the one-dimensional model due to the fact that the N-motion decouples from the l-motion.

We would like to remark that even though in this paper we have limited our discussion to the interaction of microwave fields with highly excited states, the phenomena we have described here are of a quite general nature and should also be observed in the interaction of laser fields with initially excited lower levels. It is our belief that the study of the manifestations of classical chaos in quantum mechanics will open new exciting possibilities for our understanding of the radiation matter interaction.

REFERENCES

1. "Quantum Chaos and Statistical Nuclear Physics" Lectures Notes in Physics Vol. 263, Springer Verlag, 1986.

2. G. Casati, B.V. Chirikov, F.M. Izrailev and J. Ford, 1979, Lectures Notes in Physics Vol. 93.

3. G. Casati, I. Guarneri, J. Ford, F. Vivaldi, 1986, Phys. Rev. A 34, 1413.

4. B.V. Chirikov, F.M. Izrailev, and D.L. Shepelyansky, (1981) Sov. Sci. Rev. Sec. C 2, 209.

5. S. Fishman, D.R. Grempel and R.E. Prange, 1982, Phys. Rev. Lett. 49, 509; and 1984, Phys. Rev. A 29, 1639.

6. F.M. Izrailev and D.L. Shepelyansky, 1980, Teor Mat. Fiz. 43, 417.

7. G. Casati and I. Guarneri, 1984, Commun. Math. Phys. 95, 121.

8. J.E. Bayfield and P.M. Koch, Phys. Rev. Lett. 33 (1974) 258.

9. J.E. Bayfield, L.D. Gardner and P.M. Koch, Phys. Rev. Lett. 39 (1977) 76.

10. D.L. Shepelyansky, Inst. of Nuclear Physics, Novosibirsk, Report No. 83-61 1983 (unpublished), and in "Chaotic Behaviour in Quantum stems" edited by Giulio Casati (Plenum, New York, 1985), p. 187.

11. G. Casati, B.V. Chirikov, D.L. Shepelyansky, Phys. Rev. Lett. 53, (1984) 2525.

12. J.E. Bayfield and L.A. Pinnaduwage, Phys. Rev. Lett. 54, (1985) 313, and J. Phys. B 118, (1985) 449.

13. G. Casati, B.V. Chirikov, I. Guarneri, D.L. Shepelyansky Preprint INP 87-30 (Feb. 1987), Two-Dimensional Localization of Diffusive Excitation in the Hydrogen Atom in a Monochromatic Field. (to appear in Phys. Rev. Lett.)

14. A clear presentation of the classical treatment is given by R.V. Jensen, Phys. Rev. Lett. 49 (1982), 1365; Phys. Rev. A 30, (1984), 386.

15. G. Casati, B.V. Chirikov, I. Guarneri, D.L. Shepelyansky Phys. Rev. A, Rapid Communications 36, 1987, 3501.

16. G. Casati, B.V. Chirikov, I. Guarneri, D.L. Shepelyansky, "Hydrogen Atom in a Monochromatic Field: Chaos and Dynamical Photonic Localization", to appear on a special issue of the IEEE Journal of Quantum Electronics.

17. B.V. Chirikov, D.L. Shepelyansky Radiofizika 29 (1986), 1041.

18. D.L. Shepelyansky, Phys. Rev. Lett. 56 (1986), 677.

19. G. Casati, B.V. Chirikov, I. Guarneri, D.L. Shepelyansky Physics Report 154 (1987) 77.

20. K.A.H. van Leeuwen, G.V. Oppen, G.B. Bowling, P.M. Koch, R.V. Jensen, O. Rath, D. Richards and J.G. Leopold, Phys. Rev. Lett. 55, (1985) 2231.

21. J.E. Bayfield and P.M. Koch, private communication.

DYNAMICS OF AUTOMATA, SPIN GLASSES AND NEURAL NETWORK MODELS

B. Derrida
Service de Physique Théorique
CEN-Saclay
91191 Gif-sur-Yvette Cedex, France

I - INTRODUCTION

These lectures describe a few dynamical properties of some statistical mechanics systems such as spin models, automata, neural networks. Two aspects will be mostly discussed: the multivalley structure of phase space (sections II, IV, V, VI, VII, VIII) and the distance between two configurations which evolve according to the same rules (sections III, IX, X).

Let me start with a few examples:

Example 1: the Ising model[1]

The system consists of N Ising spins S_i ($S_i = \pm 1$) which are located on a regular lattice. These spins interact through the following Hamiltonian $\mathcal{H}$:

$$\mathcal{H} = - \sum_{ij} J_{ij} \; S_i S_j \tag{1}$$

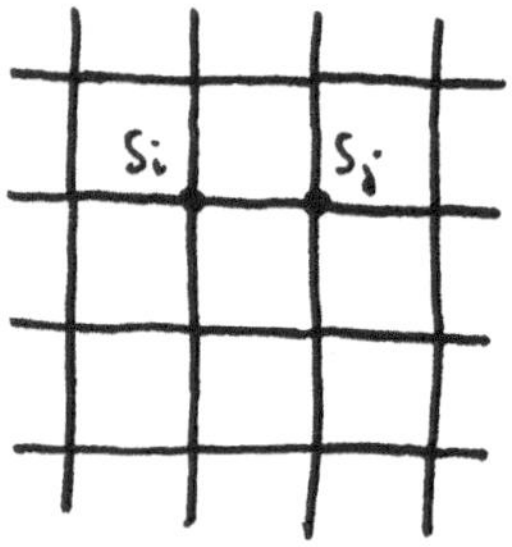

J_{ij} is the interaction between spin i and spin j. Since each spin can be either +1 or -1, the system has 2^N possible configurations. The equilibrium properties of such a system, at temperature T, can be computed by considering that each configuration $\mathcal{C} = \{S_i\}$ is visited with a probability $P_{eq}(\mathcal{C})$:

$$P_{eq}(\mathcal{C}) = \exp\left(- \frac{E(\mathcal{C})}{T}\right)/Z \tag{2}$$

where

$$E(\mathcal{C}) = - \sum_{ij} J_{ij} \; S_i S_j \tag{3}$$

and

$$Z = \sum_{\mathcal{C}} \exp\left(- \frac{E(\mathcal{C})}{T}\right) \tag{4}$$

Using (2), all the properties of the system at equilibrium can be computed. For example the average energy $\langle E \rangle$ is

$$\langle E \rangle = \sum_{\mathcal{C}} P(\mathcal{C}) \; E(\mathcal{C}) \tag{5}$$

There are several ways of introducing a time dependence for this model. Let me first describe the <u>parallel heat bath algorithm:</u> assume that at time t, the system is in a certain configuration $\mathcal{C}_t = \{S_i(t)\}$. One can compute all the fields $h_i(t)$ defined by

$$h_i(t) = \sum_j J_{ij} \; S_j(t) \tag{6}$$

and the parallel heat bath algorithm consists in updating the spins with the following rule

$$\begin{aligned} S_i(t+1) &= +1 \text{ with probability } p_i = \left(1+e^{-2h_i(t)/T}\right)^{-1} \\ S_i(t+1) &= -1 \text{ with probability } 1-p_i = \left(1+e^{2h_i(t)/T}\right)^{-1} \end{aligned} \tag{7}$$

To do so one chooses N independent random numbers $z_i(t)$ uniformly distributed between 0 and 1 and one updates the spins

$$S_i(t+1) = \text{sign}[p_i - z_i(t)] \tag{8}$$

These dynamics are stochastic. The $\{S_i(t+1)\}$ are not entirely determined by the $\{S_i(t)\}$ since the numbers $z_i(t)$ are random. If $P_t(\mathcal{C})$ is the probability of finding the system in configuration $\mathcal{C}$ at time t, one can show that for the parallel heat bath dynamics $P_t(\mathcal{C}) \rightarrow P_{eq}(\mathcal{C})$ in the limit $t \rightarrow \infty$ where the equilibrium distribution is slightly different from (2) and is given by[2]

$$P_{eq}(\mathcal{C}) = c^{te} \prod_{i=1}^{N} \cosh \left(\sum_j \frac{J_{ij}}{T} S_j \right) \tag{9}$$

There exist several other ways of introducing dynamics on Ising models: for example <u>the random sequential heat bath:</u> at each time step, one updates only one spin chosen at random among the N spins. So one chooses one site i, one computes $h_i(t)$ using (6) and then one updates this spin i according to (8). For a finite system, one can show that in the limit $t \rightarrow \infty$, $P_t(\mathcal{C}) \rightarrow P_{eq}(\mathcal{C})$ given by (2).

One can also choose other updating rules. For example the heat bath algorithm by the Metropolis algorithm: at each time step, one $h_i(t)$ has been computed (6), the rule (7), (8) is replaced by

$$S_i(t+1) = -\,\text{sign}\,(p_i - z_i(t))\; S_i(t) \tag{10}$$

where

$$p_i = \exp\left(-\frac{2h_i(t)\ S_i(t)}{T}\right) \qquad (11)$$

and $z_i(t)$ is uniform between 0 and 1.

Like for the heat bath, the updating can be either parallel or sequential. Again in the limit $t \to \infty$, $P_t(\mathcal{C}) \to P_{eq}(\mathcal{C})$ given by (2) for the sequential case.

In the zero temperature limit $(T = 0)$, the probabilities p_i become either 0 or 1 and the dynamics become deterministic

$$S_i(t+1) = \text{sign}\left(\sum_j J_{ij}\ S_j(t)\right) \qquad (12)$$

Example 2: Non-symmetric spin glasses - Neural networks[3]

In the previous example, the matrix J_{ij} of interactions was symmetric $(J_{ij} = J_{ji})$. However the stochastic dynamics described in the example 1 can be extended to cases where the matrix J_{ij} is non-symmetric $(J_{ij} \neq J_{ji})$. One can still compute the fields $h_i(t)$ using (6) and update the spins according to the rules (8) or (10). The only difference is that $P_t(\mathcal{C})$ has a limit when $t \to \infty$ which has not a simple expression in terms of the J_{ij} ($P_{eq}(\mathcal{C})$ is no longer given by the Boltzmann factor (2)).

Systems with non-symmetric J_{ij} are not very common in statistical mechanics. They are nevertheless useful to describe the dynamics of neurons, genes, etc...

Example 3: Network of automata[3]

Consider a system of N spins $S_i = \pm 1$. For each spin S_i, one gives K input sites $j_1(i), j_2(i) \ldots j_K(i)$ and a Boolean function f_i of K variables. Then the dynamics of this system are defined by

$$S_i(t+1) = f_i\left(S_{j_1(i)}(t), \ldots S_{j_K(i)}(t)\right) \qquad (13)$$

(An example of a Boolean function of 2 variables:

S_1	S_2	$f(S_1,S_2)$
+	+	+
+	−	−
−	+	+
−	−	+

A Boolean function of K variables is completely characterized by 2^K bits).

The functions f_i and the inputs $j_1(i) \ldots j_K(i)$ of each site i are chosen at time t = 0 and remain fixed. The dynamics are deterministic.

These three examples show two kinds of dynamics: at finite temperature ($T > 0$) for the first two examples, the dynamics are stochastic. The configuration $\mathcal{C}_{t+1}$ at time t+1 is not entirely determined by its configuration $\mathcal{C}_t$ at time t. It depends also on the random numbers $z_i(t)$ (and also on the random choice of the updated site i for sequential dynamics). Therefore

$$\mathcal{C}_{t+1} = F(\mathcal{C}_t, \text{Noise}_t) \tag{14}$$

where

$$\text{Noise}_t = \{z_i(t)\} \tag{15}$$

At zero temperature ($T = 0$) for parallel dynamics for the first two examples and for the 3rd example, the $S_i(t+1)$ are entirely given by the $S_i(t)$. The dynamics are deterministic and $\mathcal{C}_{t+1}$ is a function of $\mathcal{C}_t$ only.

$$\mathcal{C}_{t+1} = F(\mathcal{C}_t) \tag{16}$$

II - VALLEYS

For deterministic dynamics (eq. (16)), the time evolution ends up by becoming periodic for any initial configuration $\mathcal{C}_0$.

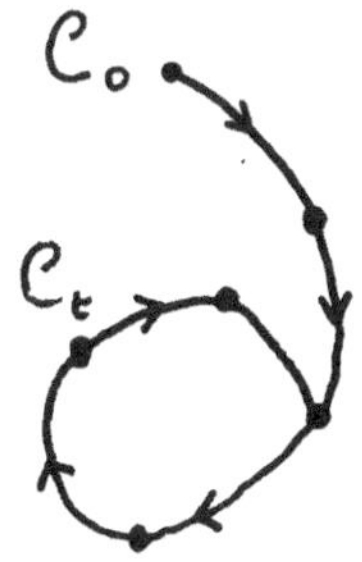

This is because phase space is finite (2^N different configurations). After a time $t > 2^N$, the system must have visited twice the same configuration, say at times t_1 and t_2

$$\mathcal{C}_{t_1} = \mathcal{C}_{t_2} \tag{17}$$

Since the dynamics are deterministic, one has for

any later time τ

$$\mathcal{C}_{t_1+\tau} = F^{\tau}\left(\mathcal{C}_{t_1}\right) = F^{\tau}\left(\mathcal{C}_{t_2}\right) = \mathcal{C}_{t_2+\tau} \tag{18}$$

There are in general (and even for finite N) several periodic attractors: phase space is broken into the basins of attraction (that we will call valleys) of these different attractors.

Two randomly chosen configurations can either belong to the same valley ($\mathcal{C}_0$ and $\mathcal{C}_0'$) or to two distinct basins ($\mathcal{C}_0$ and $\mathcal{C}_0''$)

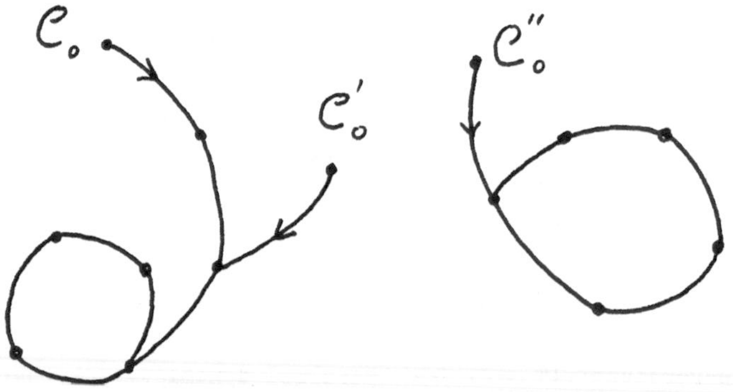

For a given measure on the set of initial conditions, i.e. if one chooses the initial condition according to a given probability distribution, one can define the weight W_s of an attractor S as the probability that a randomly chosen initial condition belongs to the basin of the attractor s.

For stochastic dynamics (at finite temperature T) it is more difficult to define valleys and they are usually well defined only in the thermodynamic limit ($N \longrightarrow \infty$). This can be illustrated by the example of the ferromagnetic Ising model below T_c.

A large but finite system spends $\frac{1-\varepsilon}{2}$ of its time in the + phase (A), $\frac{1-\varepsilon}{2}$ of its time in the - phase (C) and the rest of the time ε transiting (B) between the + and the - phase

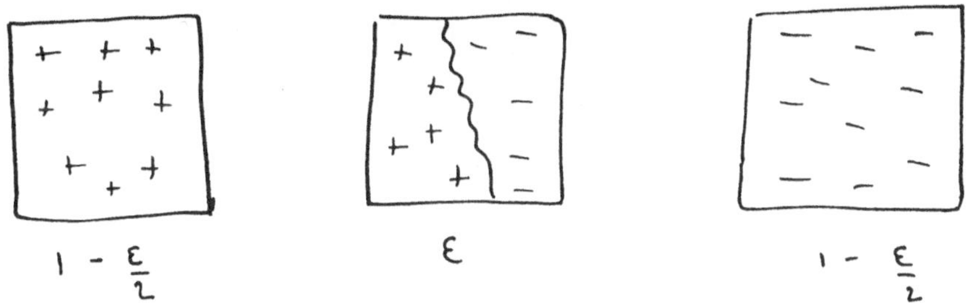

A B C

An order of magnitude of ε for a system of linear size L in dimension d is

$$\varepsilon \sim \exp\left[L^{d-1} \frac{\sigma(T)}{T}\right] \tag{19}$$

where $\sigma(T)$ is the interface free energy.

For finite L, one cannot define two distinct valleys. It is only in the limit $L \longrightarrow \infty$ that ε vanishes and that the + and the - phases become distinct valleys.

III - DISTANCES IN THE KAUFFMAN MODEL

The Kauffman model was introduced as a model for cell differentiation[4,5]. It describes a system of N genes $S_i = \pm 1$ (+1 if the gene is on and -1 if it is off). The activity of a gene i is influenced by K other genes $(j_1(i), j_2(i) \ldots j_K(i))$. The time evolution of the $S_i(t)$ is given by the parallel dynamics of N automata

$$S_i(t+1) = f_i\left(S_{j_1(i)}(t), \ldots S_{j_K(i)}(t)\right) \tag{20}$$

By definition of the model:

- each site i has exactly K inputs $j_1(i), \ldots j_K(i)$

- for each site i, these inputs are chosen at random among the N sites

- for each i, the function f_i is a random Boolean function of K variables: the function f_i is chosen at random among the 2^{2^K} possible Boolean functions of K variables

- the inputs $j_1(i), \ldots j_K(i)$ and the function f_i remain fixed at all times. The disorder is <u>quenched</u>.

Example: for K = 1, there are 4 possible Boolean functions g_1, g_2, g_3, g_4 of 1 variable

σ	$g_1(\sigma)$	$g_2(\sigma)$	$g_3(\sigma)$	$g_4(\sigma)$
-	+	+	-	-
+	+	-	+	-

Example of the connections in the Kauffman model for N = 5 sites and K = 3

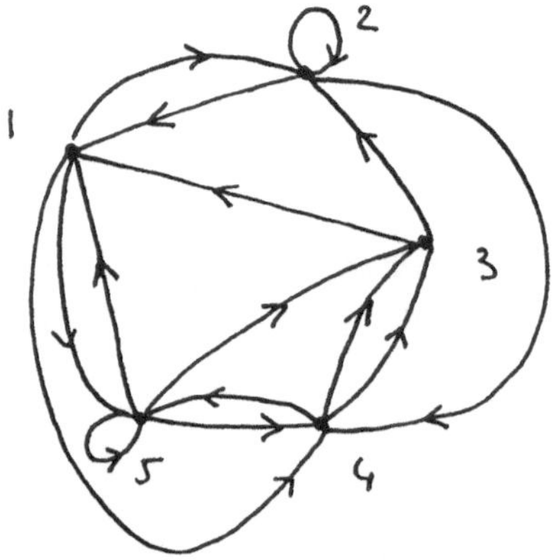

i	$j_1(i)$	$j_2(i)$	$j_3(i)$
1	2	3	5
2	3	2	1
3	4	4	5
4	2	5	1
5	1	5	4

The first results[4-6] obtained by Kauffman concerned the periods of the attractors. Kauffman studied numerically how the period P of an attractor depends on the system size N. He found two regimes: for $K \leqslant 2$

$$P \sim N^x \tag{21}$$

whereas for $K > 2$, the period P increases exponentially with N

$$P \sim e^{N\alpha} \tag{22}$$

Up to now, there does not exist any exact analytical calculation of the period P except for K = 1[7,8]. In particular the value of α (in eq. (22)) as a function of K has not yet been calculated analytically[9]. There is however another quantity, <u>the distance between two configurations</u> which has a different behaviour in the two regimes $K \leqslant 2$ and $K > 2$ and which can be calculated exactly.

Consider for the same sample, i.e. the same choice of the input sites and of the functions f_i, the time evolution of 2 different configurations $\mathcal{C}_t$ and $\mathcal{C}'_t$

$$\mathcal{C}_t = \{S_i(t)\} \qquad \text{and} \qquad \mathcal{C}'_t = \left\{S'_i(t)\right\} \tag{23}$$

We start with two different initial conditions $\mathcal{C}_0$ and $\mathcal{C}'_0$ and we let them evolve with the same rules. One can define the distance $D(t)$ between the two configurations by

$$D(t) = \frac{1}{4N}\sum_i \left(S_i(t) - S'_i(t)\right)^2 \tag{24}$$

$D(t)$ counts the fraction of spins i which are different (i.e. such that $S_i(t) = - S'_i(t)$).One can prove[10,11] that when $N \to \infty$, the time evolution of $D(t)$ is given for the Kauffman model by

$$D(t+1) = \frac{1 - (1-D(t))^K}{2} \tag{25}$$

To establish (25), one needs two steps. The first step consists in showing that in the limit $N \to \infty$, the distance D(t) is the same as the distance D(t) of the annealed model (the annealed model is defined as the model for which at each time step, the inputs $j_1(i)\ldots j_K(i)$ and the functions f_i are changed). Secondly one calculates $D(t)$ for the annealed model.

If one wants to calculate the value of a spin S_i at time t, one needs to know the value of its K inputs at time t-1, of the inputs of his inputs at time t-2, etc... up to time t = 0. So one needs to know the whole tree of its ancestors which contains a priori X sites:

$$X = 1 + K + \ldots + K^t = (K^{t+1}-1)/(K-1) \tag{26}$$

Since for each site i, the inputs $j_1(i)\ldots j_K(i)$ are chosen at random among the N sites, the same site may be present several times in this tree of ancestors. On can calculate the probability P that all the sites in a given tree are different:

$$P = \left(1 - \frac{1}{N}\right)\left(1 - \frac{2}{N}\right)\ldots\left(1 - \frac{X-1}{N}\right) \tag{27}$$

This formula is easy to understand. The probability that the 2nd site is different from the 1st is 1-1/N, that the 3rd differs from the 1st and the 2nd is 1-2/N, and so on. On can rewrite P

$$P = \exp\left[\sum_{i=1}^{X-1} \log\left(1 - \frac{i}{N}\right)\right] \simeq \exp - \frac{X^2}{2N} \qquad \text{if} \qquad X \ll N \tag{28}$$

We see that

$$\text{if} \quad X \ll \sqrt{N} \quad \Rightarrow \quad P \simeq 1 \tag{29}$$

So in the thermodynamic limit ($N \rightarrow \infty$), there is no repetition in the tree with probability 1. Since each site in the tree of the ancestors of a given site i is used only once, it does not matter if the functions f_i and the input sites are changed at each time step (annealed model) or remain fixed (quenched). Therefore one can compute the time evolution of D(t) for the annealed model.

If the distance is D(t) at time t, this means that N D(t) spins are different ($S_i(t) = - S_i'(t)$) and N(1-D(t)) spins are identical. There are then $N(1-D(t))^K$ sites i which have their K input sites identical at time t. These sites are of course identical at time t+1. There are also $N[1 - (1-D(t))^K]$ sites i which have at least one of their K inputs different. These sites have a probability 1/2 of becoming identical at time t+1 and 1/2 of becoming different. Therefore

$$D(t+1) = \frac{1}{2} \left[1 - (1-D(t))^K\right] \tag{30}$$

So the time evolution of D(t) is completely given by (25). We see that for $K \leqslant 2$,

$$\lim_{t \to \infty} D(t) = 0 \qquad \text{for any} \qquad D(0) \tag{31}$$

whereas for $K > 2$

$$\lim_{t \to \infty} D(t) = D^* \qquad \text{for any} \qquad D(0) \neq 0 \tag{32}$$

where D^* is the attractive fixed point of (25).

So for $K \leqslant 2$, the fraction of different spins vanishes in the long time limit. This means that $\mathcal{C}_t$ and $\mathcal{C}_t'$ become identical except may be for a non extensive number of sites. We called this phase the frozen phase[12]. For $K > 2$, it is remarkable that the final distance $D(\infty) = D^*$ does not depend on the initial distance D(0). The phase is chaotic[12] since two very close initial configurations $\mathcal{C}_0$ and $\mathcal{C}_0'$ end up by becoming different.

The analogy between the annealed and the quenched model allows the calculation of many other properties of the Kauffman model: one can compare more than two configurations[10], one can calculate the stable core which is the fraction of spins which never move in the limit $t \rightarrow \infty$[13], one can calculate the effect of thermal noise (section IX). It can also be used to obtain exact results for other models like non-symmetric spin glasses[14] and neural network models[15,16,9].

The Kauffman model as defined up to now is an infinite dimensional model. The number of different ancestors of a given site grows exponentially with time (K^t).

The Kauffman model has also been studied on regular lattices (square, cubic) in finite dimension[12,17-20]. The function f_i remain random Boolean functions but the inputs $j_1(i),\ldots j_K(i)$ are the neighbours of the site i on the lattice ($K = 4$ on the square lattice, $K = 6$ on the cubic lattice). The value of a spin i at time t depends on a number of spins at time $t = 0$ which increases like t^d where d is the dimension of the lattice.

In the finite dimensional case, one can also see the two phases: chaotic and frozen. The main difference with the infinite dimensional case is that $D(\infty) = \lim_{t\to\infty} D(t)$ depends on $D(0)$.

In the frozen phase, $D(\infty)$ vanishes if $D(0) \to 0$ whereas in the chaotic phase $D(\infty)$ has a non zero limit if $D(0) \to 0$.

IV - VALLEYS IN SPIN GLASSES

Most of the interesting properties of disordered systems and of spin glasses are due to the fact that phase space is broken into many valleys. I am going to describe briefly in this section some of the statistical properties of the multivalley structure of spin glasses in the mean field limit. We will see in the next section that the basins of attraction in the Kauffman model have very similar features.

Mezard, Parisi, Sourlas, Toulouse and Virasoro have studied the multivalley structure[21-22] of the Sherrington Kirkpatrick (S.K) model[23]. The SK model is a system of N Ising spins $S_i = \pm 1$ with a random interaction J_{ij} for each pair ij. So the Hamiltonian $\mathcal{H}$ of this system is

$$\mathcal{H} = - \sum_{i<j} J_{ij} \, S_i S_j \qquad (33)$$

One of the simplest quantities that Mezard et al considered is Y defined by

$$Y = \sum_S W_S^2 \qquad (34)$$

where W_S is the weight of the valley S. For a spin glass at thermal equilibrium, W_s is given by

$$W_s = \frac{e^{-f_s/T}}{\displaystyle\sum_{S'} e^{-f_{s'}/T}} \tag{35}$$

where f_s is the free energy of the valley S. One has of course

$$\sum_S W_s = 1 \tag{36}$$

It is not easy (and I will not try) to explain how the free energies f_s are defined in the SK model. However all the properties of Y for the SK model which will be discussed below can also be recovered for the random energy model[24,25]. The random energy model is defined as a system of 2^N energy levels E_s which are independent random variables distributed according to

$$P(E_s) = \frac{1}{\sqrt{N\pi J^2}} \exp\left(-\frac{E_s^2}{NJ^2}\right) \tag{37}$$

The partition function $Z(T)$ of this system at temperature T is

$$Z(T) = \sum_{S=1}^{2^N} \exp\left(-\frac{E_s}{T}\right) \tag{38}$$

and one can define the weight W_s of each configurations as

$$W_s = \frac{e^{-E_s/T}}{Z(T)} \tag{39}$$

Then for the random energy model, each configuration can be seen as a valley and Y is simply given by

$$Y = \sum_S W_s^2 = \frac{Z(T/2)}{Z^2(T)} \tag{40}$$

Y is an interesting quantity to consider because it tells us whether there are big or small valleys.

- If Y = 1, this means that there is only one big valley with weight W = 1

- If $Y \simeq 0$, there are many valleys and each of these valleys has a small weight

- If $0 < Y < 1$, there are a few big valleys plus may be many small ones.

For the SK model[21-23] and for the random energy model[25], one can show that in the spin glass phase and for $N \longrightarrow \infty$:

1) $\langle Y \rangle \neq 0$ and $\neq 1$. So there are a few big valleys. (Here $\langle\ \rangle$ means the average over disorder: the interactions J_{ij} in the SK model and the energies E_S in the random energy model).

2) $\langle Y^2 \rangle - \langle Y \rangle^2 \neq 0$. So the weights of the big valleys are sample dependent even in the thermodynamic limit $N \longrightarrow \infty$. The multivalley structure changes from sample to sample. Y is not self averaging.

3) In the limit $N \longrightarrow \infty$, there exist simple relations between the moments of Y or the Y_P defined by

$$Y_P = \sum_S W_S^P \tag{41}$$

$(Y = Y_2)$

For example

$$\langle Y^2 \rangle = \frac{1}{3} \langle Y \rangle + \frac{2}{3} \langle Y \rangle^2 \tag{42}$$

$$\langle Y_3 \rangle = \frac{1}{2} \langle Y \rangle + \frac{1}{2} \langle Y \rangle^2 \tag{43}$$

$$\langle Y_4 \rangle = \frac{1}{6} \langle Y \rangle \ (\langle Y \rangle + 1) \ (\langle Y \rangle + 2) \tag{44}$$

This means that even if $\langle Y \rangle$ depends in a complicated way on the spin glass model, temperature, magnetic field,..., all the moments $\left\langle (Y_P)^n \right\rangle$ are related to $\langle Y \rangle$ by simple expressions (42-44) which do not depend on the physical parameters.

4) Since Y is not self averaging, the meaningful quantity to consider is the probability distribution $\pi(Y)$ of Y. This probability distribution has a shape[26] which depends on $\langle Y \rangle$ only.

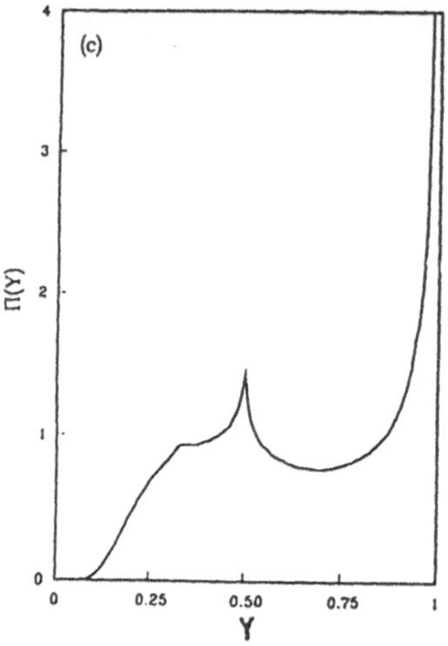

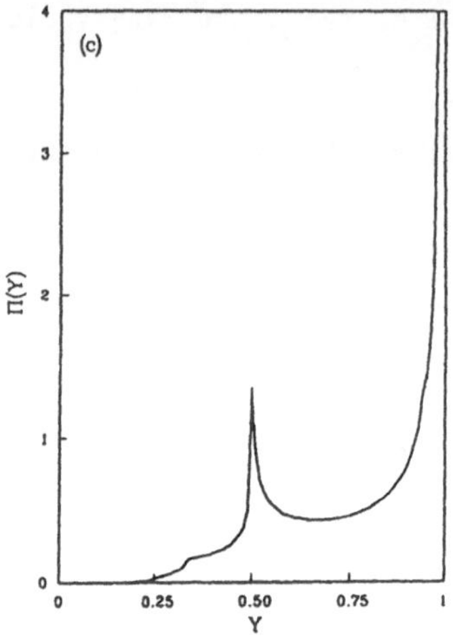

$\langle Y \rangle = .7$ $\langle Y \rangle = .9$

One should notice that $\pi(Y)$ has visible singularities at $Y = \frac{1}{2}$ and $Y = \frac{1}{3}$.

V - VALLEY STRUCTURE IN THE KAUFFMAN MODEL[27]

With Henrik Flyvbjerg we investigated numerically the multivalley structure of the Kauffman model by studying Y and the Y_P. Because the dynamics are deterministic (eq. (20)), phase space is composed of several basins of periodic attractors. We decided to define the weight of the Sth basin by

$$W_S = \frac{\Omega_S}{2^N} \tag{45}$$

where Ω_S is the number of configurations in the basin of the Sth attractor. So W_S is the probability that an initial configuration chosen at random falls on the attractor S. Since $Y = \sum_S W_S^2$, <u>Y is the probability that two randomly chosen initial conditions fall on the same attractor</u>, $\langle Y \rangle$, $\langle Y^2 \rangle$, $\langle Y_3 \rangle \ldots \langle Y_P \rangle \ldots$ can be measured in numerical simulations: for example, since $\langle Y_P \rangle$ is the probability that P configurations fall on the same attractor, one just generates P random configurations many times and one counts the number of times they belong to the same basin.

To measure $\langle Y^2 \rangle$, one can choose 4 configurations $\mathcal{C}_1, \mathcal{C}_2, \mathcal{C}_3$ and $\mathcal{C}_4$ at random and count the average number of times that $\mathcal{C}_1$ and $\mathcal{C}_2$ fall on the same attractor and that $\mathcal{C}_3$ and $\mathcal{C}_4$ fall on the same attractor.

To average Y over disorder, we generated typically 10^4 samples. For each sample we chose[27] 2 random initial configurations $\mathcal{C}_1$ and $\mathcal{C}_2$. $\langle Y \rangle$ is the average number of times that $\mathcal{C}_1$ and $\mathcal{C}_2$ belong to the same basin. A similar procedure was used to measure $\langle Y^2 \rangle$, $\langle Y_3 \rangle$ and $\langle Y_4 \rangle$.

1) The first result we have obtained is that in the limit $N \to \infty$, $\langle Y \rangle \neq 0$ and $\langle Y \rangle \neq 1$ (at least for K = 1,3,4 and ∞). So like in the SK model there are a few big basins even in the thermodynamic limit.

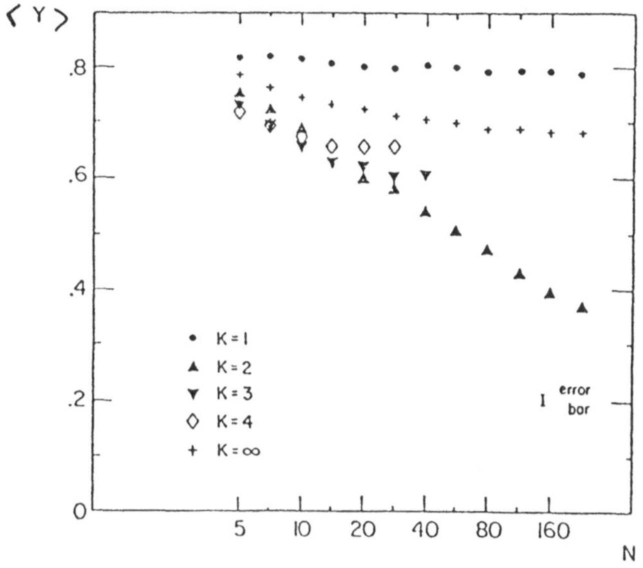

2) In the limit $N \to \infty$, we found[27] also that $\langle Y^2 \rangle - \langle Y \rangle^2$ does not vanish. So Y is not self averaging like in the SK model.

Since the multivalley structure is qualitatively the same 1) and 2) for the Kauffman model and for the mean field spin glass (SK model or random energy model), we tried to see whether the relations (42-44) could also be valid for the Kauffman model. In the following figures $\langle Y_4 \rangle$ and $\langle Y^2 \rangle$ are plotted versus $\langle Y \rangle$.

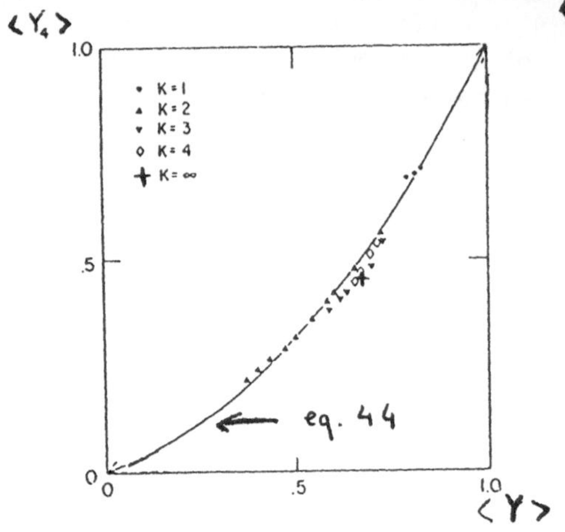

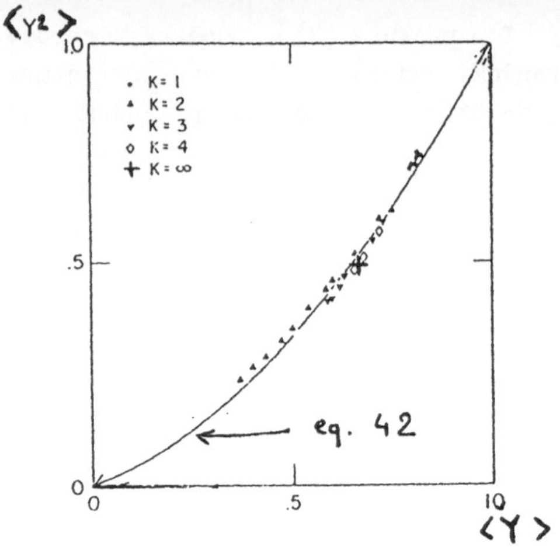

For each K, the different points correspond to different sizes N. The relations (42-44) seem to be satisfied within the statistical errors. It was then tempting to see whether these relations are exact in the Kauffman model or not. It turns out that there are two cases which can be solved exactly i.e. for which $\langle Y \rangle$, $\langle Y^2 \rangle$, $\langle Y_p \rangle$... can be calculated exactly: $K = 1$ and $K = \infty$. In both cases, one can show that the relations are not exactly satisfied for the Kauffman model. We will see however that there is a strong resemblance between spin glasses and the Kauffman model.

VI - THE RANDOM MAP MODEL[28]

The case $K = 1$ has been studied analytically in great detail[7]. I will discuss here only the case $K \longrightarrow \infty$ for which the dynamics become a random map of phase space into itself[28]. The random map model is defined as follows.

Consider a system which has M ($= 2^N$) possible configurations $\mathcal{C}$. The dynamics of this model are given by a random map F:

$$\mathcal{C}_{t+1} = F(\mathcal{C}_t) \tag{46}$$

1) For each $\mathcal{C}$, $F(\mathcal{C})$ is chosen at random among the M configurations

2) The $F(\mathcal{C})$ are independent random variables; in particular $F(\mathcal{C})$ and $F(\mathcal{C}')$ are not correlated for $\mathcal{C} \neq \mathcal{C}'$.

One can show[28] that the Kauffman model reduces to the random map model in the limit $K \longrightarrow \infty$. Let me just give here a simple argument. The time evolution of the distance D(t) is (see eq. (25))

$$D(t+1) = \frac{1 - [1-D(t)]^K}{2} \quad (47)$$

In the limit $K \to \infty$, we see that

$$D(t+1) = \frac{1}{2} \quad (48)$$

for any $D(t)$. This is due to the fact that $F(\mathcal{C})$ and $F(\mathcal{C}')$ become independent random variables in the limit $K \to \infty$.

For a given map F, one can determine the different attractors and the weights W of the different attractors. For example, the following random map F:

$\mathcal{C}$	1	2	3	4	5	6	7	8
$F(\mathcal{C})$	2	5	4	6	7	4	8	2

has 2 basins (1,2,5,7,8) and (3,4,6) so $W_1 = 5/8$ and $W_2 = 3/8$.

For the random map model, one can calculate[28] analytically $\langle Y \rangle$, $\langle Y^2 \rangle$, $\langle Y_p \rangle$... in the limit $M \to \infty$. Let us see as an example how $\langle Y \rangle$ can be calculated:

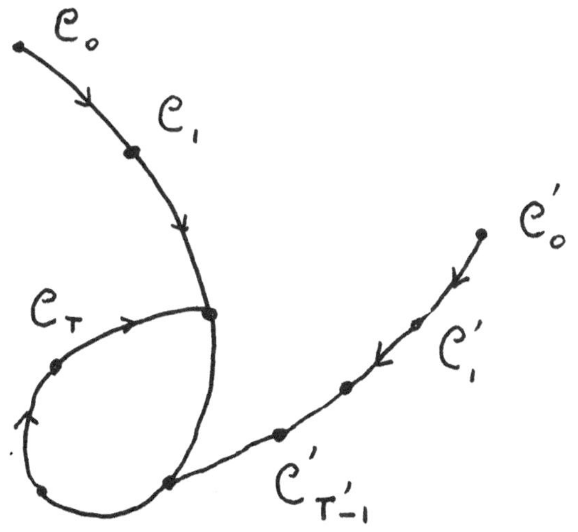

$\langle Y \rangle$ is the probability that two configurations fall on the same attractor. Let us call P(T) the probability that the first configuration has visited at times 0,1,2,...T different points of phase space and at time T+1 falls on a point which had been previously visited. One has:

$$P(T) = \left(1 - \frac{1}{M}\right)\left(1 - \frac{2}{M}\right) \cdots \left(1 - \frac{T}{M}\right)\frac{T+1}{M} \tag{49}$$

P(T) is the probability that the first configuration visits exactly T+1 different points of the basin of its attractor. Let us call Q(T,T') the probability that the first configuration visits T+1 different points and that the second configuration visits T' points before it falls on a point already visited by the first configuration

$$Q(T,T') = P(T)\left(1 - \frac{T+1}{M}\right)\left(1 - \frac{T+2}{M}\right) \cdots \left(1 - \frac{T'+T}{M}\right)\frac{T+1}{M} \tag{50}$$

Then $\langle Y \rangle$ is given by

$$\langle Y \rangle = \sum_{T,T'} Q(T,T') \tag{51}$$

For large N, one can use continuous variables and write

$$T = \sqrt{M}\, x \qquad ; \qquad T' = \sqrt{M}\, y$$

$$1 = \sqrt{M}\, dx = \sqrt{M}\, dy$$

$$Q(T,T') = \frac{x^2}{M} \exp\left(- \frac{(x+y)^2}{2}\right) \tag{52}$$

Then

$$\langle Y \rangle = \int_0^\infty dx \int_0^\infty dy\; x^2 \exp\left(- \frac{(x+y)^2}{2}\right) = \frac{2}{3} \tag{53}$$

So one finds

$$\langle Y \rangle = \frac{2}{3} \tag{54}$$

One can obtain $\langle Y^2 \rangle$, the $\langle Y_p \rangle$ by similar calculations

$$\langle Y^2 \rangle = \frac{32}{105} \tag{55}$$

$$\langle Y_P \rangle = \frac{4^{P-1}\ [(P-1)!]^2}{(2P-1)!} \tag{56}$$

We see that (54-56) do not satisfy (42-44). Therefore the statistical properties of the valleys are different in the Kauffman model and in the mean field theory of spin glasses. They have nevertheless very similar properties: if one defines f(W) as the average number of valleys with weight W:

$$f(W) = \left\langle \sum_S \delta(W-W_S) \right\rangle \tag{57}$$

One can deduce f(W) from the knowledge of the $\langle Y_P \rangle$

$$\langle Y_P \rangle = \int_0^1 W^P\ f(W)\ dW \tag{58}$$

For the random map model, one finds

$$f_{RM}(W) = \frac{1}{2}\ W^{-1}\ (1-W)^{-1/2} \tag{59}$$

whereas for the mean field spin glasses one has

$$f_{SG}(W) = \text{cste}\ W^{y-2}\ (1-W)^{-y} \tag{60}$$

where $y = \langle Y \rangle$ contains the whole dependence on the physical parameters (temperature, magnetic field, etc...). We see that the shapes (59) and (60) of f(W) are very similar although there is no choice of y in (60) which gives (59).

VII - DISTRIBUTION $\pi(Y)$ FOR THE RANDOM MAP MODEL

Since Y is not self averating ($\langle Y^2 \rangle \neq \langle Y \rangle^2$) one has to study the probability distribution $\pi(Y)$ of Y. All the moments $\langle Y^n \rangle$ of Y can be calculated by generalizing the calculation of section VI. However there is another way of calculating these moments which I am going to describe now.

Consider a random map F of M points (M is large). The probability P(W)dW that a given configuration $\mathcal{C}_0$ falls on an attractor S_0 of weight between W and W+dW is given by

$$P(W) = W\, f(W) = \frac{1}{2}\,(1-W)^{-1/2} \qquad (61)$$

If one removes the MW points of the basin of this attractor S_0, it remains a random map $\tilde{F}$ of a set of $M(1-W)$ points into itself. If the map $\tilde{F}$ has attractors of weight $\tilde{W}_1, \tilde{W}_2 \ldots$, the weights of the attractors of F are $W, W_1, W_2 \ldots$ with

$$W_i = (1-W)\,\tilde{W}_i \qquad (62)$$

This implies a simple relation between the Y of F and the $\tilde{Y}$ of $\tilde{F}$

$$Y = W^2 + \sum_i W_i^2 \qquad (63)$$

$$\tilde{Y} = \sum_i \tilde{W}_i^2 \qquad (64)$$

$$Y = W^2 + (1-W)^2\,\tilde{Y} \qquad (65)$$

When $M \rightarrow \infty$, Y and $\tilde{Y}$ have the same probability distribution $\pi(Y)$ and since $\tilde{Y}$ and W are not correlated, one can write an integral equation for $\pi(Y)$

$$\pi(Y) = \int \pi(\tilde{Y})\, d\tilde{Y} \int P(W)\, dW\, \delta(Y-W^2-(1-W)^2\tilde{Y}) \qquad (66)$$

From (63), one can calculate all the moments of Y

$$\langle Y^p \rangle = \sum_{q=0}^{p} \frac{p!}{q!(p-q)!} \langle Y^q \rangle \int_0^1 P(W)\, W^{2p-2q}\, (1-W)^{2q}\, dW \qquad (67)$$

Because the support of $\pi(Y)$ is [0,1], these moments $\langle Y^p \rangle$ determine the distribution $\pi(Y)$. However it is not easy to deduce the shape of $\pi(Y)$ from (67).

In order to draw $\pi(Y)$, we solved the integral equation (66) using a Monte Carlo method[28]. We generated a sequence W_n of 10^7 random numbers distributed according to P(W) and then we constructed the sequence Y_n by the following recurrence (Y_0 can be anything)

$$Y_{n+1} = W_n^2 + (1-W_n)^2\, Y_n \qquad (68)$$

The histogram of the Y_n is $\pi(Y)$.

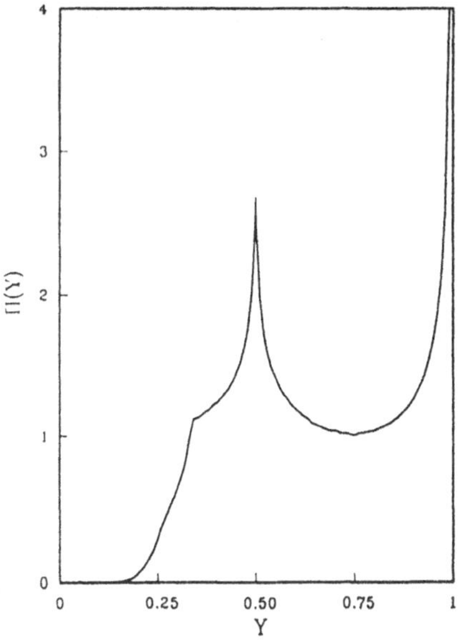

We see that $\pi(Y)$ for the random map model has a shape very similar to its shapes in the mean field spin glass. Again we see that $\pi(Y)$ is singular at $Y = 1, \frac{1}{2}, \frac{1}{3}$.

VIII - RANDOMLY BROKEN OBJECTS

We have seen in section IV and VII that $\pi(Y)$ has singularities at $Y = 1, \frac{1}{2}, \frac{1}{3}$ in the infinite ranged spin glass and also in the random map model. One can calculate the nature of these singularities analytically[26] and show that there are other singularities at all the points $Y = 1/n$.

In both problems, phase space is broken into many valleys in a random manner. The presence of these singularities (at $Y = 1/n$) is a more general phenomenon and should be present in a large class of models of randomly broken objects.

If one considers an object (glass, dish, chalk, phase space,...) of total weight 1 and one breaks it randomly into pieces of weights $W_1, W_2, \ldots,$ one can define Y by

$$Y = \sum_S W_s^2 \tag{69}$$

Since the breaking process is random, Y is a random variable. For a large class of breaking processes, one can show[26] that the probability distribution $\pi(Y)$ has singularities at all the points

$$Y = 1, \frac{1}{2}, \frac{1}{3}, \ldots, \frac{1}{n}, \ldots \tag{70}$$

This shows that the way phase space is broken into many valleys in spin glasses or in the random map is very similar to the way an object is broken into pieces.

One can build easily models of randomly broken objects for which all the moments of Y can be calculated. Let me give here an example[26]. Consider an interval of length 1. We start by breaking this interval into two pieces, one of weight W_1 and the other of weight $W' = 1-W_1$. We keep the piece of weight W_1 as the first piece of our sample. At the second step we break the piece W' into two pieces, one of weight W_2 which we keep and another piece of weight $W'' = W-W_2$ that we are going to break at the 3rd step and so on. After n steps, we have n pieces of our final sample: $W_1, W_2, \ldots, W_n$ and a piece $W^{(n)}$ which will be broken at the next step. To make the model simple, we consider only processes which are self-similar in the following sense: the probability that $W^{(n)}$ is broken into W_{n+1} and $W^{(n)}-W_{n+1}$ depends only on the ratio $W_{n+1}/W^{(n)}$. Therefore after an infinite number of steps the system consists of an infinite number of pieces which can be described in the following way:

$$\begin{aligned} W_1 &= x_1 \\ W_2 &= (1-x_1)\, x_2 \\ W_3 &= (1-x_1)\,(1-x_2)\, x_3 \\ &\vdots \\ W_n &= (1-x_1)\ldots(1-x_{n-1})\, x_n \end{aligned} \tag{71}$$

where all the numbers $x_1, x_2, \ldots, x_n$ are randomly distributed according to the same probability distribution $\rho(x)$.

Clearly since the piece of weight W' has statistically the same breaking process as the initial interval, one can write

$$Y = x^2 + (1-x)^2\, Y' \tag{72}$$

where x is randomly chosen according to the distribution $\rho(x)$. Then $\pi(Y)$ is solution of

$$\pi(Y) = \int \rho(x)\, dx \int \pi(Y')\, dY'\, \delta(Y-x^2-(1-x)^2 Y') \tag{73}$$

This integral equation is identical to eq. (66) which was obtained for the random map model. One can show[26] that for a large class of $\rho(x)$ (probably for any $\rho(x)$ analytic on $]0,1[$, $\pi(Y)$ has singularities at $Y = 1, \frac{1}{2}, \ldots, \frac{1}{n} \ldots$

IX - THERMAL NOISE AND DYNAMICS

We have seen in section I how the thermal noise appears in the dynamics of the Ising model:

Examples 1 and 2

The N Ising spins S_i evolve according to the following rule (eq. (10))

$$S_i(t+1) = \text{sign}[p_i(t) - z_i(t)] \tag{74}$$

where the $z_i(t)$ are random numbers uniformly distributed between 0 and 1 and

$$p_i(t) = (1+e^{-2h_i/T})^{-1} = \left[1+\exp\left(-\frac{2}{T}\sum_j J_{ij}\, S_j(t)\right)\right]^{-1} \tag{75}$$

The numbers $z_i(t)$ represent the thermal noise. At low temperature T, p_i is close to 0 and 1 and $S_i(t+1) = \text{sign}(h_i(t))$ with a high probability. At high temperature, p_i is close to 1/2 and the effect of the $z_i(t)$ is stronger.

Example 3

One can introduce a temperature in automata problems in a similar way. If one considers a system of N automata $S_i = \pm 1$ with for each automaton i, a Boolean function f_i of K variables and K inputs $j_1(i) \ldots j_K(i)$, one can update the $S_i(t)$ like in the examples 1 and 2:

$$S_i(t+1) = \text{sign}[p_i(t) - z_i(t)] \tag{76}$$

where $z_i(t)$ are random numbers and the probabilities $p_i(t)$ are given by

$$p_i(t) = \left(1 + \exp - \frac{2h_i(t)}{T}\right)^{-1} \tag{77}$$

where

$$h_i(t) = f_i\left(S_{j_1}(t), S_{j_2}(t), \ldots S_{j_K}(t)\right) \tag{78}$$

Of course in the limit $T \to 0$, one recovers the deterministic dynamics discussed in the example 3 of section I.

For all these examples at finite temperature, the configuration $\mathcal{C}_{t+1}$ of the system at time t+1 depends on $\mathcal{C}_t$ and also on the noise (eq. (14-15))

$$\mathcal{C}_{t+1} = F(\mathcal{C}_t, \text{Noise}_t) \quad \text{where} \quad \text{Noise}_t = \{z_i(t)\} \tag{79}$$

Therefore a configuration $\mathcal{C}_t$ at time t depends on its initial configuration $\mathcal{C}_0$ and on the noise at times 0,1,2,...t-1. In the limit $t \to \infty$, one can then ask the following question: Does $\mathcal{C}_t$ depend on $\mathcal{C}_0$?

To answer this question, one can compare the time evolution of two different configurations $\mathcal{C}_t$ and $\mathcal{C}'_t$ which are submitted to the same thermal noise. If they become identical in the limit $t \to \infty$, this means that in the long time limit there is no memory of the initial condition.

In all the examples we have considered (spin models, automata) one can easily compare two different configurations $\mathcal{C}_t = \{S_i(t)\}$ and $\mathcal{C}'_t = \left\{S'_i(t)\right\}$ submitted to the same thermal noise. At time t, one computes the two fields $h_i(t)$ and $h'_i(t)$, and the probabilities $p_i(t)$ and $p'_i(t)$ (74-78). Then one updates the spins using the same random numbers $z_i(t)$ for the two configurations

$$S_i(t+1) = \text{sign}\ (p_i(t) - z_i(t))$$
$$S'_i(t+1) = \text{sign}\ (p'_i(t) - z_i(t)) \tag{80}$$

One can then measure the distance D(t) between the two configurations

$$D(t) = \frac{1}{4N} \sum_i (S_i(t) - S'_i(t))^2 \tag{81}$$

In the high temperature regime, the thermal noise is strong, the system forgets faster its initial condition and one expects that D(t) vanishes when $t \to \infty$. In the low temperature regime, the effect of thermal noise is weak and D(t) remains finite as $t \to \infty$.

For the Kauffman model (section III), one can introduce a temperature as explained in the example 3 (eq. (76-78)). For the same reasons as in section III, the time evolution of D(t) is the same in the annealed and in the quenched model. By extending the calculation of section III, one finds

$$D(t+1) = \frac{1}{2} \tanh\left(\frac{1}{T}\right) (1-(1-D(t))^K) \tag{82}$$

There is a critical temperature T_c given by

$$\frac{K}{2} \tanh\left(\frac{1}{T_c}\right) = 1 \tag{83}$$

For $T > T_c$, $D(t) \longrightarrow 0$ as $t \longrightarrow \infty$. The thermal noise is strong enough to make two configurations become identical in the long time limit.

For $T < T_c$, $D(t) \longrightarrow D^* \neq 0$ as $t \longrightarrow \infty$. D^* is the fixed point of (83). It is interesting to notice that the limit of D(t) when $t \longrightarrow \infty$ does not depend on D(0).

One can generalize this calculation at finite temperature to other models like diluted non symmetric spin glasses[14] and one finds a similar transition.

X - THE 3 DIMENSIONAL SPIN GLASS

One can study the time evolution of the distance between pairs of configurations submitted to the same thermal noise in all kinds of systems. I will describe here the case of the 3d ± spin glass that we have studied recently with G. Weisbuch[29]. We considered Ising spins on a cube of size L(= 8 or 12) with nearest neighbour interactions J_{ij}:

$$\begin{aligned} J_{ij} &= +J \quad \text{with probability} \quad 1/2 \\ J_{ij} &= -J \quad \text{with probability} \quad 1/2 \end{aligned} \tag{84}$$

At each time step, we updated the spins $S_i(t)$ and $S_i'(t)$ of two configurations according to eq. (80).

To compare the time evolution of these two configurations submitted to the same thermal noise, we considered the three following situations.

a) The two initial configuration $\mathcal{C}_0$ and $\mathcal{C}_0'$ are opposite, $\mathcal{C}_0$ being random ($S_i(0) = -S_i'(0)$ for all i). Therefore D(0)= 1. (The data corresponding to this case will be represented by triangles in the figures).

b) The two initial configurations $\mathcal{C}_0$ and $\mathcal{C}_0'$ are random and independent: D(0) = 1/2 (squares on the figures).

c) The two initial configurations $\mathcal{C}_0$ and $\mathcal{C}_0'$ differ by a single spin: D(0) = 1/N (diamonds on the figures).

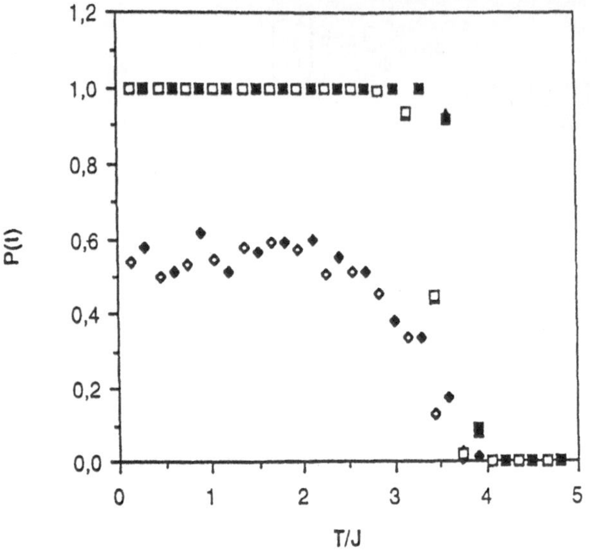

The first quantity one can consider is the survival probability P(t): P(t) is the probability that $\mathcal{C}_t$ and $\mathcal{C}_t'$ are still different after t time steps (t = 500). We repeated the calculation for N = 256 spins (white symbols) and N = 864 (black symbols) and averaged over hundreds of samples. The triangles are everywhere masked by the squares. We see a rather sharp transition at $T_c/J \simeq 3.5$ or 4. For $T > T_c$, the survival probability vanishes. For $T < T_c$, we can notice that even in case c) where only one spin was changed initially, the survival probability is high.

The next figure shows the average distance D(t) (for t = 500) in the 3 cases as a function of temperature T. D(t) is averaged only over the samples for which $\mathcal{C}_t$ and $\mathcal{C}_t'$ remain different. (The average over all samples would be given by P(t)D(t)).

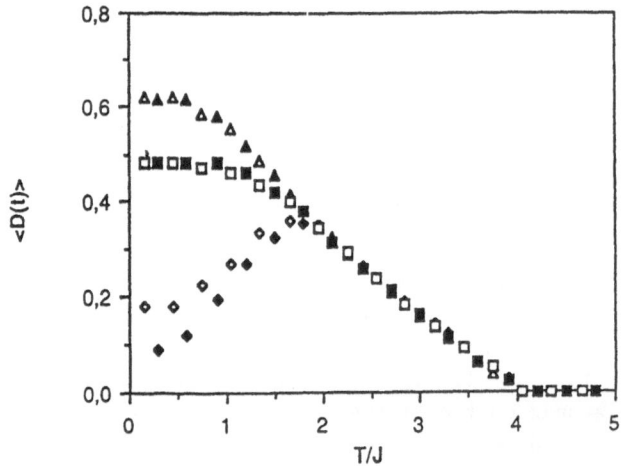

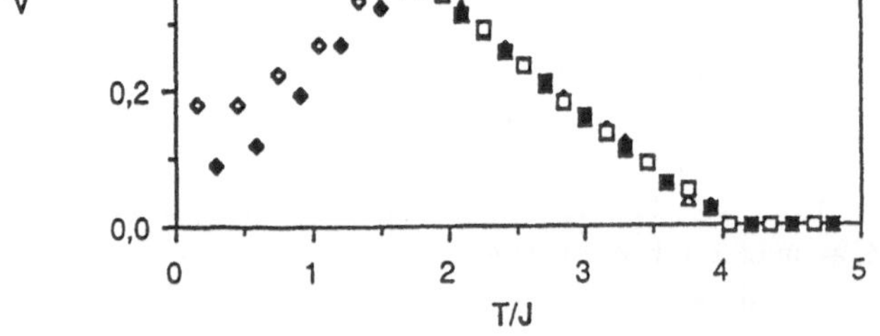

Like in the example discussed in section IX, the distance D(t) vanishes above a certain temperature $T_1 \simeq 4.J$. We see also a second important temperature $T_2 \simeq 1.8$ J below which D(t) depends on D(0).

The region $T_2 < T < T_1$ is very similar to the low temperature phase of the Kauffman model (section IX). The low temperature region $T < T_2$ is new since D(t) depends on D(0). Therefore D(t) can no longer be given by a recursion similar to (82). This region $T < T_2$ seems to coincide with the region where the spin glass phase appears[30].

It would of course be interesting to study in more detail the way the temperatures T_1 and T_2 depend on the system size N and on the time t in order to know if these apparent phase transitions persist in the limit $N \to \infty$ and $t \to \infty$. It would also be interesting to see how the time evolution of other quantities like the remanent magnetization, the autocorrelation function could be related to the time dependence of D(t). In particular, it has been shown[30] that the long time decay of the autocorrelation function q(t)

$$q(t) = \lim_{\tau \to \infty} \frac{1}{N} \sum_{i=1}^{N} \langle S_i(\tau)\ S_i(\tau+t) \rangle \tag{85}$$

is a stretched exponential in a range of temperatures which is roughly $T_2 < T < T_1$. It would be of course interesting to know whether there is a relation between the stretched exponential decay of q(t) and the non zero limit of D(t).

In this section, I have only discussed a recent but very partial aspect of the dynamics of spin glasses. In the last few years, much important progress has been done in the dynamics of spin glasses, in particular in the mean field limit[31-34]. I suggest the following recent reviews[35,36] for a detailed discussion of the present knowledge on spin glasses.

Lastly, I would like just to mention that an alternative way of submitting two configurations to the same thermal noise has been proposed[37-38] to study how the effect of changing one spin can spread over a sample.

XI - A NEURAL NETWORK MODEL

In the last few years, a lot of work has been done to describe neural networks as particular cases of spin glass models. I will not describe here all the interesting results[3,40,41,42] which have been obtained recently on neural networks. I will just talk about a simple model of neural network for which the dynamics can be solved exactly[15,16,43]. The idea which leads to the solution of this neural network model is very similar to the one which was already used in section III to calculate the distance in the Kauffman model.

A simple way of describing neural networks consists in considering a system of N neurons S_i which interact through synapses J_{ij}. The neurons are represented by Ising variables

$$\begin{aligned} S_i(t) &= +1 \quad \text{if the neuron is on} \\ S_i(t) &= -1 \quad \text{if the neuron is off} \end{aligned} \tag{86}$$

and the time evolution of this network is given by

$$S_i(t+1) = \text{sign}\left(\sum_j J_{ij} \; S_j(t)\right) \tag{87}$$

Assume that p patterns $\left\{\xi_i^{(\mu)}\right\}$ are given:

$$\begin{aligned} &1^{st} \text{ pattern} \qquad \xi_i^{(1)} = \pm 1 \qquad \text{for} \qquad 1 \leqslant i \leqslant N \\ &\mu^{th} \text{ pattern} \qquad \xi_i^{(\mu)} = \pm 1 \qquad \text{for} \qquad 1 \leqslant i \leqslant N \end{aligned} \tag{88}$$

One would like the neural network to memorize these patterns in the following sense: by definition a pattern $\left\{\xi_i^{(\mu)}\right\}$ is memorized if for the dynamics (87) there is an attractor near this pattern. So the problem is to choose the J_{ij} such that there are attractors as close as possible to the patterns.

At a given time t, one can easily measure the distance of the configuration $\{S_i(t)\}$ to a pattern $\left\{\xi_i^{(\mu)}\right\}$ by considering the projection

$$m_\mu(t) = \frac{1}{N} \sum_i S_i(t) \; \xi_i^{(\mu)} \tag{89}$$

There exist several choices[44] of the J_{ij} which give attractors in the neighbourhood of the p stored patterns. Some of these choices lead to interesting effects like short or long term memory[16,45], forgetting. However the rule which has been mostly studied for neural network models is the "Hebb rule" for which the J_{ij} have the following expression

$$J_{ij} = \sum_{\mu=1}^{P} \xi_i^{(\mu)} \; \xi_j^{(\mu)} \tag{90}$$

The model[15,16,3] for which the dynamics can be solved exactly is the following: consider a system of N neurons $S_i = \pm 1$. For each neuron, one chooses K input neurons $j_1(i)\ldots j_K(i)$ at random among the N neurons. The time evolution of the network is given by

$$S_i(t+1) = \text{sign}\left(J_{ij_1(i)}S_{j_1(i)}(t)+\ldots+J_{ij_K(i)}S_{j_K(i)}(t)\right) \tag{91}$$

where the J_{ij} are given by (90).

For the same reasons as in section III, one can show that the time evolution of the annealed and of the quenched model are the same in the thermodynamic limit. This allows one to calculate the time evolution of the $m_\mu(t)$. The result is the following: if the p patterns $\left\{\xi_i^{(\mu)}\right\}$ are random, i.e. the $\xi_i^{(\mu)} = 1$ with probability 1/2 and -1 with probability 1/2 and are independent, one can show that in the limit $p \gg 1$ and $K \gg 1$ with

$$\alpha = \frac{p}{K} \tag{92}$$

$$m_\mu(t+1) = \frac{1}{\sqrt{\pi}}\int_{-\infty}^{+\infty} dy\; e^{-y^2}\, \text{sign}(m_\mu(t) - y\sqrt{2\alpha}) \tag{93}$$

One sees from (93) that there is a critical value α_c of α

$$\alpha_c = 2/\pi \tag{94}$$

If $\alpha > \alpha_c$, the number of patterns memorized is too large and the only attractive fixed point of (93) is $m_\mu^* = 0$. The system does not remember anything.

If $\alpha < \alpha_c$, there appears an attractive fixed point $m_\mu^* \neq 0$ of (93) corresponding to the attractor near the pattern μ. One should notice that the retrieval is not perfect since $m_\mu^* \neq 1$ (the fraction of wrong bits is given by $(1-m_\mu^*)/2$).

The calculation which leads to (93) can be generalized to many other situations[15,16,46]. Let me just finish by giving the example of correlated patterns.

In the above calculation, the typical projection of one pattern μ on another pattern ν was $N^{-1/2}$:

$$\frac{1}{N}\sum_i \xi_i^{(\mu)}\,\xi_i^{(\nu)} \sim N^{-1/2} \tag{95}$$

for all pairs μ and ν.

If one considers that the p patterns are random but that two of them (patterns 1 and 2 are correlated)

$$\frac{1}{N} \sum_i \xi_i^{(1)} \xi_i^{(2)} = Q \tag{96}$$

Then one finds that there are two critical values of α:

$$\alpha_1 = \frac{2}{\pi} (1-Q)^2$$
$$\alpha_2 = \frac{2}{\pi} (1+Q)^2 \tag{97}$$

For $\alpha > \alpha_2$, the only fixed point in $m_1^* = m_2^* = 0$. Too many patterns have been stored. The system does not remember anything.

For $\alpha_1 < \alpha < \alpha_2$, there is an attractive fixed point $m_1^* = m_2^* \neq 0$. The system remembers patterns 1 and 2 but cannot distinguish them.

For $\alpha < \alpha_1$, there is an attractive fixed point $m_1^* > m_2^*$. The system can distinguish the two patterns.

There are some limiting cases which can be easily understood.

If $Q \to 0$, the patterns become uncorrelated and α_1 and $\alpha_2 \to \alpha_c$.

If $Q \to 1$, $\alpha_1 \to 0$. If the 2 patterns become identical, it is impossible to distinguish them.

ACKNOWLEDGEMENTS

The content of these lectures is based upon works done in collaboration with H. Flyvbjerg, E. Gardner, J.P. Nadal, Y. Pomeau, D. Stauffer, G. Weisbuch and A. Zippelius. It is a pleasure to thank them as well as C. de Dominicis, H.J. Hilhorst, K. Kürten, S. Kauffman and G. Toulouse for useful discussions.

REFERENCES

1. K. Binder, Applications of the Monte Carlo Method in Statistical Physics (Berlin: Springer Verlag) (1984).
2. J.F. Fontanari and R. Köberle, preprint (1986).
3. E. Bienenstock, F. Fogelman Soulié and G. Weisbuch, Disordered systems and biological organisation (Springer Verlag, Heidelberg) (1986).
4. S.A. Kauffman, J. Theor. Biol. 22, 437 (1969); 44, 167 (1974).

5. S.A. Kauffman, Math. Life Sci. 3, 63 (1970).
6. A.E. Gelfand and C.C. Walker, Ensemble Modelling (M. Dekker) (1984).
7. H. Flyvbjerg and N.J. Kjaer, preprint (1987), submitted J. Phys. A.
8. H.J. Hilhorst and M. Nijmeijer, J. Physique 48, 185 (1987).
9. K.E. Kürten, proceedings of the San Diego Conference (1987).
10. B. Derrida and G. Weisbuch, J. Physique 47, 1297 (1986).
11. B. Derrida and Y. Pomeau, Europhys. Lett. 1, 45 (1986).
12. B. Derrida and D. Stauffer, Europhys. Lett. 2, 739 (1986).
13. H. Flyvbjerg, Lectures given at the Niels Bohr Institute and Nordita (1987), to be published in Physica Scripta.
14. B. Derrida, J. Phys. A20 (1987) to appear.
15. B. Derrida, E. Gardner, A. Zippelius, Europhys. Lett. 4, 167 (1987).
16. B. Derrida, J.P. Nadal, preprint (1987).
17. G. Weisbuch, D. Stauffer, J. Physique 48, 11 (1987).
18. L. de Arcangelis, J. Phys. A20, L369 (1987).
19. L. de Arcangelis, D. Stauffer, preprint (1987).
20. B. Derrida, Les Houches (1986) "Chance and Matter", J. Souletie, J. Vannimenus and R. Stora eds, in press.
21. M. Mezard, G. Parisi, N. Sourlas, G. Toulouse, M. Virasoro, Phys. Rev. Lett. 52 1146 (1984); J. Physique 45, 843 (1984).
22. M. Mezard, G. Parisi, M. Virasoro, J. Physique Lett. 46, 217 (1985).
23. D. Sherrington, S. Kirkpatrick, Phys. Rev. Lett. 35, 1972 (1975); Phys. Rev. B 17, 4384 (1978).
24. B. Derrida, Phys. Rev. Lett. 45, 79 (1980); Phys. Rev. B24, 2613 (1981).
25. B. Derrida, G. Toulouse, J. Physique Lett. 46, 223 (1985).
26. B. Derrida, H. Flyvbjerg, J. Phys. A20 (1987) to appear.
27. B. Derrida, H. Flyvbjerg, J. Phys. A19, L1003 (1986).
28. B. Derrida, H. Flyvbjerg, J. Physique 48, 971 (1987).
29. B. Derrida, G. Weisbuch, Europhys. Lett. 4 (1987) to appear.
30. A. Ogielsky, Phys. Rev. B32, 7384 (1985).
31. H. Sompolinsky, Phys. Rev. Lett. 47, 935 (1981).
32. H. Sompolinsky, A. Zippelius, Phys. Rev. Lett. 47, 359 (1981); Phys. Rev. B25, 6860 (1982).
33. C. de Dominicis, H. Orland, F. Lainée, J. Physique Lett. 46, L463 (1985).
34. C. de Dominicis, J. of Magnetism and Magnetic Materials 54, 17 (1986).
35. K. Binder, A.P. Young, Rev. Mod. Phys. 58, 801 (1986).
36. D. Chowdhury, Spin glasses and other frustrated systems (1986), World Scientific and Co.
37. D. Stauffer, proceedings of the Antigonish Workshop (1987).
38. V. Costa, J. Phys. A20, L583 (1987).
39. H.E. Stanley, D. Stauffer, J. Kertesz, H.J. Herrmann, preprint (1987).
40. J.J. Hopfield, Proc. Natl. Aca. Sci. (USA) 79, 2554 (1982).
41. P. Peretto, Biol. Cybern. 50, 51 (1984).
42. D.J. Amit, H. Gutfreund, H. Sompolinsky, Ann. of Phys. 173, 30 (1987).
43. R. Kree, A. Zippelius, preprint (1987).

44. E. Gardner, preprints (1987).
45. J.P. Nadal, G. Toulouse, J.P. Changeux, S. Dehaene, Europhys. Lett. 1, 535 (1986).
46. R. Meir, E. Domany, preprints (1987).

ON THE WEAKLY COUPLED LORENTZ GAS

Detlef Dürr
Fachbereich Mathematik , Ruhr Universität Bochum
und BiBoS, Universität Bielefeld
4800 Bielefeld, West-Germany

BEGINNING

One of the main concerns of rigorous nonequilibrium statistical mechanics is to establish diffusive behaviour of testparticles in fluids. We call diffusive behavior the Brownian motion like wandering out to infinity of the initially localised testparticle, i.e. its mean square displacement grows asymptotically like time. There are only few results on this; we are concerned here with the Lorentz gas where the fluid consists of an infinite scatterer field, which does not change with time, i.e. the fluid particles are at rest and have infinite mass. The testparticle is a point particle moving according to the dynamical laws in the scatterer field.

The field may be a realisation of a random process - that is the choice I talk about here - or it may be deterministic. Then the testparticle's initial position and velocity are random. In the latter case the only rigorous results on testparticle diffusions are for scatterer fields which arise from periodic arrays of scatterers, the periodic Lorentz gas. For hard core scatterers this is a Sinai Billiard, for which the stationary measure for the particle flow is known. The hyperbolicity of the (non differentiable) flow allows for a construction of an (infinite) Markovpartition and a symbolic dynamics, such that under the isomorphism the velocity process of the testparticle exhibits good mixing properties |1|. It seems to be well accepted by now, that the assumed mixing property rests on a faulty argument; I recommend Levy's short note on this |2|. I also wish to draw your attention to the nice work of Knauf |3| on the smooth periodic Lorentz gas, where he constructs in a two dimensional fundamental domain a field of potentials with Coulomb like singularities, such that the flow is geodesic on a negatively curved manifold. Being thus hyperbolic and smooth he quickly gets the Central Limit Theorem, i.e. the diffusive behavior for the position of the testparticle, but the diffusion matrix (2×2,since the motion takes place in R^2) might have one zero eigenvalue. Such is the case also for the result of Sinai and Bunimovich |1| ,where the argument for the nondegeneracy of the diffusion matrix is incomplete.

The situation of the random Lorentz gas is somewhat peculiar.In rigorous studies the quantum mecanical flow drew most attention, partly because there is the phenomenon of localisation (the testparticle stays in a finite region) for large disorder in higher dimensions, whereas diffusive behaviour has not been established as yet |4|.

Somewhat simpler is the Lorentz gas in weak coupling situations, the van Hove limits. Unfortunately one needs some arguing to describe diffusive behaviour in weak coupling limits, where one naturally studies the velocity process of the testparticle.

I wish to describe the van Hove limit and some mathematics to go with it for Newtonian mechanics. To complete however my list of compliments, I mention |5,6,7| as works on the quantum evolution, and I recommend to solve the open two dimensional problem and to clarify the fate of the position process in the limit |7|.

SOME HEURISTICS

Of course I could take any random potential field with some mild mixing properties but then being asked what potentials this really are, I think of uniformly distributed finite range potentials (scatterers);and the point I eventually want to make is most naturally for this choice. For ease of drawings we take a two dimensional system: A Poisson field (Ω,F,P,N) with density ρ, i.e. for the random point measures $N(d^2q)$, $P(N(d^2q)) = \rho d^2q$. Think of $\omega = (q_i)_{i\ Z}\varepsilon\Omega$ as a configuration of points in R^2, and $N_\omega(A) = |\omega\ A|$ =number of points in A , then $N(A)$, $N(B)$ are independent when A, B are disjoint.

Next we take a nice symmetric finite range potential $V(x)$ and build the random potential field.

(1) $$U_\omega(x) = \int N_\omega(d^2q)\, V(x-q) = \sum_{q_i \in \omega} V(x-q_i) \,.$$

If you draw a picture of this the scatterers should overlap a lot. The forcefield is of course

$$K(x) = -\int N(d^2q)\, \nabla V(x-q)$$

and

(2) $$P(K(x)) = -\, P(\int N(d^2q)\nabla V(x-q)\,) = -\int \rho d^2q \nabla V(x-q) = 0$$

and because of that

(3) $$P(K(x)K(y)) = P(\int N(d^2q)\nabla V(x-q)\int N(d^2q')\nabla V(y-q')) =$$

$$= \int \rho d^2q \nabla V(x-q)\nabla V(y-q) = P(K(x-y)K(0))$$

also since $N(d^2q)^2 = N(d^2q)$.

Now comes the particle with Newton's equations of motion

(4) $$\dot{x} = v \qquad x(0) = x_o$$
$$m\dot{v} = \varepsilon K(x) \quad v(0) = v_o$$

Clearly $x(t;x_o,v_o)$, $v(t;x_o,v_o)$ inheret the randomness from K. Hence we may talk about the processes $x_t, v_t, t \geq 0$, once their existence is proven, which we take for granted. There is also the ε.Letting $\varepsilon \to 0$ we get a particle moving with constant velocity v_o, but only for t in a compact interval. We need to find the right scale for t to see something intersting happening. To find this, one does the same I do next- perturbation to second order- to show what to expect.But I say already that the right scale is t/ε^2. Hence $x^\varepsilon(t) = x(t/\varepsilon^2)$, $v^\varepsilon(t) = v(t/\varepsilon^2)$ and for (4) comes

(5) $$\dot{x}^\varepsilon = \frac{1}{\varepsilon^2} v^\varepsilon$$
$$\dot{v}^\varepsilon = \frac{1}{\varepsilon} K(x^\varepsilon) \qquad , \quad m=1$$

Define for a function $f(v)$, $u^\varepsilon(t,x,v) = f(v^\varepsilon(t;x,v))$, $u^\varepsilon(0,x,v) = f(v)$, and get from (5) that $u^\varepsilon(t,x,v)$ satisfies the adjoint Liouville equation

(6) $$\frac{\partial u^\varepsilon(t,x,v)}{\partial t} = \left(\frac{v}{\varepsilon^2}\cdot\frac{\partial}{\partial x} + \frac{1}{\varepsilon} K(x)\cdot\frac{\partial}{\partial v}\right) u^\varepsilon(t,x,v,)$$

and then starting the game with

(7a) $$u\,(t,x,v) = u_o(t,x,v) + \varepsilon u_1(t,x,v) + \varepsilon^2 u_2(t,x,v) \ldots$$

we get

(7b) $$v\cdot\frac{\partial}{\partial x}\, u_o(t,x,v) = 0$$

(7c) $v\cdot\frac{\partial}{\partial x}u_1(t,x,v) + K(x)\cdot\frac{\partial}{\partial v}u_o(t,x,v) = 0$

(7d) $\frac{\partial}{\partial t}u_o(t,x,v) = v\cdot\frac{\partial}{\partial x}u_2(t,x,v) + K(x)\cdot\frac{\partial}{\partial v}u_1(t,x,v)$.

(7b) is satisfied when $u_o(t,x,v) = u_o(t,v)$, $u_o(0,v) = f(v)$, in particular u_o is nonrandom.(Averaging is like spatial averaging).(7c) says then that

$$u_1(t,x,v) = \chi(x,v)\cdot\frac{\partial}{\partial v}u_o(t,v)$$

with
$$v\cdot\frac{\partial}{\partial x}\chi(x,v) + K(x) = 0 ,$$

i.e.
$$\chi(x,v) = \int_0^\infty K(x+vt)dt$$

which is a particular pleasant expression for infinity, but now (7d) becomes

$$\frac{\partial}{\partial t}u_o(t,v) = K(x)\cdot\frac{\partial}{\partial v}\int_0^\infty K(x+vt)dt\cdot\frac{\partial}{\partial v}u_o(t,v) = v\cdot\frac{\partial}{\partial x}u_2(t,x,v) .$$

K is random, hence we average the equation, and because of (3) the left side will not depend on x; so why not putting the right side equal to zero (spatial average)? So

(7e) $\frac{\partial}{\partial t}u_o(t,v) = Lu_o(t,v)$

with the generator

(8) $$L = \frac{\partial}{\partial v}\cdot\int_0^\infty P(K(x)\ K(x+vt)dt)\cdot\frac{\partial}{\partial v} = \frac{1}{2}\ D\ \nabla\cdot(I - \frac{v\ v}{|v|^2})\cdot\nabla$$

$$= \frac{\rho}{2|v|}\int dk|\hat{V}(k)|^2k^2\frac{\partial^2}{\partial\phi^2}$$

in our units, Fourier transform and polar coordinates.
So we should prove that $v^\varepsilon(t) = v(t/\varepsilon^2)$ (with an intrinsic ε coming from the force) converges in distribution to a diffusion process on the circle of radius the speed $|v_o|$(of course,the potential energy goes to zero). Clear, that all it should take is that the diffusion matrix in (8) with the force correlation funtion is finite. In dimensions >2 all this has been shown in particular also for non conservative force fields in |8|.

I give now a second argument, which I think goes more to the heart of the problem and which does not stretch the weak coupling character.At the same time we learn a new way to look upon the scaling. We give the Poisson field $(\Omega,F,P^\varepsilon,N)$ the index

ε, by setting $\rho^{\varepsilon} = \rho\varepsilon^{-4}$ $(=\rho\varepsilon^{-2d})$. Then one convinces oneself that the process $\varepsilon^2 x^{\varepsilon}(t)$, $v^{\varepsilon}(t);t\geq 0$,solving (5) is equivalent to the process x_t^{ε}, v_t^{ε} on $(\Omega,F,P^{\varepsilon},N)$ solving

$$(9) \qquad \dot{x}_t^{\varepsilon} = v_t^{\varepsilon} \qquad\qquad x_0^{\varepsilon} = \varepsilon^2 x_o$$
$$\dot{v}_t^{\varepsilon} = K^{\varepsilon}(x_t^{\varepsilon}) \qquad\qquad v_0^{\varepsilon} = v_o$$

where

$$(10) \quad K^{\varepsilon}(x) = -\nabla U^{\varepsilon}(x)$$

and

$$(10a) \quad U^{\varepsilon}(x) = \int N(d^2q)\ V^{\varepsilon}(x-q) = \int N(d^2q)\ \varepsilon V(\tfrac{x-q}{\varepsilon^2}) \ .$$

Now we give an "order of magnitude" argument: Since the potential gets weak, the speed of the particle is $|v_o|$ and we only have to worry about the direction ϕ. The time for the particle to cross one scatterer is $\Delta t \sim \varepsilon^2$, since the range of V^{ε} is $\sim\varepsilon^2$. Thus the change in v due to one scatterer is

$$\Delta\phi \sim \Delta v = \nabla V^{\varepsilon}\ \Delta t \ \sim \varepsilon^{-1}\varepsilon^2 = \varepsilon \ .$$

Now assume that the effects of the scatterers are independent, (the Poisson distribution hints at that, but don't forget that they overlap).Then the total variance of the direction in time t is the sum of the single variances $\Delta\phi^2 \sim \varepsilon^2$, and we have as many as

t×density×crosssection of one scatterer×speed of the particle

$$\sim t\rho^{\varepsilon}\ \varepsilon^2\ |v_o| \sim t/\varepsilon^2.$$

Hence the total variance equals tD, D the right constant. So what we got is a process on the circle, rotationally invariant with an angle variable behaving like Brownian motion.

SOME MATHEMATICS

Hence we wish to prove that the solution of (9) converges in law to (x_t,v_t) , $x_t = \int_0^t v_s ds$, and v_t is the Wiener process on the circle of radius $|v_o| = 1$. Clear, but something to reflect upon, we need only look at the velocity. We represent $(v_t^{\varepsilon})_{t\geq 0}$ canonically on pathspace, say $C(0,T)$, some T, by the induced measure ν^{ε} on path space. The same we do with $(v_t)_{t\geq 0}$ with the induced measure ν. Convergence in law means that $\nu^{\varepsilon}(f)\to\nu(f)$ as $\varepsilon\to 0$, for all bounded and continuous functions on C.

$(\nu^{\varepsilon})_{\varepsilon}$ is a sequence of measures which might be sequentially compact (in the usual sense), then we like to have a unique limit point namely ν. So assume we have compactness (which is

not easy to establish) how do we get uniqueness?
ν corresponds to the diffusion process with generator L, i.e. letting F_t be a family of increasing σ-algebras to $F(C(0,T))$ of histories of the paths $w(\cdot)$, we have that for good functions g

$$Lg(w(t))\Delta t = \nu(g(w(t+\Delta t))/F_t) - g(w(t))$$

with $\nu(\cdot/F_t)$ conditional expectation , so that

$$\nu(\{g(w(t+\Delta t)) - g(w(t)) - Lg(w(t))\}/F_t) = 0 \quad ,$$

and one calls $\{\cdot\}$ martingale difference ΔM_t, whose sum yields a martingale M_t: $\nu(M_t/F_s) = M_s$, and this is a martingale <u>only</u> under ν. So we might understand that $\nu^\varepsilon \to \nu$ if

$$(11) \quad \nu^\varepsilon(g(w(t)) - g(w(s)) - \int_s^t Lg(w(u))du \;/\; F_s) \to 0 \ , \text{ as } \varepsilon \to 0 \ .$$

We may express (11) also in terms of the process v_t^ε :

$$(11a) \quad P^\varepsilon(g(v_t^\varepsilon) - g(v_s^\varepsilon) - \int_s^t Lg(v_u^\varepsilon)du \;/\; F_s^{(\varepsilon)}) \to 0 \quad .$$

To get a feeling let us take $g(x) = x$, then (11a) says that in v_t^ε hides a martingale and a systematic part which is very much like $\int Lvdu$.

The essence of all classical weak coupling limits is an underlying martingale structure.
We are going to find one now. We go back to (9): By integration

$$(12) \quad v_t^\varepsilon - v_0^\varepsilon = \int_0^t K^\varepsilon(x_u) = \int_0^t N(d^2q)\, \frac{1}{\varepsilon}\, \nabla V(\frac{x_t^\varepsilon - q}{\varepsilon^2}) = \int_0^t \sum_{q_i \in \omega} \frac{1}{\varepsilon}\, \nabla V(\frac{x_t - q}{\varepsilon^2})$$

Now we follow the particle with a clock through the realisation ω and register the times t_i at which the particle enters the scatterer q_i for the first time and also the place of first entry: $\sigma_i \in S^\varepsilon = \partial(\text{supp } V^\varepsilon)$
For each ω we get a sequence $(t_i, \sigma_i)_i$. These are points in $[0,\infty) \times S^\varepsilon$, they are random according to P^ε, hence we may build a point process $N^\varepsilon(d\sigma, dt)$ by

$$(13) \qquad \int_{I\times A} f(\sigma,t) N^\varepsilon(d\sigma,dt) = \sum f(\sigma_i, t_i) \qquad , I\times A \cdot [0,\infty)\times S^\varepsilon .$$

With $N^\varepsilon(A,[0,t])$ comes a natural σ-algebra F_t and it is no surprise that at least morally

(14) $P^{\varepsilon}(N^{\varepsilon}(d\sigma,dt)/F_t) \sim \rho^{\varepsilon} d\sigma \cdot v_t^{\varepsilon}\, dt = r_t^{\varepsilon}(d\sigma)dt$

where $r_t^{\varepsilon}(d\sigma)$ is the collision rate in $d\sigma$.

Accordingly $N^{\varepsilon}(d\sigma,dt) - r_t^{\varepsilon}(d\sigma)dt$ is nearly a martingale difference and

(15) $\int_A \int_0 f(\sigma,s)\, N^{\varepsilon}(d\sigma,ds) - \int_A \int_0 f(\sigma,s) r_s^{\varepsilon}(d\sigma)ds \sim M_t^{\varepsilon}$

ought to be a martingale for every F_t measurable function f. So all we have to do is to express the right of (12) in terms of N^{ε}.

Unfortunately also for low moral standards (14) needs careful thinking, as it expresses in fact independence of future increments of the point process (future scatterers to come) from the past (the ones the particle has already gone through). So what happens when the particle trajectory crosses itself? Then for some amount of time the particle does not enter new scatterers so definitely no independence there. ($N^{\varepsilon} = 0$ there , since we only registered first entries!) But think of our second argument and think how a trajectory crosses itself: Transversally of course (means at least not tangentially) so the region of selfcrossing is $\sim$area of a scatterer $\sim \varepsilon^4$. In that area we find $\rho^{\varepsilon}\varepsilon^4 = O(1)$, that means many many scatterers, but not enough to influence the motion of the particle (and thus the structure we are after), since every scatterer has merely an effect of order ε.

Hence we stop worrying and rewrite (12):

(16) $v_t - v_o = \int_{S^{\varepsilon}} \int_0^t N^{\varepsilon}(d\sigma,du) \int_u^t \nabla V^{\varepsilon}(x_{\sigma,u}^{\varepsilon}(s))ds$

+ effect from scatterers initially overlapping x_0^{ε}

where

$$x_{\sigma,u}^{\varepsilon}(t) = x_t^{\varepsilon} - x_u^{\varepsilon} + \sigma \quad .$$

Perhaps one has to think a bit to see (16) come out. One might like to reinsure oneself about the hiding place of the martingale. It is for sure in $N^{\varepsilon}(d\sigma,du)$ and if it were integrating a F_u- measurable function..., but it does not. The function it integrates ($\int_u^t \ldots$) depends clearly on the future after u.

Clear, that we couldn't do without perturbation and estimates: We need to approximate $\int_u^t \nabla V^{\varepsilon}(x_{\sigma,u}^{\varepsilon}(s))$ by a function which depends only scatterers before and until time u. A hint is here the straight line path through a scatterer which appears in the diffusion constant in L. The details are in |9|.

CONCLUSION

Now we know that in the van Hove limit the velocity of the test-particle describes a Brownian motion on the sphere of radius v_o. The convergence of v_t^ε as a process is for compact time intervals.

So when we look now for the long time behaviour of x_t, we took the limit $\varepsilon \to 0$ first, and that is in general not the same as studying the long time behaviour simultaneously with $\varepsilon \to 0$. One might then look at $a(\varepsilon)x^\varepsilon(t/a(\varepsilon)^2)$, $a(\varepsilon) \to 0$ as $\varepsilon \to 0$ to prove directly for a class of $a(\varepsilon)$ what I say next. It is certainly interesting since $a(\varepsilon)$ may interpolate between the weak coupling situation (when $a(\varepsilon)$ varies slowly with ε) and the Lorentz diffusion ($a(\varepsilon) = a$, $a \to 0$, ε fixed.)

In the weak coupling regime we expect the following:

In Itô- differential form v_t obeys

$$dv_t = -D\frac{d-1}{|v|^2}v_t dt + \sqrt{D}\,\left(I - \frac{v_t v_t}{|v|^2}\right)\cdot dW_t \quad , \text{ in d dimensions}$$

and since $dx_t = v_t dt$

$$x_t - x_o = \frac{v^2}{D(d-1)}(v_t - v_o) \quad + \frac{v^2}{(d-1)\sqrt{D}}\int_0^t \left(I - \frac{v_s v_s}{v^2}\right)\cdot dW_s \quad .$$

The stochastic integral is a martingale (that's always so) and its mean square is $(d-1)t$ (Brownian motion on the sphere!). Since $|v_t| = |v_o|$ we have by the martingale invariance principle that

$$\tilde{\varepsilon}x(t/\tilde{\varepsilon}^2) \to \sqrt{D}_{wc}\, W_t \;, \quad \tilde{\varepsilon}\to 0 \quad ,\to = \text{in law, with}$$

$D_{wc} = v^4/D$ and W_t a standard Wiener process. This is so simple because we took the limit $\varepsilon \to 0$ first!

REFERENCES

|1| L.A. Bunimovich, Ya.G. Sinai: Statistical Properties of the Lorentz Gas with Periodic Configurations of Scatterers. Commun. Math. Physics 78, 479-497,1981

|2| Y.E. Levy : A Note On Sinai and Bunimovich's Markov Partition for the Billiard. Preprint CNRS LP 014

|3| A. Knauf : Ergodic and Topological Properties of Hamiltonian Systems with two Degrees of Freedom. Dissertation, to be published in Commun. Math. Physics.

|4| F. Martinelli, E. Scoppola: Introduction to the Mathematical Theory of Anderson Localisation. Preprint Universitá di Roma.

|5| G. Emch, P. MArtin, Helv.Phys. Acta,t.48,59,1975.

|6| H. Spohn : J. Stat. Physics, 17, 6, 1977.

|7| G.F. Dell' Antonio: Large Time, Small Couling Behaviour of a Quantum Particle in a Random Field. Ann.Inst. Henri Poincare XXXIX , 339-384 ,1983

|8| H. Kesten, G. Papanicolaou : A Limit Theorem for Stochastic Acceleration. Commun. Math. Physics,78, 19-63 ,1980

|9| D.Dürr, S. Goldstein, J.L. Lebowitz : Asymptotic Motion of a Classical Particle in A Random Potential in Two Dimensions: Landau Model. To appear in Commun. Math. Physics .

DYNAMICAL SYSTEM DESCRIBING THE LOW TEMPERATURE PHASE OF A DRIVEN LATTICE GAS

Joachim Krug and Herbert Spohn

Theoretische Physik, Universität München
Theresienstr. 37, 8 München 2, Germany

1. The Problem

We want to understand the structure of the steady state of a driven diffusive system [1-9]. Such a system is modelled as a lattice gas on a square lattice (for spatial dimension two). In the y-direction particle jumps are biased downwards due to the driving field, cf. Figure 1. In the x-direction the jump rates satisfy detailed balance with respect to an attractive nearest neighbor energy. The strength of the interaction energy introduces a temperature-like parameter into the model.

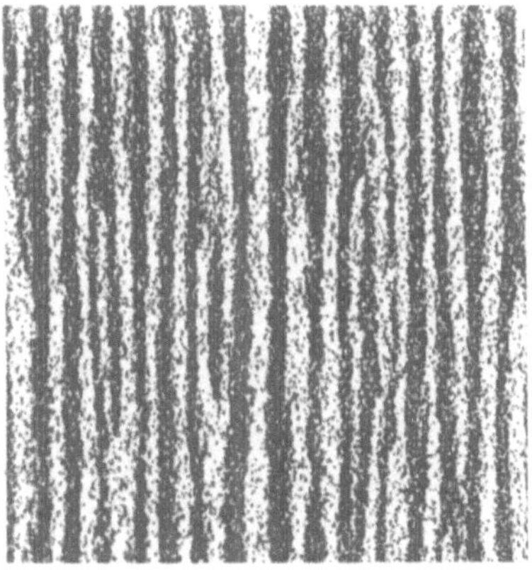

Figure 1. Typical low-temperature configuration of a driven lattice gas in two dimensions. Snapshot from a simulation by J. Marro and J.L. Vallès, performed on a 300×300 lattice. The driving force is directed downwards.

Let $\rho_j(t)$ be the (actual) density of column j at time t, $0 \leq \rho_j(t) \leq 1$. The densities satisfy the mass conservation

$$\frac{d}{dt}\rho_j(t) = J_{j-1} - J_j \quad . \tag{1.1}$$

J_j is the current from column j to column j+1. To simplify somewhat we assume that the driving field mixes each column very rapidly [4-5]. Then $\rho_j(t)$ is deterministic up to a small noise of order $N^{-1/2}$, where N is the column height. Since the interaction between particles is nearest neighbor, the current from j to j+1 depends on the densities ρ_j, ρ_{j+1} and on the densities in the neighboring columns j-1, j+2. It is the difference of the number of jumps per unit time from j to j+1 and those from j+1 to j. Therefore we have

$$J_j(\underline{\rho}) = R(\rho_{j-1},\rho_j,\rho_{j+1},\rho_{j+2}) - R(\rho_{j+2},\rho_{j+1},\rho_j,\rho_{j-1}) + \text{noise of } O(N^{-1/2}) \quad . \tag{1.2}$$

The noise will be ignored ($N \to \infty$). A specific choice for R will be discussed below.

In Monte-Carlo simulations [1-3] one observes that at sufficiently strong interaction (= sufficiently low temperature) (i) the system segregates into a high and low density phase, (ii) the system stays for a long time in metastable configurations with a periodic variation in the density. Figure 1 shows such a configuration on a 300×300 lattice. If one imposes initially a configuration with a single interface a spontaneous break-up into periodic patterns is never observed. On the basis of (1.1) and (1.2) we would like to understand then (i) the phase diagram, i.e. the high (low) density as a function of the temperature, (ii) the structure of metastable states.

2. The discrete dynamical system

The basic strategy is straightforward. Stationary solutions satisfy $\frac{d}{dt}\rho_j(t) = 0$. Since the jumps orthogonal to the field are governed by thermal rates, $J_j = 0$ in the steady state and the physical stationarity condition is

$$R(\rho_{j-1},\rho_j,\rho_{j+1},\rho_{j+2}) = R(\rho_{j+2},\rho_{j+1},\rho_j,\rho_{j-1}) \quad , \tag{2.1}$$

$j = 0,\pm 1,\ldots$. We read (2.1) as a recurrence equation: given ρ_{j-1}, ρ_j, ρ_{j+1} the value of ρ_{j+2} is determined. (2.1) defines a dynamical system with a three-dimensional phase space. Every orbit of (2.1) is a stationary, zero current solution to (1.1), (1.2), provided that $0 \leq \rho_j \leq 1$. Of physical interest are only those orbits leading to stationary profiles of (1.1), (1.2) which are dynamically stable. Among them there should be in particular a solution which represents the coexistence of a high and low density phase. This kink solution rapidly interpolates between two half-lines of essentially constant density. From our numerical evidence we expect also stable stationary density profiles which are spatially periodic.

The function R in (2.1) is obtained by averaging the rate of jumps, $\Gamma_\perp$, perpendicular to the driving force with respect to the steady state of the fast dynamics parallel to the driving field. $\Gamma_\perp$ is a function only of the energy difference in the jump. We write schematically

$$R(\rho_{j-1},\rho_j,\rho_{j+1},\rho_{j+2}) = \langle\Gamma_\perp(\beta\Delta H_{j,j+1})\rangle_\parallel \quad . \tag{2.2}$$

β denotes the inverse temperature. Since the jumps perpendicular to the driving field satisfy detailed balance,

$$\Gamma_\perp(\lambda) = e^{-\lambda}\,\Gamma_\perp(-\lambda) \quad . \tag{2.3}$$

For systems in thermal equilibrium (2.3) ensures the relaxation into the Gibbs state $Z^{-1}\exp[-\beta H]$ irrespective of the precise choice of $\Gamma_\perp$. By contrast, the nonequilibrium steady state of our model depends explicitely on the jump rates. Still some general features, such as the topology of the phase diagram and the values of the critical exponents, were shown to be rate independent [5]. We expect then the low temperature dynamics discussed below to be qualitatively correct for a large class of jump rates.

A particularly convenient choice is $\Gamma_\perp(\lambda) = \exp[-\lambda/2]$ because it implies the factorization of R. With this choice, in terms of $v_j = 2\rho_j-1$, the recursion relation (2.1) reads

$$v_j = K\,\frac{1 - F(v_{j-3},v_{j-2},v_{j-1})}{1 + F(v_{j-3},v_{j-2},v_{j-1})} \equiv G(v_{j-3},v_{j-2},v_{j-1}) \quad , \tag{2.4}$$

$$F(x,y,z) = \frac{(1+y)(1-z)(K-x)(K-y)^2(K+z)^2}{(1-y)(1+z)(K+x)(K+y)^2(K-z)^2} \quad . \tag{2.5}$$

K is the temperature parameter

$$K = \coth(\beta J) \tag{2.6}$$

with $-J$, $J > 0$, the attractive coupling. We regard(2.4) as a discrete map, T, defined by

$$T(x,y,z) = (y,z,G(x,y,z)) \quad . \qquad (2.7)$$

T as a map on R^3 has singularities. Orbits of physical relevance have to stay in $[-1,1]^3$. In this domain T is smooth, but not onto. T is reversible. It does not preserve volume.

Let us briefly discuss the fixed point structure of T. The diagonal $\{x = y = z\}$ is a line of fixed points and no other fixed points exist. The linearization of T ~~at~~ (v^*,v^*,v^*), $DT(v^*)$, is volume preserving, $\det DT(v^*) = 1$. One eigenvector of DT points along the diagonal. Its eigenvalue is marginal, $\lambda_0 = 1$. In the remaining two-dimensional eigenspace DT is either hyperbolic or elliptic. For $K > 5$ all fixed points are hyperbolic. In terms of the lattice gas this means that the only stationary configurations have a uniform density. This is the high temperature situation and we identify $K_c = 5$ as the critical temperature. For $K < K_c$, (v^*,v^*,v^*) is hyperbolic if $(v^*)^2 > v_s(K)^2$ and elliptic if $(v^*)^2 < v_s(K)^2$, where

$$v_s(K)^2 = K \frac{K_c - K}{5K - 1} \quad . \qquad (2.8)$$

$v_s(K)$ defines the <u>spinodal</u>. If (1.1) with currents given through (2.2) is linearized around a configuration of uniform density ρ, then the linearization is unstable for $(2\rho-1)^2 < v_s(K)^2$.

The fairly complex fixed point structure of T might lead the reader to the suspicion that no intelligible pattern to the global dynamics can be obtained. Surprisingly, T "foliates" into <u>two-dimensional</u> dynamics. Let us first discuss then a two-dimensional dynamical system isomorphic to T restricted to one leaf of the foliation. Afterwards we will return to the dynamical system of interest and show how the foliation is achieved.

3. Discrete dynamics in an inverted double well potential

We consider the equilibrium dynamics of a two-dimensional lattice gas. The fast parallel rate in the driven system has as correspondance a mean-field interaction in the y-direction. Let ρ_j, $0 \le \rho_j \le 1$, be the density of the j-th column. Then the free energy is given by

$$F(\underline{\rho}) = \sum_j \{(\rho_j \log(\rho_j) + (1-\rho_j)\log(1-\rho_j)) - 4\beta J(\rho_j\rho_{j+1} + \rho_j^2)\} \quad . \tag{3.1}$$

The first term is (minus) the entropy per column and the second term is the interaction energy between columns. We assume that the current J_j in (1.2) is proportional to the gradient of the local chemical potential,

$$J_j = -\left(\frac{\partial F}{\partial \rho_{j+1}} - \frac{\partial F}{\partial \rho_j}\right) \quad . \tag{3.2}$$

The Onsager coefficient has been set to unity. Then the stationarity condition reads

$$\frac{\partial F}{\partial \rho_j} = \mu \quad , \tag{3.3}$$

μ being the (global) chemical potential, i.e.

$$(\Delta\rho)_j = V'(\rho_j) \tag{3.4}$$

with

$$V'(\rho) = \frac{1}{4\beta J}\log(\rho/(1-\rho)) - 4\rho + \mu \quad . \tag{3.5}$$

Δ is the discrete Laplacian. (3.4) describes the motion of a classical particle in an <u>inverted</u> double well potential - in discrete time. (3.4) has been studied by Pandit and Wortis [10] in the context of wetting. Qualitative properties do not change, if the potential is expanded at $\rho = 1/2$. Let $\rho_j = \frac{1}{2} + \phi_j$. Then the potential is

$$V(\phi) = \frac{1}{2}\tau\phi^2 + \frac{1}{4!}\phi^4 + \mu\phi \quad . \tag{3.6}$$

In this approximation the equations of motion (1.1) and (3.2) read

$$\frac{d}{dt}\phi_j(t) = [\,\Delta(-\Delta\underline{\phi}(t) + V'(\underline{\phi}(t)))\,]_j \tag{3.7}$$

with $V'(\underline{\phi})_j \equiv V'(\phi_j)$. $\phi_j(t)$ takes arbitrary real values. The stationarity condition is

$$\Delta\underline{\phi} = V'(\underline{\phi}) \tag{3.8}$$

corresponding to the dynamical system

$$\left.\begin{aligned} x' &= V'(x) + 2x - y \\ y' &= x \end{aligned}\right\} = T_\mu(x,y) \quad . \tag{3.9}$$

T_μ is area-preserving.

Let us describe the fixed point structure for T_μ, cf. Figure 2. In the symmetric case, $\mu = 0$, (0,0) is the only fixed point of T_o for $\tau > 0$. It is hyperbolic and corresponds to the stable homogeneous phase $\underline{\phi} = 0$. The critical coupling is $\tau_c = 0$. For $\tau < 0$, T_o has the three fixed points (0,0) and $\pm(\sqrt{-6\tau}, \sqrt{-6\tau})$. The outer fixed points are hyperbolic. They correspond to the pure phases of low and high density. (0,0) is elliptic for $-4 < \tau < 0$. For $\tau < -4$ it period doubles as studied by Janssen and Tjon [11]. Since no period doubling occurs for the potential (3.1), we require $\tau \geq -4$. We are interested in the low temperature phase, i.e. $\tau < 0$. In the asymmetric case, $\mu \neq 0$, the picture is "tilted". The high temperature fixed point does not bifurcate. For a particular $\tau^* = \tau^*(\mu) < 0$ a pair of fixed points appears, the inner being elliptic, the outer hyperbolic.

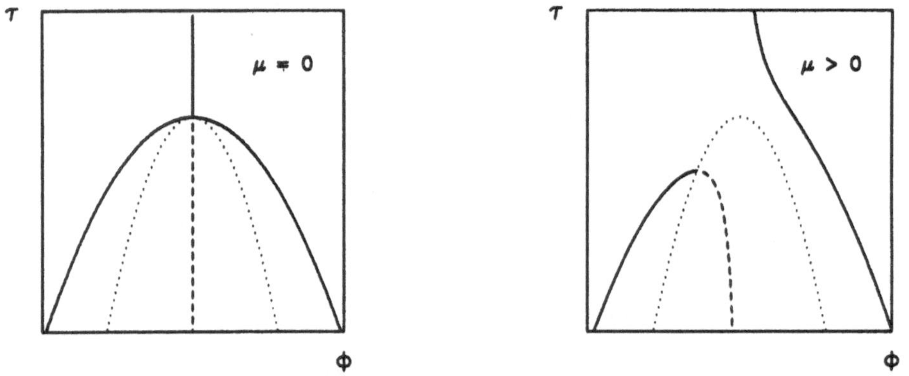

Figure 2. Phase diagram of the equilibrium model (3.1). Full lines mark the (meta-) stable phases, corresponding to hyberbolic fixed points of (3.9), dashed lines mark the unstable phase (elliptic fixed point). The dotted line gives the spinodal.

$|\tau|$ should be thought of as the strength of the non-linearity. If τ is close to zero, T_μ is essentially integrable and the continuum approximation

$$\frac{d^2}{dt^2} x(t) = V'(x(t)) \tag{3.10}$$

is fairly accurate. For the lattice gas this corresponds to the critical region, where the correlation length ξ is large on the scale of the lattice spacing. As usual in mean field theories, ξ diverges as $|\tau|^{-1/2}$ for $\tau \to 0$. As τ decreases a structure familiar from the KAM theorem and Chirikov's standard map appears between the outer and central fixed points. The role of τ is displayed in a more canonical form upon rescaling ϕ such that the hyperbolic fixed points are located at $\pm(1,1)$.

Any orbit of the dynamical system T_μ represents a stationary density configuration, provided $|\phi_j|$ remains bounded. Of physical interest are only those configurations which are dynamically stable in the sense of (3.7). Therefore we linearize (3.7) around the stationary solution $\underline{\phi}$ as $\phi_j + \xi_j$. Then

$$\frac{d}{dt} \xi_j(t) = [\Delta(-\Delta + V''(\underline{\phi}))\, \underline{\xi}(t)]_j \quad . \tag{3.11}$$

The first Laplacian results from the conservation of mass. It slows down relaxation, but does not influence the character of stability. We conclude that $\underline{\phi}$ is stable, if

$$- \Delta + V''(\underline{\phi}) \equiv H(\underline{\phi}) \geq 0 \tag{3.12}$$

as an operator on $\ell^2(Z)$. $H(\underline{\phi})$ is a discrete Schrödinger operator (= Jacobi matrix) with potential $W_j = V''(\phi_j)$. W_j is bounded from below. We are not aware of a "simple" criterion on $\underline{\phi}$ which would guarantee the positivity of $H(\underline{\phi})$. There are however two helpful general observations.

(i) $V''(\phi) = 0$ defines the spinodal, cf. Figure 2. For ϕ inside the spinodal $V''(\phi) < 0$. Therefore orbits $\underline{\phi}$ with ϕ_j inside the spinodal region for all j have to be dynamically unstable. On the other hand, if ϕ_j is outside the phase coexistence curve, then the repulsive part of the potential dominates the motion and the orbit diverges. We conclude, that dynamically stable orbits must have densities which "mostly" lie between the phase coexistence curve and the spinodal.

(ii) $\underline{\phi}$ is an orbit of the dynamical system T_μ. Linearizing T_μ along that orbit as $\underline{\phi} + \underline{\xi}$ yields the linear recursion relation $(H(\underline{\phi})\underline{\xi})_j = 0$. In fact, the two-dimensional set of initial conditions (ξ_o,ξ_1) generates two linearly independent $\underline{\xi}$'s. If $\underline{\xi}$ grows exponentially, then zero does not belong to the spectrum of $H(\underline{\phi})$. On the other hand, if $\underline{\xi}$ stays bounded, then zero is in the spectrum of $H(\underline{\phi})$ [16]. In particular, let $\underline{\phi}$ be an N-cycle, $\phi_j = \phi_{j+N}$. Then W_j is a periodic potential. If $\underline{\phi}$ corresponds to an elliptic fixed point of T_μ^N with eigenvalues $e^{\pm i2\pi k}$, then $H(\underline{\phi})$ has an eigenvector of the form

$$\xi_j = u(j)e^{i2\pi kj/N} \quad \text{with } u(j) = u(j+N) \quad . \tag{3.13}$$

Therefore zero lies in a band. Since the lowest band has its minimum at $k = 0$, $H(\underline{\phi})$ must have spectrum below zero provided $k \neq 0$. $\underline{\phi}$ is dynamically unstable. If $\underline{\phi}$ corresponds to a hyperbolic fixed point of T_μ^N, then zero is in a band gap. Without further information we cannot decide on the dynamical stability of $\underline{\phi}$. As for the other systems [12-15] the orbits of physical interest are unstable for T_μ and, hence, hard to find numerically.

Periodic orbits with small periods can be followed analytically. Linearizing, for $\mu = 0$, $T_o^N(x,y) = (x,y)$ at the origin yields a period N provided $\tau = \tau_N$ with

$$\cos(2\pi/N) = 1 + (\tau_N/2) \ . \tag{3.14}$$

At τ_N, (0,0) "ejects" a pair of an elliptic and a hyperbolic N-cycle. The hyperbolic N-cycle is always dynamically unstable. The elliptic N-cycle turns hyperbolic approximately upon crossing the spinodal. There the bottom of the spectrum of $H(\underline{\phi})$ moves through zero. The hyperbolic N-cycle is then dynamically stable and remains so all the way to $\tau = -4$. The N-cycle corresponds to a metastable density configuration with densities inside the phase coexistence curve. Of course, the elliptic N-cycles eject again cycles of higher period, ad infinitum. Therefore we should expect the existence of (dynamically stable) quasiperiodic density configurations. Presumably there are also (dynamically stable) aperiodic density configurations.

The kink solution is discussed in [10]. It consists of points of intersection of the unstable manifold of $-(\sqrt{-6\tau}, \sqrt{-6\tau})$ with the stable manifold of $(\sqrt{-6\tau}, \sqrt{-6\tau})$. Among the kink orbits there is exactly one which crosses $\phi = 0$ only once. This is the dynamically stable density configuration. All others are unstable.

For the dynamical system of interest, given by the map (2.7), the analogue of the fixed points $\pm(\sqrt{-6\tau}, \sqrt{-6\tau})$ are not known a priori. Therefore a direct determination of the stable and unstable manifold does not seem to be feasable. Another numerical approach is to approximate the kink solution through periodic orbits. We choose the initial condition $(-\kappa,\kappa)$. As κ increases the period of the orbit becomes longer up to a point κ^*, where the orbit diverges. Close to $\tau = 0$ the density of the orbit at κ^* gives a good approximation to the true phase coexistence curve, $\tau = -\frac{1}{6}\phi^2$. For smaller τ the appearance of the kink is masked by the chaotic layer close to the separatrix. In fact, our numerical procedure is essentially identical to the nonlinear transmission problem as studied by Delyon, Levy and Souillard [17]. In their context the transmission coefficient labels a line of initial points ($\hat{=}\ \kappa$), whereas the wave number is the parameter controlling the nonlinearity ($\hat{=}\ \tau$). Quasi-periodic orbits correspond to transmission and diverging orbits to no transmission. The "borderline of transmission" corresponds to the numerical approximation to the phase coexistence curve. Its fractal shape due to the disappearance of periodic orbits [17,18] shows that this approximation is reasonable only close to integrability, i.e. to $\tau = 0$.

4. The foliation

We return to the dynamical system given by the map (2.7). We would like to partition the three-dimensional phase space for the mapping T into two-dimensional manifolds corresponding to a constant chemical potential μ. On each of these submanifolds T should operate as a two-dimensional map similar to T_μ. Although we cannot explicitely construct such an isomorphism, we give some evidence that a foliation of phase space actually exists and that, moreover, it has a simple structure.

Let us fix some temperature $K < K_c$ and some $v_o \geq 0$ such that $v_o^2 < v_s(K)^2$. Consider then the one-parameter set of initial conditions

$$\vec{x}_o(\lambda) = (v_o - \lambda, v_o, v_o + \lambda) \quad , \tag{4.1}$$

where $\lambda \in [v_o-1, 1-v_o]$. (4.1) defines a line in phase space intersecting the diagonal at the fixed point $\vec{v}_o = (v_o,v_o,v_o)$, which is elliptic by construction. Thus for small values of λ, the orbits $\{T^n(\vec{x}_o(\lambda))\,|\,n = 0,1,\dots\}$ trace out ellipses in a plane transverse to the diagonal. The corresponding configurations oscillate around v_o and have a mean density $(1+v_o)/2$. Let $\tilde{\Sigma}(K,v_o)$ be the closure of the set of all trajectories $\{T^n(\vec{x}_o(\lambda))\}$. In general, we must expect $\tilde{\Sigma}(K,v_o)$ to have holes and to be rough. Numerically it appears that $\tilde{\Sigma}(K,v_o)$ can be embedded into a smooth surface $\Sigma(K,v_o)$. In Figure 3 we show

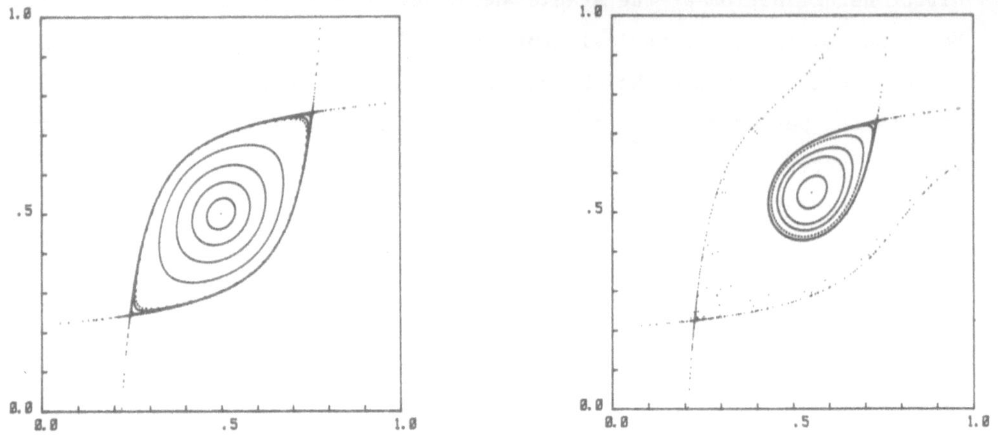

Figure 3. Projection of orbits generated by the set of initial conditions (4.1), with a) $K = 4.545$, $v_o = 0$, b) $K = 4.545$, $v_o = 0.1$.

some orbits of T with initial conditions (4.1) projected onto the x-y-plane. They show no self-intersection. This is a highly non-trivial property of T. Indeed, it indicates that there are manifolds $\Sigma(K,v_o)$ defined through $z = f_{K,v_o}(x,y)$ with single valued functions f_{K,v_o}. On $\Sigma(K,v_o)$ the phase portrait is indistinguishable from the one of T_μ, cf. [5] for a comparison.

Let us discuss the phase portrait in more detail, beginning with the case $v_o = 0$, cf. Figure 3a. In addition to the elliptic fixed point at the origin, two hyperbolic fixed points, $(v_\pm, v_\pm)$, together with their stable and unstable manifolds are clearly visible. As explained in Sec.3, the densities $(1+v_\pm)/2$ define the phase coexistence line of the lattice gas. Since K is close to K_c, the mapping is nearly integrable and the stochastic layer around the separatrix is narrow. The hyperbolic points can then be approached very closely by following quasiperiodic orbits around the origin. This technique was employed in [5] to determine the phase coexistence line. It was found to coincide with the results of a nonequilibrium generalization of the Maxwell construction [4,5].

To understand the case $v_o \neq 0$, cf. Figure 3b, we imagine a nonlinear coordinate transformation of the three-dimensional phase space such that the projection of the $\Sigma(K,v_o)$ at fixed K onto, say, the x'-z'-plane, gives a family of one-dimensional curves. Since $\Sigma(K,v_o=0)$ contains three fixed points, it intersects the diagonal $x' = y' = z'$ (which we assume to be left invariant by the transformation) transversally at three points. The resulting topology in the x'-z'- projection is shown in Fig.4. We shift now v_o away from zero.

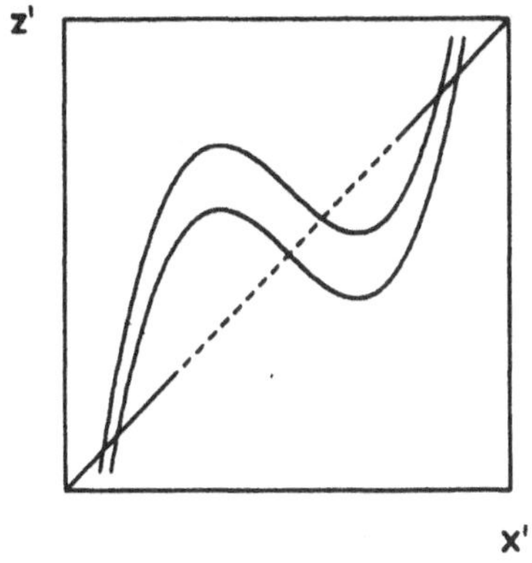

Figure 4. Topology of the surface $\Sigma(K,v_o)$ for $K < K_c$ and $v_o = 0$ (lower curve) and $v_o > 0$ (upper curve). The dashed part of the diagonal contains the elliptic fixed points inside the spinodal.

Since the $\Sigma(K,v_o)$ for different v_o cannot intersect, the whole surface is shifted upwards. Therefore one of the hyperbolic points moves away from the origin, while the other one comes closer to the elliptic point. This is the situation shown in Figure 3b. If we further increase v_o, then the elliptic and the innermost hyperbolic point further approach each other, until at $v_o = v_s(K)$ they coalesce and disappear, leaving only one hyperbolic fixed point. The resulting phase diagram is identical to that shown in Figure 2, with v_o playing the role of the chemical potential μ.

At least numerically, it seems to be possible to partition the three-dimensional phase space for T into manifolds of constant chemical potential. Except for some gentle curvature, these manifolds appear to be simply stacked one on top of the other. As further evidence the analytically determined N-cycles (up to N = 8) follow the pattern found for T_o. The pair of N-cycles, one elliptic and one hyperbolic, appears at the temperature defined through

$$K_N = 3 + 2\cos(2\pi/N) \quad . \tag{4.2}$$

As $N \to \infty$, K_N tends to K_c as $1/N^2$. The elliptic cycle turns hyperbolic upon crossing the spinodal and yields then a dynamically stable, periodic density configuration. We have finally identified stationary solutions of (1.1) which describe the long-lived periodic configurations of the lattice gas shown in Figure 1.

5. Conclusion

The stability of a stationary solution to (1.1), (1.2) yields information only about the dynamics in its close neighborhood. Metastability and "thermodynamic" stability are global concepts. They reflect the time it takes for a solution to escape from a stationary point under the influence of small noise. In principle, the Wentzell-Freidlin theory [19] can be used to study small stochastic perturbations. In practice, it is difficult to obtain explicitely the stationary density configurations and to determine the quasi-potential. For a small system of eight columns such kind of analysis has been carried through [5,18]. As main conclusion we found that the exponential of differences in energy for stationary density configurations provides a reasonable measure of its life-time. As expected physically, for average density 1/2, the kink solution has the longest life-time and in this sense is the most stable configuration.

If the current is proportional to the gradient of a chemical potential, as in (3.2), then the steady states of the lattice gas are Gibbs measures. They can be investigated by means of equilibrium statistical mechanics. A driven lattice gas does not satisfy detailed balance. At present the dynamical system approach seems to be the only method to study the low temperature behavior.

Acknowledgements. We thank M.Scheucher for plotting Figures 2 and 3, and J.Marro for his permission to publish the typical configuration of Figure 1 taken from his Monte-Carlo simulation.

References

[1] S. Katz, J.L. Lebowitz, and H. Spohn, J. Stat. Phys. 34, 497 (1984)

[2] J. Marro, J.L. Lebowitz, H. Spohn, and M.H. Kalos, J. Stat. Phys. 38, 725 (1985)

[3] J.L. Vallès and J. Marro, J. Stat. Phys. 43, 441 (1986)

[4] H. van Beijeren and L.S. Schulman, Phys. Rev. Lett. 53, 806 (1984)

[5] J. Krug, J.L. Lebowitz, H. Spohn, and M.Q. Zhang, J. Stat. Phys. 44, 535 (1986)

[6] K. Gawedzki and A. Kupiainen, Nucl. Phys. B269, 45 (1986)

[7] K.T. Leung and J.L. Cardy, J. Stat. Phys. 44, 567 (1986)

[8] H.K. Janssen and B. Schmittmann, Z.Phys. B64, 503 (1986)

[9] H. Spohn, N.Y. Academy of Sciences Vol. 491, 157 (1987)

[10] R. Pandit and M. Wortis, Phys. Rev. B25, 3226 (1982)

[11] T. Janssen and J.A. Tjon, Phys. Lett. 87A, 139 (1982)

[12] T. Janssen and J.A. Tjon, J. Phys. A16, 673 (1983)

[13] M.H. Jensen and P. Bak, Phys. Rev. B27, 6853 (1983)

[14] S. Aubry and P.Y. Le Daeron, Physica 8D, 381 (1983)

[15] P.I. Belobrov, V.V. Beloshapkin, G.M. Zaslavskii, and A.G. Tretyakov, Sov. Phys. JETP 60, 180 (1984)

[16] H.L. Cycon, R.G. Groese, W. Kirsch, and B. Simon, Schrödinger Operators (Springer, Berlin 1987)

[17] F. Delyon, Y.E. Lévy, and B. Souillard, Phys. Rev. Lett. 57, 2010 (1986)

[18] J. Krug, Diplomarbeit, Universität München 1985 (unpublished)

[19] M.I. Freidlin and A.D. Wentzell, Random Perturbations of Dynamical Systems (Springer, Berlin 1984)

COMPUTING BOUNDS ON CRITICAL INDICES

Hans Koch [1]
University of Texas at Austin, Department of Mathematics
Austin, TX 78712

Peter Wittwer [2]
Rutgers University, Department of Mathematics
New Brunswick, NJ 08903

1. INTRODUCTION

In this paper we give an outline of a computer assisted proof in which we use an extended version of the Wilson-Kadanoff renormalization group scheme to get rigorous bounds on a critical exponent that is universal for a class of one parameter families F_μ of hierarchical lattice systems in $d = 3$ dimensions. The parameter μ will be referred to as temperature. Assume that a given family F_μ undergoes a phase transition at $\mu = \mu_c$, then we define κ to be the exponent that describes the scaling behavior of the free energy as the temperature approaches the critical value μ_c. More precisely, if $U_n(\mu)$ is the free energy density coresponding to the system F_μ confined to a cube of volume 2^{dn} at temperature μ, then

$$\kappa = \lim_{\mu\to\mu_c} \frac{1}{\log(|\mu-\mu_c|)} \log\big(\lim_{n\to\infty} U_n(\mu)\big). \tag{1.1}$$

From a renormalization group analysis this quantity is hard to compute since it requires global bounds on the renormalization group transformation. A possible strategy is to show that the double limit in (1.1) can be replaced by a scaling limit along certain trajectories $n \mapsto \mu_n$, i.e. that

$$\kappa = \kappa_s \equiv \lim_{n\to\infty} \frac{1}{\log(|\mu_n-\mu_c|)} \log(U_n(\mu_n)). \tag{1.2}$$

The limit κ_s can be computed from local quantities only. We assume here the validity of this exchange of limits, i.e. we assume that $\kappa = \kappa_s$ for our systems, and concentrate on the task of getting sharp bounds for κ_s. For comparison with existing results we will use the more common index ν [5], assuming the correctness of the scaling relation

$$\nu = \kappa/d. \tag{1.3}$$

[1] Supported by the National Science Foundation under Grant No. DMR-85-40879.

[2] Supported by the National Science Foundation under Grants No. DMS-85-18622 and DMS-87-03539.

We obtain our results by an application of the method which has been developed in [2] to the class of hierarchical models which is discussed in [1]. We refer to [1] for the definition of the models and for motivations, and restrict ourselves here to explain the techniques which have been used to get the desired bounds.

Let C be the set of bounded continuous functions on $\mathbb{R}^+ = \mathbb{R} \setminus (-\infty, 0)$. In [1] we have studied for $f \in C$, $f(0) \neq 0$, the renormalization group map $\mathcal{N}_{\alpha\beta}$, which is defined through the equation

$$(\mathcal{N}_{\alpha\beta} f)(t^2) = \frac{1}{c} \int_{-\infty}^{\infty} ds \exp(-\beta^2 s^2)[f((\alpha t + s)^2)]^\ell, \tag{1.4}$$

and we have been interested in finding fixed points for this map. The normalization constant $c = c(f)$ in (1.4) is chosen such that $(\mathcal{N}_{\alpha\beta} f)(0) = 1$, i.e.

$$c = \int_{-\infty}^{\infty} ds \exp(-\beta^2 s^2)[f(s^2)]^\ell. \tag{1.5}$$

The definition of α, β and ℓ in (1.4) is as follows. If L is the side length of blocks in $\mathbb{Z}^d$ over which the spins are averaged in order to get the renormalization group map (1.4), then $\ell = L^d$ is the number of spins in a block and α is the rescaling of the spin variables associated with such a block spin transformation. For a field with canonical dimension one has $\alpha = L^{-(d-2)/2}$. We choose here $L = 2$, and therefore $\ell = 8$ and $\alpha = 1/\sqrt{2}$. β is just a normalization constant: if f is a fixed point of $\mathcal{N}_{\alpha\beta}$, then the function $f_\beta = f(\beta\, \cdot)$ is a fixed point of $\mathcal{N}_{\alpha 1}$. Without loss of generality we can therefore choose $\beta = 1$. 1)

It is easy to see that the two functions $f_{tr}(t^2) = 1$ and $f_{ht}(t^2) = \exp(-\frac{3}{8}t^2)$ are fixed points of $\mathcal{N}_{\alpha 1}$. f_{tr} is the trivial fixed point. It corresponds to the case of a noninteracting massless lattice field theory. f_{ht} is a massive high temperature fixed point. The fact that its mass is finite and not infinite is a particularity of the renormalization group map (1.4) and is not shared by other maps. The existence of such an additional massive gaussian fixed point is essential for our analysis: we use it to get an a priory bound on the large field behavior of the nontrivial fixed point of (1.4). Assume therefore that $f^* = \mathcal{N}_{\alpha 1} f^*$ and that $f^*(t)$ decays faster than $\exp(-\varepsilon t)$

1) The recursion relation (1.4) can also be studied for noninteger values of L and for fields with non-canonical dimension. A numerical study shows, that the critical index ν depends, as expected, on the dimension of the field. More careful studies show, that it also depends on the block size L, for fixed dimension of the field. This result seems to contradict the universality dogma at first sight. However, one can understand this effect in terms of the formulation which we have given in [1]. There, the recursion relation (1.4) is shown to be exact for spin systems on a cubic lattice with a L-dependent hierarchical symmetry, and models with different symmetries usually do have different critical behavior.

It is interesting to compare our results with those obtained in [3,4], where Wilsons hierarchical model — which differs from the one we study here — has been analyzed. In these references the value $L=2^{1/3}$ is used. We find for this case numerically that ν=0.6495704... which is in very good agreement with the values ν_S=0.649... and ν_E=0.6492... given in [3] and [4]. (As a comparison: the best numerical value for the case of the Ising model is ν_I=0.638....) Wilson studied originally his model for L=2 [6]. For this case we prove here rigorously that ν=0.6501625.... This value differs considerably from the number ν_W=0.609... given by Wilson [6] and by Golner [7]. We have no explanation for such a disagreement.

as $t \to +\infty$, for some $\varepsilon > 0$, then by iterating (1.4) one gets for every $\delta > 0$ the stability bound

$$C_1 \exp(-(\frac{3}{8}+\delta)\, t^2) \leq f^*(t^2) \leq C_2 \exp(-(\frac{3}{8}-\delta)\, t^2), \tag{1.6}$$

with $0 < C_1 \leq C_2 < \infty$.

In [1] we have already proved that there is a nontrivial fixed point $f^* \in C$, and we have shown that f^* extends to a function which is entire analytic. Here, we improve now our analysis and show that the fixed point f^* is hyperbolic in a appropriately chosen Banach space of analytic functions, with a splitting of the tangent space into a one dimensional unstable direction and a codimension one stable direction. As mentioned above, we apply the method which has been developed in [2] to prove the theorems of this paper. The idea is to use an extended renormalization group transformation, which has a whole piece of the one dimensional unstable manifold of f^* as its fixed point. One aspect of this particular way of doing the analysis is — and this might not have been stressed enough in [2] — that very precise bounds for the critical exponent which is associated with this unstable direction can be obtained.

By comparing the methods in this paper with those of [1] the reader will find that we have been improving considerably our method of dealing with the large field problem. The method is formulated now in such a way that a generalization to the physically interesting case of translation invariant models seems feasible.

Before we define the extended renormalization group map we recall some more details about the operator $\mathcal{N}_{\alpha 1}$. An important technical step is to choose coordinates in function space which take advantage of the a priory bound (1.6). Define C^+ to be the space of continuous functions h of the form:

$$h(t^2) = (\mathcal{H}f)(t^2) \equiv \exp(\frac{3}{8}t^2) f(t^2) \tag{1.7}$$

for $f \in C$. Clearly $\mathcal{H} : C \to C^+$ is one–to–one. $\mathcal{N}_{\alpha 1}$ induces on C^+ the map $\hat{\mathcal{N}} = \mathcal{H}\mathcal{N}_{\alpha 1}\mathcal{H}^{-1}$. An easy computation shows that, for $h \in C^+$, $h(0) \neq 0$, $\hat{\mathcal{N}}$ is given by the equation

$$(\hat{\mathcal{N}}h)(t) = \frac{1}{c}\int_{-\infty}^{\infty} ds \exp(-L^2 s^2)[h((\frac{\alpha}{L^2}t + s)^2)]^{\ell}. \tag{1.8}$$

where $c = c(h)$ is chosen such that $(\hat{\mathcal{N}}h)(0) = 1$. A comparision with (1.4) shows that $\hat{\mathcal{N}} = \mathcal{N}_{\alpha/L^2 L}$, i.e. (1.4) is form invariant under the change of coordinates (1.7). In particular the replacement of $\beta = 1$ by $\beta = L$ is just a change of a normalization as we have shown above, and is therefore irrelevant. The change from α in $\mathcal{N}_{\alpha 1}$ to α/L^2 in $\hat{\mathcal{N}} = \mathcal{N}_{\alpha/L^2 L}$ is however crucial. The reason is, that the maps $\mathcal{N}_{\theta\beta}$ behave in a certain sense pretty much like the linear substitution operator $L_\theta : \zeta(t) \mapsto \zeta(\theta t)$. (This can be seen by replacing in (1.4) and (1.8) the convolution with the exponential function by a convolution with a delta function.) On functions ζ that are analytic near zero L_θ is analyticity improving for $\theta < 1$. For θ small enough this property carries over to the linearization of $\mathcal{N}_{\theta\beta}$, and makes an application of the contraction mapping principle possible. We will use that method below to prove our main theorem.

An other advantage of the new coordinates is, that the stability bound (1.6) becomes more transparent. If $h^* = \mathcal{H}f^*$ is the nontrivial fixed point of $\hat{\mathcal{N}}$, then for every $\delta > 0$ there is a constant $C_0 > 0$ such that for all $t \in \mathbb{R}$,

$$h^*(t^2) \leq C_0 \exp(\delta\, t^2). \tag{1.9}$$

This property has been guiding us in our choice of function space (see below).

2. RESULTS

The extended renormalization group map $\mathcal{R}$ acts on functions Φ that are jointly analytic in a spin variable t and in the parameter μ that characterizes the temperature of the system. The choice of $\mathcal{R}$ is dictated by the following considerations. Let $h^* = \hat{\mathcal{N}} h^*$ be the nontrivial fixed point of $\hat{\mathcal{N}} = \mathcal{N}_{\alpha/L^2 L}$ discussed above, and let h^*_μ be a parametrization of the unstable manifold of h^*, such that the action of $\hat{\mathcal{N}}$ becomes linear, i.e. $\hat{\mathcal{N}} h^*_\mu = h^*_{\delta^* \mu}$, where $\delta^* = 2.904071\ldots$ is the eigenvalue associated with the unstable direction of h^*. It is then easy to write down an operator $\mathcal{R}_0$ which has the whole one parameter family h^*_μ as its fixed point. Namely, one could define $\mathcal{R}_0 h^*_\mu = \hat{\mathcal{N}} h^*_{\mu/\delta^*}$. The situation is slightly more complicated if the unstable manifold h^*_μ is known, but not the fact that $h^*_{\mu^*}$ is a fixed point for $\mu^* = 0$. The action of $\hat{\mathcal{N}}$ would then be $\hat{\mathcal{N}} h^*_\mu = h^*_{\mu^* + \delta^*(\mu-\mu^*)}$, with μ^* do be determined. This motivates to define the operator $\mathcal{R}$ by the equation

$$(\mathcal{R}\Phi)(t,\mu) = \frac{1}{c(\mu/\delta)} \int_{-\infty}^{\infty} ds \exp(-L^2 s^2)\, [\Phi((\frac{\alpha}{L^2} t + s)^2, \mu_0 + \mu/\delta)]^\ell. \tag{2.1}$$

Here $c(\mu) = c(\mu, \Phi)$, $\delta = \delta(\Phi)$ and $\mu_0 = \mu_0(\Phi)$ are to be determined from normalization conditions which will be given below. For convenience later on we decompose now $\mathcal{R}$ into a product of simpler operators. We define for these matters for a given function Ψ, $\Psi(t,\mu) = \sum_{ij \geq 0} \Psi_{ij} t^i \mu^j$, the real function m by the equation

$$m(\Psi) = \frac{1}{2}(\Psi_{10} + \Psi_{20}), \tag{2.2}$$

and for $\mu_0 \in \mathbb{R}$, μ_0 small, the operators S_{μ_0} by the equations

$$(S_{\mu_0}\Psi)(t,\mu) = \Psi(t, \mu_0 + \mu), \tag{2.3}$$

and finally the extension $\mathcal{N}$ of the operator $\hat{\mathcal{N}}$ by the equation

$$(\mathcal{N}\Psi)(t,\mu) = \frac{1}{c(\mu)} \int_{-\infty}^{\infty} ds \exp(-L^2 s^2)\, [\Psi((\frac{\alpha}{L^2} t + s)^2, \mu)]^\ell. \tag{2.4}$$

Here,

$$c(\mu) = c(\mu, \Psi) = \int_{-\infty}^{\infty} ds \exp(-L^2 s^2)\, [\Psi(s^2, \mu)]^\ell, \tag{2.5}$$

so that $(\mathcal{N}\Psi)(0,\mu) = 1$.

$\mathcal{R}$ is now given by the following sequence of steps.
a) Define μ_0^* as the solution of the equation

$$m(\mathcal{N} S_{\mu_0^*} \Phi) = m(S_{\mu_0^*} \Phi). \tag{2.6}$$

b) Write S^* for $S_{\mu_0^*(.)}$ and $\mathcal{N}^*$ for $\mathcal{N} S^*$. (2.7)
c) Define, for $\Psi = \mathcal{N}^* \Phi$, δ by the equation $\delta = \Psi_{11}$. (2.8)

d) Define $(\mathcal{R}\Phi)(t,\mu) = (\mathcal{N}^*\Phi)(t,\mu/\delta)$ (2.9)

Note, that with the definition of δ given in (2.8), and with the normalization (2.5) for $\mathcal{N}$, we get for any Φ on which $\mathcal{R}$ is defined, the following normalizations for $\Psi = \mathcal{R}\Phi$.

$$\Psi(0,\mu) = 1, \tag{2.10}$$

$$(\partial_1\partial_2\Psi)(0,0) = 1. \tag{2.11}$$

For completeness we also specify here the action of the tangent map $D\mathcal{R}_\Phi$ of $\mathcal{R}$ evaluated at Φ and acting on $\delta\Phi$. ($^\bullet$ denotes the partial derivative with respect to the second argument.)

$$\begin{aligned}(D\mathcal{R}_\Phi\delta\Phi)(t,\mu) &= (D\mathcal{N}^*_\Phi\delta\Phi)(t,\frac{\mu}{\delta}) + (\mathcal{N}^*\Phi)^\bullet(t,\frac{\mu}{\delta})\frac{-\mu}{\delta^2}\delta\delta \\ &= (D\mathcal{N}^*_\Phi\delta\Phi)(t,\frac{\mu}{\delta}) + (\mathcal{R}\Phi)^\bullet(t,\mu)\frac{-\mu}{\delta}\delta\delta,\end{aligned} \tag{2.12}$$

$$(D\mathcal{N}^*_\Phi\delta\Phi)(t,\mu) = (D\mathcal{N}_{S^*\Phi}DS^*_\Phi\delta\Phi)(t,\mu), \tag{2.13}$$

$$\begin{aligned}(DS^*_\Phi\delta\Phi)(t,\mu) &= (S^*\delta\Phi)(t,\mu) + ((\partial_{\mu_0}S_{\mu_0})\big|_{\mu_0=\mu_0^*(\Phi)}\Phi)(t,\mu)\delta\mu_0^* \\ &= (S^*\delta\Phi)(t,\mu) + (S^*\Phi^\bullet)(t,\mu)\delta\mu_0^*,\end{aligned} \tag{2.14}$$

$$\begin{aligned}(D\mathcal{N}_\Phi\delta\Phi)(t,\mu) =& \frac{\ell}{c(\mu)}\int_{-\infty}^{\infty}\exp(-L^2s^2)[\Phi((\frac{\alpha}{L^2}t+s)^2,\mu)]^{\ell-1}\delta\Phi((\frac{\alpha}{L^2}t+s)^2,\mu)\,ds \\ &- \frac{\delta c(\mu)}{c(\mu)}(\mathcal{N}\Phi)(t,\mu).\end{aligned} \tag{2.15}$$

Here, $\delta c = \delta c(\mu,\delta\Phi)$ and $\delta\delta = \delta\delta(\delta\Phi)$ are defined such that the normalizations (2.10) and (2.11) remain unchanged by the variation $\delta\Phi$. It follows that $\varphi = D\mathcal{R}_\Phi\delta\Phi$ is normalized according to

$$\varphi(0,\mu) = 0, \tag{2.16}$$

$$(\partial_1\partial_2\varphi)(0,0) = 0. \tag{2.17}$$

Furthermore, in (2.14), $\delta\mu_0^* = \delta\mu_0^*(\delta\Phi)$ is given by the equation

$$\delta\mu_0^* = -\frac{m(D\mathcal{N}_{S^*\Phi}S^*\delta\Phi - S^*\delta\Phi)}{m(D\mathcal{N}_{S^*\Phi}S^*\,\Phi^\bullet - S^*\,\Phi^\bullet)}. \tag{2.18}$$

In order to complete the definition of the operator $\mathcal{R}$ we define now the function spaces with which we will work below.

Definition 2.1. *Given two positive numbers ρ_1 and ρ_2, we define $\mathcal{A}_{\rho_1\rho_2}$ to be the Banach space of all functions Φ, $\Phi(t,\mu) = \sum_{ij\geq 0}\Phi_{ij}t^i\mu^j$ which are analytic on the domain $D_{\rho_2}\subset\mathbb{C}^2$,*

$$D_{\rho_2} = \mathbb{C}\times\{|\mu| < \rho_2\}, \tag{2.19}$$

which take real values for real arguments t and μ, and for which the norm

$$\|\Phi\|_{\rho_1\rho_2} = \sum_{ij\geq 0}|\Phi_{ij}|\cdot(i!\cdot\rho_1^i)\cdot\rho_2^j \tag{2.20}$$

is finite. Furthermore, we define $\mathcal{A} = \mathcal{A}_{\rho_t \rho_\mu}$*, where* $\rho_t = 3$ *and* $\rho_\mu = 0.01$.

Let $\Phi \in \mathcal{A}_{\rho_1 \rho_2}$, then we can write $\Phi(t, \mu) = \sum_{j \geq 0} h_j(t) \mu^j$, where the functions h_j are elements of the spaces $\mathcal{E}_{\rho_1}$ which we now define.

Definition 2.2. *For each* $\rho > 0$*, we define* $\mathcal{E}_\rho$ *to be the Banach space of all functions* h, $h(t) = \sum_{i \geq 0} h_i t^i$ *which are entire analytic, which take real values for real arguments* t*, and for which the norm*

$$\|h\|_\rho^{\mathcal{E}} = \sum_{i \geq 0} |h_i| \cdot (i! \cdot \rho^i) \tag{2.21}$$

is finite. Furthermore, we define $\mathcal{E} = \mathcal{E}_{\rho_t}$.

To understand the choice of the norms (2.20) and (2.21) consider a function $h(t) = \sum_{i \geq 0} h_i t^i$, for which $\|h\|_\rho^{\mathcal{E}}$ is finite. Such a function is entire analytic, and it satisfies, for $t \in \mathbb{C}$ the inequality

$$|h(t)| \leq \sum_{i \geq 0} |h_i| \cdot |t|^i \leq \sum_{i \geq 0} (|h_i| \cdot i! \cdot \rho^i) \frac{(|t|/\rho)^i}{i!} \leq \|h\|_\rho^{\mathcal{E}} \cdot \exp(|t|/\rho). \tag{2.22}$$

This shows that the growth rate of functions for which the norm (2.21) is finite is bounded by $1/\rho$. In our definition of $\mathcal{A}$ we have chosen the inverse growth rate ρ_t such, that for a function $\Phi \in \mathcal{A}$, and fixed μ, the function $f = \mathcal{H}^{-1}\Phi(\,.\,, \mu)$ is exponentially decreasing along the positive real axis, and is therefore within the class of functions for which we have the stability bound (1.6).

The following theorem is our main result.

Theorem 2.3. *There is a ball* $B \subset \mathcal{A}$ *on which* $\mathcal{R}$ *is defined and once continuously differentiable. For each* $\Phi \in B$*, the Fréchet derivative* $D\mathcal{R}_\Phi$ *of* $\mathcal{R}$ *at* Φ *is a compact operator on* $\mathcal{A}$*, whose spectrum is contained inside the unit disk.* $\mathcal{R}$ *maps* B *into itself and has a fixed point* $\Phi^* \in B$*. The number* $\delta^* = \delta(\Phi^*)$ *satisfies the bound* $\delta^* \in [2.9040714905322, 2.9040715072204]$.

The proof of this theorem is outlined in the next section.

Consider now the one parameter family of maps $h_\mu^* = \Phi^*(\,.\,, \mu)$. For $-\rho_\mu < \mu < \rho_\mu$ we have that $h_\mu^* \in \mathcal{E}$. We have the following corollary.

Corollary 2.4. *There is a ball* $\hat{B} \subset \mathcal{E}$ *on which the operator* $\hat{\mathcal{N}}$ *is defined and once continuously differentiable. For* $|\mu| < \rho_\mu/\delta^*$ *the action of* $\hat{\mathcal{N}}$ *is given by* $\hat{\mathcal{N}} h_\mu^* = h_{\delta^* \mu}^*$. $h^* = h_0^*$ *is a fixed point of* $\hat{\mathcal{N}}$*. The Fréchet derivative* $D\hat{\mathcal{N}}_{h^*}$*, of* $\hat{\mathcal{N}}$ *at* h^* *is a compact operator on* $\mathcal{E}$*. Its spectrum is contained inside the unit disk with the exception of the simple eigenvalue* δ^**.* h^* *is hyperbolic with a splitting of its tangent space into a one dimensional expanding and a codimension one contracting direction. The function* $\delta h^* = \partial_\mu \Phi^*(\,.\,, 0)$ *is an eigenvector of* $D\hat{\mathcal{N}}_{h^*}$ *with eigenvalue* δ^*.

Proof. All the statements of this corollary are immediate consequences of the definition of the operator $\mathcal{R}$ and of Theorem 2.3 with the exception of the fact, that h_0^* is a fixed point of $\hat{\mathcal{N}}$, i.e. that μ_0^*, as defined through (2.6), is zero at the fixed point Φ^*. To see this we have to discuss briefly how (2.6) is solved. We define a function $\Delta : \mathbb{R} \to \mathbb{R}$

$$\Delta(\mu_0) = m(\mathcal{N}S_{\mu_0}\Phi) - m(S_{\mu_0}\Phi), \tag{2.23}$$

and we prove that this function is zero for some value of its argument. This proof is given by using Newtons method: we show that the map $F : \mathbb{R} \to \mathbb{R}$

$$F(\mu_0) = \mu_0 - \frac{\Delta(\mu_0)}{\Delta'(\mu_0)} \tag{2.24}$$

is a contraction of a small interval into itself, which proves that F has a unique fixed point there. Equivalently we conclude, that the function Δ has a unique zero in this interval, or equivalently that (2.6) has a unique solution in this interval. In our computer proof this interval contains $\mu_0 = 0$. Now, from (2.6) and the definition (2.2) of the function $m(.)$ it follows, that

$$m(\mathcal{R}\Phi) = m(S^*\Phi), \tag{2.25}$$

for any function Φ on which $\mathcal{R}$ is defined. In particular at the fixed point $\Phi^* = \mathcal{R}(\Phi^*)$ of $\mathcal{R}$ we get that $m(\Phi^*) = m(S^*\Phi^*)$. This equation is satisfied for $\mu_0^* = 0$. Therefore, since the solution of (2.6) is unique, we have that $\mu_0^*(\Phi^*) = 0$.

Corollary 2.5. *Each of the one parameter families $h_\mu = \Phi(\,.\,,\mu)$ in the domain of $\mathcal{R}$ intersects the stable manifold of h^* transversally for some critical value μ_c of the temperature. If ν is the index defined in (1.3), evaluated in the scaling limit $\mu_n = \mu_c + \mu_c + k(\delta^*)^{-n}$, then $\nu = \log(L)/\log(\delta^*) \in [0.65016251767896, 0.65016252118341]$, for any nonzero constant k.*

Proof. For the particular one parameter family of maps $h_\mu^* = \Phi^*(\,.\,,\mu)$, for which $\mu_c = \mu_0^*(\Phi^*) = 0$, $\nu = \log(L)/\log(\delta^*)$ is obtained by using the relations

$$Z_n(\mu) = Z_{n-1}(\mu\delta^*), \tag{2.26}$$

satisfied by the partition functions Z_n associated with cubes of volume $2^{d(n+1)}$. The equation (2.26) has been proved in [1] for $\mu = 0$, but the same argument applies for arbitrary μ and n sufficiently large. In the general case one uses an equation similar to (2.26). However if Z_n is a partition function associated with a family h_μ then Z_{n-1} is associated with the family $\tilde{h}_\mu = \mathcal{R}(\Phi(\,.\,,\delta\mu))$. (See (2.8) for the definition of $\delta = \delta(\Phi)$.) The convergence of the sequence $Z_n(\mu_n)$ follows from the fact that $\mathcal{R}$ is a contraction with uniformly bounded derivative. The details of this computation are left to the reader.

3. METHOD OF PROOF

In this section we give an outline of the proof of Theorem 2.3. Details of the proof, including the listing of the computer program will appear elsewhere [8]. First, because a function Ψ which is in the range of $\mathcal{R}$ is normalized according to (2.10) and (2.11), we can restrict our search for a fixed point to such functions. A normalized function Ψ is of the form

$$\Psi(t,\mu) = 1 + t\mu + \varphi(t,\mu) \tag{3.1}$$

with φ satisfying (2.16) and (2.17).

Definition 3.1. *We define $\mathcal{A}_H$ and $\mathcal{A}_T$ to be the sets of functions $\Phi \in \mathcal{A}$ and $\varphi \in \mathcal{A}$ which satisfy the normalization conditions (2.10), (2.11) and (2.16), (2.17), respectively.*

If we want to prove Theorem 2.3 by an application of the contraction mapping principle, then we have to find a ball $\mathcal{B}_H \subset \mathcal{A}_H$ on which $\mathcal{R}$ is defined and differentiable. Then we have to study for each $\Phi \in \mathcal{B}_H$ the linearization $D\mathcal{R}_\Phi$ as a map from $\mathcal{A}_T$ to $\mathcal{A}_T$. The condition for $\mathcal{R}$ to be a contraction is, that the operator norm $\|D\mathcal{R}_\Phi\|$ is smaller than one, uniformly for Φ in such a ball $\mathcal{B}_H$. Numerical studies show, however, that this is not the case. We cure this problem brute force by a change of coordinates to a basis in which $D\mathcal{R}_\Phi$ is essentially diagonal.

Definition 3.2. *Let I_1 and I_2 be the two sets*

$$I_1 = \{(1,0),(2,0),\ldots,(9,0)\}, \tag{3.2}$$

and

$$I_2 = \{(i,j) \in \mathbb{Z}^2 \,|\, i > 0, (i,j) \neq (1,1), (i,j) \notin I_1\}, \tag{3.3}$$

then we define Λ to be the Banach space of all real sequences (λ_{ij}) indexed by $I_1 \cup I_2$,

$$\Lambda = \{(\lambda_{ij}) | (i,j) \in I_1 \cup I_2\}, \tag{3.4}$$

for which with the norm

$$\|(\lambda_{ij})\|_1 = \sum_{I_1 \cup I_2} |\lambda_{ij}| \tag{3.5}$$

is finite.

To each element $(\lambda_{ij}) \in \Lambda$ we can now associate a function $\varphi \in \mathcal{A}_T$, by the equation

$$\varphi(t,\mu) = \sum_{k=1\ldots9} \lambda_{k0} \cdot \mathbf{e}_k(t) + \sum_{I_2} \lambda_{ij} \cdot \frac{t^i}{\rho_t^i \cdot i!} \cdot \frac{\mu^j}{\rho_\mu^j}, \tag{3.6}$$

where the functions $\mathbf{e}_k$ are given by the equation

$$\mathbf{e}_k(t) = \sum_{i=1\ldots9} \sigma_{ik} \cdot \frac{t^i}{\rho_t^i \cdot i!}. \tag{3.7}$$

Here, $\sigma = (\sigma_{ij})$ is a nonsingular 9×9–matrix with inverse σ^{-1}. (See [8] for the choice of σ.) Conversely if $\varphi \in \mathcal{A}_T$, $\varphi = \sum_{I_1 \cup I_2} \varphi_{ij} t^i \mu^j$, then there is an element $(\lambda_{ij}) \in \Lambda$ associated with it by the equations

$$\lambda_{i0} = \sum_{j=1\ldots9} \sigma_{ij}^{-1} \cdot \varphi_{j0} \cdot (\rho_t^j \cdot j!), \qquad i = 1\ldots9, \tag{3.8}$$

$$\lambda_{ij} = \varphi_{ij} \cdot (\rho_t^i \cdot i!) \cdot \rho_\mu^j, \qquad (i,j) \in I_2. \tag{3.9}$$

These relations induce a new norm on $\mathcal{A}_T$. Namely, let $\varphi \in \mathcal{A}_T$ and let (λ_{ij}) be defined through (3.8) and (3.9), then we define (misusing the notation $\|\ \|_1$ slightly)

$$\|\varphi\|_1 = \|(\lambda_{ij})\|_1. \tag{3.10}$$

We now have the following theorem.

Theorem 3.3. *There is a function $\Phi_0 \in \mathcal{A}_H$, such that*

a) $\mathcal{R}$ is defined and continuously differentiable on the neighborhood $\mathcal{B}_\beta$,
$\mathcal{B}_\beta = \{\Phi \in \mathcal{A}_H |\ \|\Phi - \Phi_0\|_1 < \beta\}$, where $\beta = 5 \cdot 10^{-9}$.

b) For $\Phi \in \mathcal{B}_\beta$, the tangent map $D\mathcal{R}_\Phi$ is bounded in norm by $\|D\mathcal{R}_\Phi\|_1 \leq \Theta \leq 0.86 < 1$,

c) Φ_0 is a good approximate fixed point of $\mathcal{R}$, i.e. $\|\mathcal{R}\Phi_0 - \Phi_0\|_1 \leq \epsilon \leq 2 \cdot 10^{-11} < (1-\Theta)\beta$.

Proof. The proof of this Theorem is obtained by running our computer program. For details on the method see [1,2,9,10]. For details on the proof see [8].

As a result of a),...,c) there is a unique fixed point Φ^* of $\mathcal{R}$ in $\mathcal{B}_\beta$. Our computer program also provides the bound on δ^* in Theorem 2.3.

ACKNOWLEDGEMENTS

One of the authors (P.W.) would like to acknowledge NSF support under Grant No. DMS–85–18622 for the use of the Pittsburgh supercomputer center, where part of the numerical work as well as runs of the complete proof have been carried out on a CRAY–XMP.

REFERENCES

[1] H. Koch, P. Wittwer. A Non–Gaussian Renormalization Group Fixed Point for Hierarchical Scalar Lattice Field Theories. Comm. Math. Phys. 106, 495 (1986).

[2] J.–P. Eckmann, P. Wittwer. A Complete Proof of the Feigenbaum Conjectures. To appear in Journal of Statistical Physics.

[3] P.M. Bleher, Ja.G. Sinai. Critical indices for Dyson's asymptotically hierarchical models. Comm. Math. Phys. 45, 347 (1975).

[4] P. Collet, J.–P. Eckmann, B. Hirsbrunner. A numerical test of Borel summability in the ϵ–expansion of the hierarchical model. Physics letters 71B, 385 (1977).

[5] K.G. Wilson, J.B. Kogut. The renormalization group and the ϵ expansion. Phys. Report. 12C, 75 (1974).

[6] K. Wilson. Renormalization group and critical phenomena, II phase space cell analysis of critical behavior. Phys. Rev. B4, 3184 (1971).

[7] J. Golner. Calculation of the Critical Exponent η via Renormalization–Group Recursion Formulas.

[8] H. Koch, P. Wittwer. In preparation.

[9] J.–P. Eckmann, H. Koch, P. Wittwer. A computer–assisted proof of universality for area–preserving maps. Memoirs AMS 47, 289 (1984).

[10] J.–P. Eckmann, P. Wittwer. Computer methods and Borel summability applied to Feigenbaum's equation. Lecture Notes in Physics, Springer–Verlag, Berlin, Heidelberg, New York, Tokyo (1985).

A THERMODYNAMIC ACTIVE SYSTEM IN REGULAR AND CHAOTIC MOTION

R. Gallimbeni and L. Sertorio

Dipartimento di Fisica Teorica
Università di Torino

M. Miari

Dipartimento di Matematica
Università di Milano

A formal definition of thermodynamic active system is fairly new and therefore requires a few introductory remarks. Historically Carnot[1] was the first to evaluate the maximum ideal work that can be extracted from two heat reservoirs at different temperatures, that is in mutual disequilibrium. Later Gibbs[2] introduces explicitly the concept of maximum obtainable work as the measure, expressed in Joules, of the distance from equilibrium of a system composed of two parts one at T, P, the other at T_0, P_0.

Such ideal work, availability in modern language, is dissipated until equilibrium is reached.

Tolman[3] develops such concept further, he explicitly imposes that the availability is extracted rather than dissipated, writes a generalized efficiency equation, a particular case of which is the law for the thermoelectric effect. Landau[4] in his text book very clearly exploits the rationship between availability dissipated and entropy produced, always in the formalism of equilibrium thermodynamics, to discuss the direction of evolution from non equilibrium initial states. After 1973, the energy crisis, the doctrine of energetics[5] begins, inspired mainly by the paper of Tolman. The literature is mostly devoted to the study of a variety of cases often described by complicated mathematical models[6]. On the other hand the formulation of non equilibrium thermodynamics reached mathematical maturity in the fifties with the powerful formalism of irreversible field theory[7][8].

The body of research in energetics needed to be located in the general theory of non equilibrium thermodynamics. This logic link was badly missing not only for the above reason but also from the viewpoint of a critical modern theoretical foundation embodying the concept of work extracted already present in the first conception of Gibbs but somehow lost on the way to field theory.

This was obtained in some recent work(9).

For a homogeneous continuous system the ideal availability w is a density of mechanical power (unit: watt/meter^3) and is represented by these differential operators

$$w = w^c + w^f = -k\,\frac{(\nabla T)^2}{T} - \vec{v}\nabla p \qquad (1)$$

T = temperature, p = pressure, $\vec{v}$ = fluid velocity,

w^c = carnot availability

w^f = fluid availability

According to the definition (1) it is found that the fundamental equations for an active system (availability extracted) are

$$\begin{aligned} \dot{\rho} &= -\,\mathrm{div}\rho\vec{v} \\ (\dot{\rho\vec{v}}) &= -\,\mathrm{div}\,\Pi + \underline{\nabla p} \qquad (2) \\ (\dot{\rho e}) &= -\,\mathrm{div}\vec{e}\ \underline{-\ w} \end{aligned}$$

(the notations follow closely Landau(8)).

Erasing the underlined terms in (2) we obtain the known purely dissipative equations as they can be found in Ref. 7, 8.

The problem we are facing here is the study of the full system (2) in the same geometry and in the presence of the external gravity field as in the Bénard problem, so we study the "active Bénard system".

Concerning the physical meaning of this study we remind that the dissipative system even in the drastic truncation of Lorenz(10) has been considered, with some imagination, as a theoretical laboratory for the understanding of the atmospheric motions; with the same amount of imagination the active Bénard system can be considered for example as representing a volcano during activity (a magma fluid layer with mechanical power ejected) or an ecological model of the earth's surface ideally exploited to extract the mechanical power contained in its disequilibrium ultimately powered by the solar radiation.

We follow exactly the same approximations used by Saltzman(11) and Lorenz, essentially the Oberbeck-Boussinesq simplification and the double Fourier series expansions for vorticity and temperature fields using also their notations. Notice that the simplification of the vorticity expansion is incompatible with the term w^f, namely the velocity dependent part of the availability. So either we keep w^f and use more complicated formulae or we drop w^f; we drop it in favor of a possible direct comparison in the following with the Lorenz system. This and other explicit justifications of several statements can be found in a forthcoming paper. The Galerkin expansion which is obtained is the following

$$\dot{\psi}_1(m,n) = \frac{1}{\pi^2(m^2\ell^2+n^2)} \Big(\sum_{p,q} \pi^2(p^2\ell^2+q^2)[\psi_1(p,q)\psi_1(m-p,n-q) - \psi_2(p,q)\psi_2(m-p,n-q)]\pi^2\ell(mq-np) - \sigma\pi^4(m^2\ell^2+n^2)^2\psi_1(m,n) + \sigma m\pi\ell\theta_2(m,n)\Big)$$

$$\dot{\psi}_2(m,n) = \frac{1}{\pi^2(m^2\ell^2+n^2)} \Big(\sum_{p,q} \pi^2(p^2\ell^2+q^2)[\psi_1(p,q)\psi_2(m-p,n-q) + \psi_2(p,q)\psi_1(m-p,n-q)]\pi^2\ell(mq-np) - \sigma\pi^4(m^2\ell^2+n^2)^2\psi_2(m,n) - \sigma m\pi\ell\theta_1(m,n)\Big)$$

$$\dot{\theta}_1(m,n) = -\sum_{p,q} [\psi_1(p,q)\theta_1(m-p,n-q) - \psi_2(p,q)\theta_2(m-p,n-q)]\pi^2\ell(mq-np) - rR_o\pi m\ell\psi_2(m,n) - \pi^2(m^2\ell^2+n^2)\theta_1(m,n) + \Big\{-2\frac{r}{r+1}\pi n\theta_2(m,n) + \frac{\pi^2}{R_o}\frac{1}{r+1}\sum_{p,q}[\theta_1(p,q)\theta_1(m-p,n-q) - \theta_2(p,q)\theta_2(m-p,n-q)][q(n-q)+\ell^2 p(m-p)]\Big\}$$

$$\dot{\theta}_2(m,n) = -\sum_{p,q} [\psi_1(p,q)\theta_2(m-p,n-q) + \psi_2(p,q)\theta_1(m-p,n-q)]\pi^2\ell(mq-np) + rR_o\pi m l\psi_1(m,n) - \pi^2(m^2\ell^2+n^2)\theta_2(m,n) + \Big\{2\frac{r}{r+1}\pi n\theta_1(m,n) + \frac{\pi^2}{R_o}\frac{1}{r+1}\sum_{p,q}[\theta_1(p,q)\theta_2(m-p,n-q) + \theta_2(p,q)\theta_1(m-p,n-q)][q(n-q)+\ell^2 p(m-p)]\Big\}$$

with

$$\ell = \frac{1}{3\sqrt{2}}\ ;\ R_o = \frac{27\pi^4}{4}\ ;$$

and

rR_o the Rayleigh number

σ the Prandtl number

The terms in curly brackets express w^c. In their absence the system coincides with Ref. 11 formulae (32,35). We have to face the truncation. We remember first the truncation of Lorenz. His choice of modes is

$$X = \frac{4}{3}\psi_1(3,1)$$
$$Y = \frac{8\sqrt{2}}{27\pi^3}\theta_2(3,1)$$
$$Z = \frac{8}{27\pi^3}\theta_2(0,2)$$
$$\text{with } t \to \frac{3\pi^2}{2} t\ ; \tag{4}$$

according to (4) system (3) becomes the well known system

$$\dot{X} = -\sigma X + \sigma Y$$
$$\dot{Y} = rX - Y - XZ$$
$$\dot{Z} = -bZ + XY \tag{5}$$

For the full active system (3) the choice of modes is instead the following

$$X = \frac{4}{3}\psi_1(3,1) \qquad Y = \frac{8\sqrt{2}}{27\pi^3}\theta_2(3,1)$$
$$Z = \frac{8}{27\pi^3}\theta_2(0,2) \qquad W = \frac{8\sqrt{2}}{27\pi^3}\theta_1(3,1)\ . \tag{6}$$

The dynamical system representing the original full active system (3) is the new set of equations:

$$\dot{X} = -\sigma X + \sigma Y$$
$$\dot{Y} = rX - Y - XZ + \epsilon\frac{4}{3\pi}\frac{1}{r+1}W(r+Z)$$
$$\dot{Z} = -bZ + XY$$
$$\dot{W} = -W - \epsilon\frac{4}{3\pi}\frac{1}{r+1}Y(r+Z)\ , \qquad \text{with } b = \frac{8}{3}\ . \tag{7}$$

The present choice of modes is the simplest giving rise to (nonlinear) active terms in the truncation. The parameter ϵ is introduced to weight the availability extracted and varies obviously from 0 to 1 in the physical range.

The purpose of our work is the study of the availability

$$\overline{w^c} = -\frac{1}{T_o}\int_D (\nabla T)^2 dV =$$

$$= -6\sqrt{2}\frac{r^2}{r+1}\left[1+\frac{1}{r^2}\left(\frac{3}{4}Y^2+\frac{3}{4}W^2+4Z^2\right)\right] \tag{8}$$

where D is the domain corresponding to one roll, as a function of time and the study of the structural properties of the flow (7) which we expect to be different from the properties of its corresponding dissipative system (5) namely the Lorenz system.

The system (7) for $\epsilon=0$ coincides, apart from the inessential equation

$$\dot{w} = -w$$

with the Lorenz system (5).

The properties of (7) are the following. It is invariant under the symmetry $(x, y, z, w) \rightarrow (-z, -y, z, -w)$. It is dissipative since the divergence of the flow, namely the trace of its Jacobian is

$$\mathrm{Tr}J = -(\sigma + 2 + b)$$

The singular points of the system are: the origin, for any value of the parameters r and ϵ, and other two points, which exist for

$$r - 1 > \left(\frac{4}{3\pi}\right)^2 \epsilon^2 \frac{r^2}{(r+1)^2}$$

This inequality shows that these singularities (convective equilibria) appear for r values larger than in the Lorenz system $(r - 1 > 0)$. Let us see now what we can obtain with the linear analysis around the origin. We use for convenience the new parameter

$$e = \left(\frac{4}{3\pi}\,\epsilon\,\frac{r}{r+1}\right)^2 \quad .$$

The secular equation is

$$(\lambda+b)\{\lambda^3 + (\sigma+2)\lambda^2 + [\sigma+1-\sigma(r-1)+e]\lambda-\sigma(r-1-e)\} = 0 \quad . \tag{9}$$

We see that for

$$\mu = r - 1 - e = 0 \; .$$

one has a zero root of (9), while for

$$\nu = r - 1 - \frac{\sigma+2}{\sigma} - \frac{2e}{\sigma(\sigma+1)} = 0$$

$$\mu < 0$$

one has a couple of pure imaginary roots.

From this observation we expect, respectively, a pitchfork bifurcation for $\mu=0$ and a Hopf bifurcation for $\nu=0$ and $\mu < 0$.

With lengthy algebraic calculations one can find that another Hopf bifurcation could take place after the pitchfork bifurcation.

Now, rather than continuing with the study of the above codimension one bifurcations, we choose to study the codimension two bifurcation at the point $P = (\mu,\nu) = (0,0)$, i.e. $r-1 = e = \sigma+1/\sigma-1$, in the nonphysical range $\epsilon > 1$, where a double zero root in (9) is met.

Indeed, in a neighborhood of P in the parameter plane, we are able to find all codimension one bifurcation sets leaving this point, especially those which cannot be found by means of a local one parameter analysis.

We perform a linear transformation of coordinates which takes the Jacobian matrix at the origin for $\mu=\nu=0$ to Jordan normal form, and then, with standard techniques we reduce system (7), in a neighborhood of the critical point $(x, y, \mu, \nu) = (0, 0, 0, 0)$, on a bidimensional center manifold[12]. This manifold is evaluated up to second order in x, y, μ, ν, so we obtain a reduced system calculated up to third order in perturbation theory. The normal form theorem is applied to this system. The resulting system is

$$\dot{x} = y + 0^{(4)}$$

$$\dot{y} = \mu_1 x + \mu_2 y - x^3 - x^2 y + 0^{(4)} \qquad (10)$$

where $0^{(4)}$ means terms of degree greater than three in x, y, μ, ν; μ_1 and μ_2 are known function of μ, ν.

System (10) has been extensively studied[12,13].

In particular the Hopf bifurcation at $\nu=0$ and the pitchfork bifurcation at $\mu=0$ are supercritical; for μ, $\nu > 0$ there is a tangent bifurcation, i.e. the birth of a pair of periodic orbits of opposite stability encircling the three equilibria, at $\mu_2 = 0.752\ \mu_1$; a double homoclinic bifurcation at $\mu_2 = 4/5\ \mu_1$ and finally a subcritical double Hopf bifurcation at $\mu_2 = \mu_1$. Evaluating μ_2, μ_1, in terms of $r - 1$ and e, we obtain the following picture:

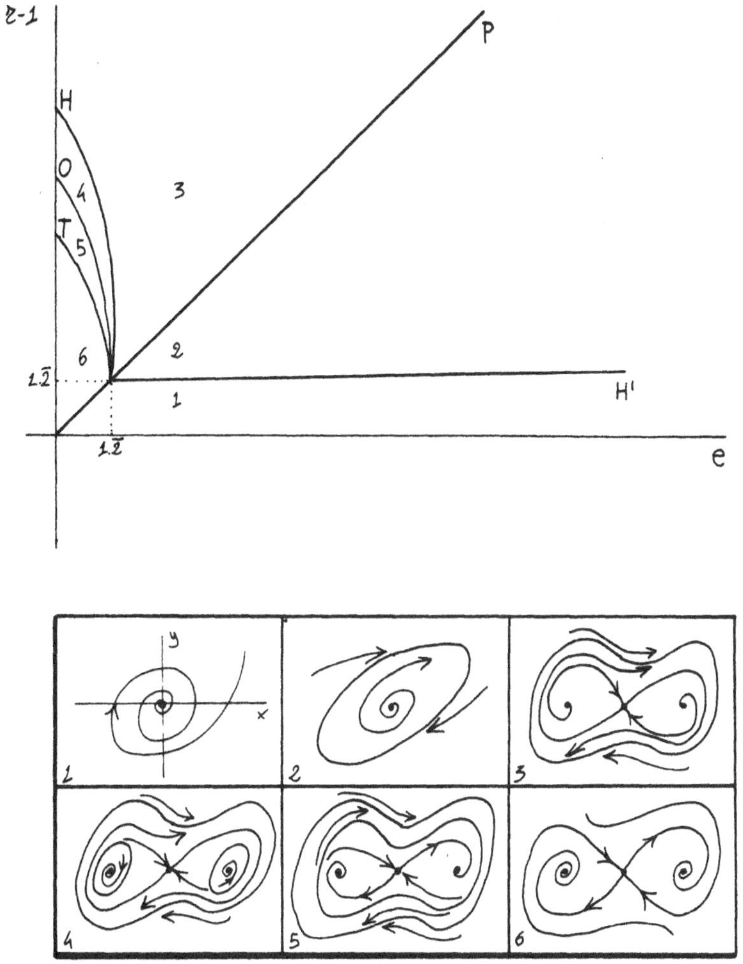

FIG. 1

The curves H, O and T intersect the axis $e = 0$ at

$$r_H = 14.21 \ , \quad r_O = 10.2 \ , \quad r_T = 8.64 \quad \text{for } \sigma = 10 \ .$$

We wish to compare these values with the corresponding values calculated for the Lorenz system: for the double subcritical Hopf bifurcation linear analysis gives

$$r_H = 24.73 \ ;$$

numerically one can find an homoclinic bifurcation at

$$r_O = 13.2 \ .$$

This comparison shows that the third order perturbative methods are qualitatively satisfactory. Concerning the tangent bifurcation, it has not been observed in the Lorenz system. A possible explanation arises from numerical investigations on the system (7), which show that the bifurcation set T intersects the curve O approximately at $r-1=5.8$, $e=0.91$. Such a global codimension two bifurcation has not yet been studied, but simple considerations of topological equivalence classes of the phase portrait in a neighborhood of this bifurcation point in the parameter plane, suggests the existence of at least another codimension one bifurcation set leaving this point, probably associated with global three dimensional effects. Obviously, such effects are not expected to be found by means of our reduction on a two dimensional center manifold.

For $\epsilon=1$ in system (7), linear analysis determines the value for subcritical Hopf bifurcation at $r = 20.887$, while by means of numerical integrations a non transverse intersection (eteroclinic) of the unstable manifold of the origin and of the stable manifold of one of the unstable periodic orbits, at $r = 19.645$, and finally a homoclinic bifurcation at $r = 12.94$ are found. The sequence of these bifurcations is quite close to the sequence of the Lorenz system, but in a smaller interval of r. No periodic orbit has been observed in this parameter range.

The most interesting difference between the dissipative system (5) and the active system (7) seems to be that a two dimensional center manifold containing the origin can be found for $e \neq 0$, while this is not possible for the Lorenz system[14].

This fact would give rise to different dynamical behaviours.

Finally we present in Fig. 2 the time behaviour of the availability $\overline{w^c}$, formula (8). The curves are normalized to the constant $\delta\sqrt{2}\,\frac{r^2}{r+1}$ which is the value of w^c in the conduction regime.

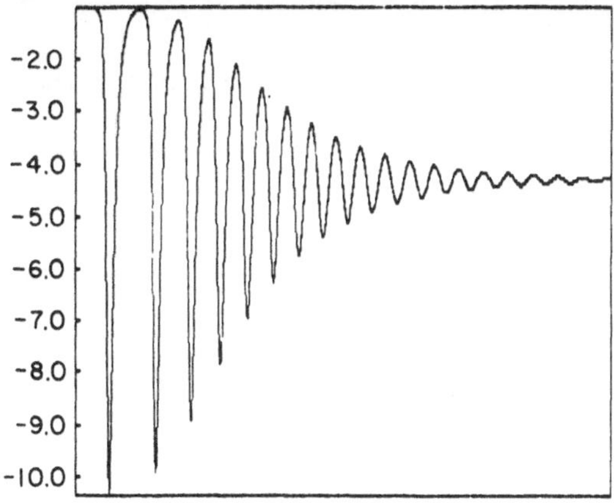

potenza estratta normalizzata ; r = 12.9 ; e = 1

FIG. 2a

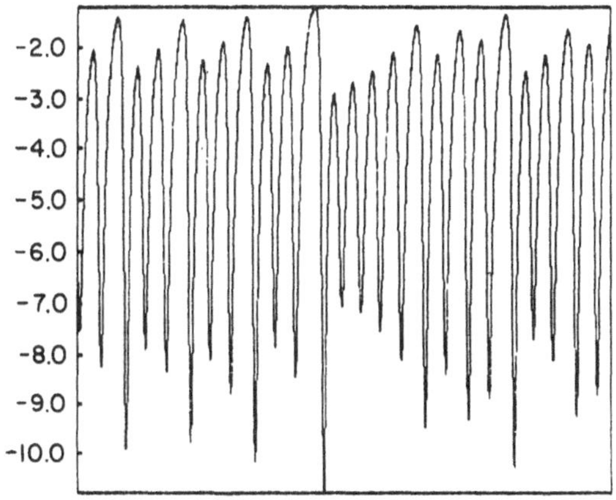

potenza estratta normalizzata; r = 25 ; e = 1

FIG. 2b

Acknowledgments

We thank G. Gallavotti and V. Franceschini for their stimulating interest in this work.

References

1. S. Carnot, "Réflexions sur la puissance motrice du feu et sur les machines propres à developper cette puissance", Paris, 1824.
2. J. W. Gibbs, "The Collected Works", New Haven, 1948.
3. R. C. Tolman, P. G. Fine, Rev. Mod. Phys. 20, 51(1948).
4. L. Landau, E. Lifshitz, "Statistical Physics", London 1958.
5. K. Ford, G. Rochlin, A. Rosenfeld, M. Ross, R. Socolow, "Efficient use of energy: a physics perspective", AIP Conf. Proc. 25, AIP, New York, 1975.
6. - Spätind Winter Institute on Irreversible thermodynamics and its applications. January 1983.
   - B. Andresen, "Finite time thermodynamics", Physics Laboratory II Press, University Copenhagen, 1984.
7. S. R. de Groot, P. Mazur, "Non equilibrium thermodynamics", North Holland, Amsterdam, 1962.
8. L. Landau, E. Lifshitz, "Fluid mechanics", London, 1959.
9. - L. Sertorio, "Energy active passively-controlled non-equilibrium systems", Preprint LA-UR-83-3590, 1983.
   - F. D'Isep, L. Sertorio, "On the relationship between entropy and availability production for active systems", Il Nuovo Cimento, 93B, 2(1986).
10. E. N. Lorenz, "Deterministic nonperiodic flow", Jour. Atm. Sci., March 1963.
11. B. Saltzman, "Finite amplitude free convention as an initial value problem", Jour. Atm. Sci., July 1962.
12. J. Carr, "Applications of center manifold theory", Springer, N. Y., Heidelberg, Berlin.
13. F. Takens, "Forced oscillations and bifurcations", Comm. Math. Inst. Rijkuniversitei Utrecht 3, 1-59(1974).
14. M. Miari, "Qualitative understanding of the Lorenz equations through a well-known second order dynamical system", Physica 21D, 1986.

EXPERIMENT AND THEORY OF THE CLASSICALLY CHAOTIC MOTION OF THE DRIVEN BOUND ELECTRON

James E. Bayfield

Department of Physics and Astronomy
University of Pittsburgh
Pittsburgh, Pennsylvania 15260 U. S. A.

Introduction

The externally driven bound electron is a quantized nonlinear system, with the nonlinearity arising from the Coulomb binding force. Laboratory experiments that probe the near classical behavior of this system involve highly excited hydrogen atoms exposed to microwave fields of ionizing strength. Ionization of atoms of initial principal quantum number n_0 between 30 and 90 and microwave frequencies between 6 and 18 GHz requires the absorption of the energy of some number k_I of microwave photons that is the order of 100. At ionizing field strengths there is rapid stimulated photon absorption and emission underlying the time evolution of the mean value of the atom's quantum number. The total number K of photon interactions during the microwave exposure time can be 10,000. As the quantities n_0, k_I, and K are all much larger than unity, it is reasonable to ask as to what extent the behavior of the system is classical.

At weak microwave field strengths F, only sharply resonant multiphoton transitions between pairs of quantal energy levels occur with observable probability. We have no classical picture of this. However, as F is increased, the quantum two-state resonances broaden and particularly the transitions of low order begin to occur over wide ranges of frequency. An intermediate regime is reached where four or five states of adjacent quantum number are all strongly coupled to the initial state. As the atom's energy levels are anharmonic, the time evolution is now quasiperiodic. We do have a classical picture of the intermediate regime that is based upon ensembles of quasiperiodic electron trajectories.

With a further increase in F, the atom anharmonicity is totally overcome by the radiative broadening of energy levels, and the bound states above the initial state and at least some of the continuum states are all strongly coupled together. Now the electron becomes delocalized, ionization becomes rapid and the classical electron motion becomes chaotic (1,2).

A crucial question here inquires about the quantal time evolution of the system at ionizing field strengths. To what extent is it classical, and to what extent can it be called chaotic? Theoretical studies of this require extensive numerical work, as Professor Casati has indicated in his lectures. Here we consider the contributions of laboratory experiments designed to establish the physical picture outlined above and to test the numerical calculations. We shall see that much of the real system's behavior is indeed adequately explained by classical physics.

Stationary Initial Quantum States and External Static Field Effects

Past experiments have used stationary states as the atom's initial state to be established before exposure to a microwave pulse. This is not a trivial point, as experiments are being devised that utilize non-stationary initial states that are spatially localized bound electron wavepackets (3,4). The stationary states correspond to ensembles of classical electron trajectories containing distributions of initial conditions.

The hydrogen atom stationary states we all know about are the states expressed in spherical coordinates that are eigenstates of the magnitude and z-component of the angular momentum. However, all experiments to date have probably involved external static electric fields large enough to produce Stark interactions larger than the relativistic spin-orbit interaction in highly excited hydrogen. In this case, a different set of stationary states of the isolated atom is useful; these are the states expressed in parabolic coordinates that are eigenstates of the z-components of the angular momentum and Runge-Lenz vectors as well as of the energy (5). Quantum numbers arising from these three quantities are m, n_1, and n. When an external static electric field is present, the z-components of angular momentum and of the generalized Runge-Lenz vector G are still conserved (6). The energy now depends upon the strength F_s of the static field and upon all three parabolic quantum numbers, the well-known Stark effect (5). In addition, the total potential energy function that the electron now sees is the sum of the Coulomb and static field interactions and contains a barrier that can be tunneled through. Thus all the bound states of the atom acquire finite rates for ionization by the static field (7), although only those with very high principal quantum number might have significant probabilities in short-time experiments. These static field effects should be included in calculations that are to be precisely compared with real experiments.

An Overview of Past Experiments

The design of a typical experiment involves the time sequence of events indicated in Figure 1. A microwave interaction of an ensemble of highly excited atoms is preceded by initial state preparation and followed by state analysis to assess the changes caused by the microwaves. Of course, one must begin with a source of atoms and end with some detector of them.

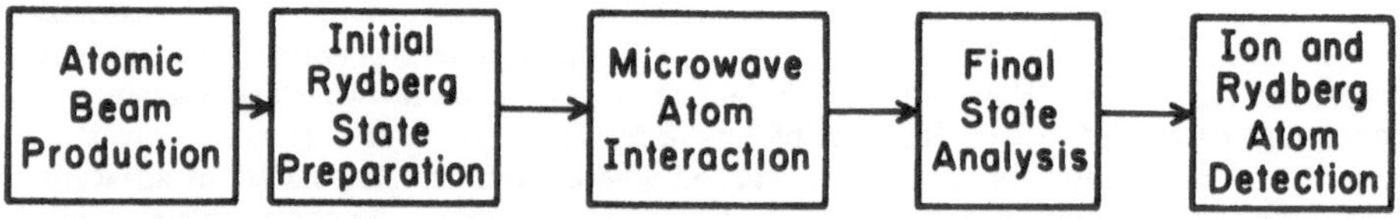

Figure 1. Schematic of an experiment for the study of highly excited (Rydberg) atoms exposed to a pulse of strong microwave radiation.

The experiments begin with the production of a beam of atoms moving within an evacuated apparatus. The motion of the atoms through a spatial sequence of regions produces a time sequence of events in the atom's frame of reference. When a fast atomic beam is used,, an atom velocity of about 10 (+8) cm/sec and a total apparatus length of a few meters means that the time sequence spans a total time of a few microseconds.

The second step of Figure 1 is to prepare some of the atoms in the beam in selected highly excited stationary states. This has usually meant selective laser excitation of atoms already in a low excited state as the result of atomic collision beam formation. In experiments working with atoms having all their quantum numbers defined, external static electric fields within the laser excitation region are necessary to remove the isolated atom energy degeneracy during the excitation.

In the microwave-atom interaction region, the atoms pass through a microwave cavity or waveguide structure, whose length has been between 1 and 70 cm. Depending upon the atom velocity and microwave frequency, the atoms are exposed to a finite number of microwave field oscillations that has ranged from 100 to 3000.

After the microwave exposure, the probabilities for occupation of the different possible stationary final states of the atoms are measured with the assistance of a final state analyzer. Final bound states can be selectively observed using differential static electric field ionization (7,8). A sum over continuum final states also can be obtained by observing the ions produced directly by the microwaves.

In the detection region, the ions made directly by the microwaves are separated from the other ions produced by the field ionization of the bound states. The intensities of these various ion fluxes are ultimately measured and compared with the intensity of the initially prepared highly excited atom flux determined with the microwaves turned off. Thus the observed quantities are the various final state fractional probabilities.

Table I lists the initial states, final observed states and nature of the microwave interaction for the principal experiments carried out to date. The work in 1974 that discovered the difference between microwave and static electric field ionization involved a crude selection of only the initial principal quantum number. Only recently has work been carried out with the initial parabolic quantum numbers n_0, n_{10}, m_0 all defined.

Table I. Past experiments. "Collinear fields" means that $\vec{F}_s$ was paralled to $\vec{F}$ in the interaction region. "One-dimensional" means that the work was with electrically polarized atoms, n_1=m=0, and that all electric fields were aligned along the direction of the atomic polarization.

Ref.	Initial State(s)	Microwave-Atom Interaction Region	Final States
9	mixed 63≤n≤69 mixed n_1 and m	weak static field 600 oscillations	ions
10	single n mixed n_1 and m	weak static field 600 oscillations	ions atoms, unknown $n>n_0$ unknown n_1 and m
11	single n mixed n_1 and m	variable static field collinear fields 300 oscillations	ions
12	single n, n_1, m	large static field collinear fields one-dimensional 3000 oscillations	ions atoms, known n, n_1,m
13	single n, n_1, m	variable static field collinear fields one-dimensional 100 oscillations	ions atoms, known n, n_1,m

Classical Behavior

One dimensional models of the electron motion are relevant to experiments using electrically polarized atoms, which are those atoms with $n_1=m=0$ (8, 12, 14). In atomic units where $\hbar=m=e=1$, one model Hamiltonian is (1):

$$H = p^2/2 + x[F_s + F\cos(\ t + \phi)] + \begin{cases} -Z/x, & x>0 \\ \infty, & x\leq 0. \end{cases} \quad (1)$$

It is useful to transform to the isolated-atom action-angle coordinate θ defined by (1):

$$\phi = \begin{cases} 2\pi[\sin^{-1}(\sqrt{x/a}) - \sqrt{x/a(1-x/a)}], & p\geq 0 \\ 2\pi - 2\pi[\ \text{"}\ ], & p<0. \end{cases} \quad (2)$$

For the isolated atom, the corresponding action I_0 is conserved, and the energy is $-Z^2/(2I_0^2)$. Numerical integration of the classical equations of motion finds that ionization occurs for initial values of action I that are above a threshold value that depends upon the initial location and direction of motion of the electron, the initial value of the microwave field, the microwave frequency ω and peak field strength F, and the strength of any external static electric field F_s. The dependence of the threshold upon these parameters does vary somewhat with how the microwave field is switched on.

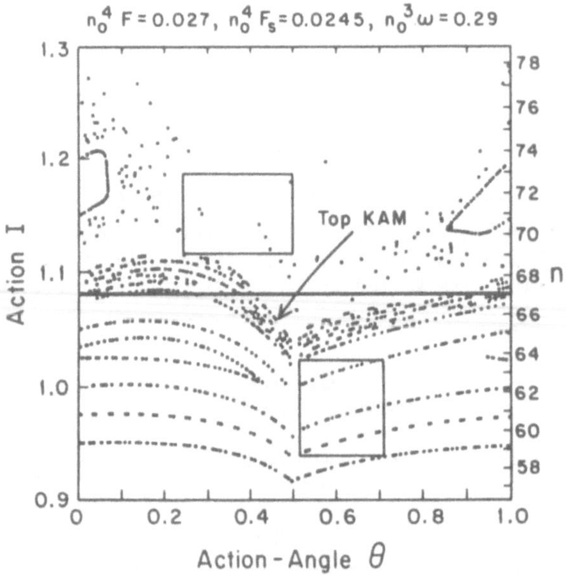

Figure 2. Time-periodic samplings of some classical electron trajectories plotted in action-angle space. The horizontal line corresponds to an initial ensemble of trajectories all having the same initial value of action. The boxes are two possible regions for future experiments limited only by the Heisenberg uncertainty principle.

Once the atom is in the field, its classical motion can be described by some set of trajectories in time-extended phase space. Suppose that we periodically sample in time, at the microwave frequency, the (I, θ) values of the trajectories. Figure 2 shows the results of a few such samplings (16, 17), for the case of the initial value of the microwave field equal to zero, i.e. the field is taken to oscillate as sin (ωt). Here classically normalized units are used, along with the isolated-atom quantum identification of the initial action with the initial quantum number: $I_0 = n_0\hbar$, $F_0=n_0^4 F$, $F_{s0}=n_0^4F_s$, and $\omega_0=n_0^3\omega$.

If the electron is initially on a trajectory below the top Kolgomorov-Arnold-Moser (KAM) surface indicated in Figure 2, then its motion is regular and usually quasiperiodic. Energy averaged over long times is then close to constant. By the KAM theorem, the electron stays on the KAM surface it was started on (2). Quantum tunneling between KAM surfaces will of course modify this (18).

If the electron initially is on a trajectory above the top KAM surface and is not in one of the classical resonance islands of quasiperiodic motion, then its motion is chaotic. The time-averaged energy now increases in time, along with the mean value of the action or quantum number. The square of the increase in I above I_0 is numerically found to be proportional to time (19), indicative of a deterministic diffusion in action space (20-21). In practice, such chaotic trajectories ultimately lead to ionization of the atom; before the electron can be infinitely far from the atomic nucleus, it certainly interacts with the experimental apparatus in some new way not contained in the Hamiltonian of equation (1). A more important mechanism for altering the microwave-induced diffusion after a finite time is the static electric field ionization already mentioned, an effect that remains to be investigated in detail.

Quantum Estimates of Some Features of the Classical Picture

When high enough in the regular region of phase space, on time-average one expects correspondences between the quantum and classical features of the absorption of energy by the atom from the microwave field. This expectation is based upon the knowledge that in the semiclassical region there exists a correspondence principle linking the classical and quantum spectral densities of the radiation field (22). These correspondences are suggested by the experimental data and are seen in the results of the numerical calculations.

First there is a correspondence between classical and quantum resonances. The classical resonance islands, see Figure 2, are regions of quasiperiodic motion containing an identifying periodic trajectory. For this trajectory the time evolution contains a single fundamental frequency plus its harmonics and subharmonics. For a periodic orbit, an integral number C_1 of microwave oscillation periods T must equal an integral number C_2 of electron orbit periods T_n in the presence of the microwave field:

$$C_1T = C_2T_n, \qquad C_i \text{ integers}$$

Thus the trajectory winding number W is rational:

$$W \equiv T_n/T = \omega/\omega_n = C_1/C_2.$$

We can easily identify a quantum quantity that also involves the same ratio of microwave and orbital frequencies. In a two-state approximation, microwave energy is resonantly absorbed when the energy of an integral number k of photons equals the energy level separation ΔE_n of the states:

$$k\omega = \Delta E_n \approx \Delta n/n^3.$$

Combining equations (4) and (5) we find that in the limit $\Delta n \ll n$, the classical and quantal labels of resonant behavior mathematically correspond according to

$$W \approx \Delta n / k. \tag{6}$$

The microwave frequency dependence of experimental data for the probability of transitions decreasing n by one unit are shown in Figure 3, for microwave field strengths where ionization probabilities are not large. One sees a field-saturated, field- broadened and field-shifted resonance at a frequency near 6.3 GHz, superimposed upon a roughly constant background probability. This background probability is obtained in the quantal numerical calculations, and is due to strong coupling of a finite number of states with quantum numbers adjacent to the initial value (15). Because of the anharmonicity of the energy levels of the hydrogen atom, each state introduces a fundamental frequency incommensurate with the others, and one has quasiperiodic time evolution and irrational winding numbers. One can numerically probe this quasiperiodic quantal time evolution using sudden switching-on of the field and evaluation of the states in the wavefunction after some finite time. The final state distribution so obtained is shown in Figure 4a. A characteristic narrow band of final states is seen. The quantal wavefunction is relatively localized in quantum number space. When the microwave frequency is away from the resonance region of Figure 3, one expects the quantal evolution to correspond to classical evolution on the non-resonance KAM surfaces of Figure 2.

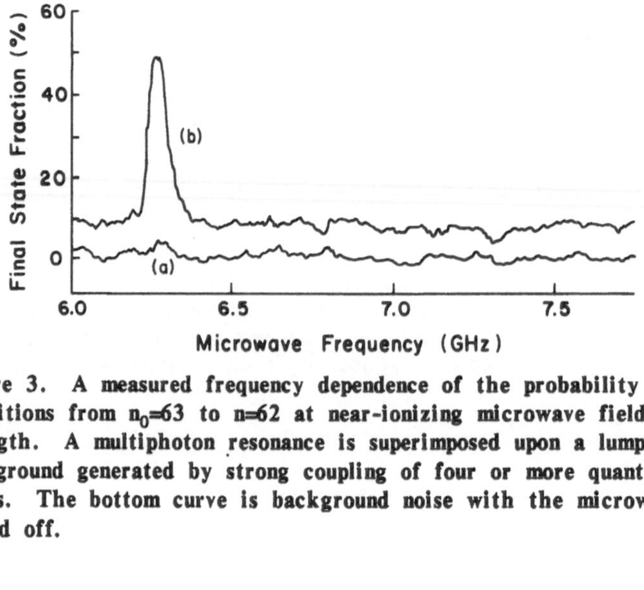

Figure 3. A measured frequency dependence of the probability for transitions from $n_0=63$ to $n=62$ at near-ionizing microwave field strength. A multiphoton resonance is superimposed upon a lumpy background generated by strong coupling of four or more quantum states. The bottom curve is background noise with the microwaves turned off.

If at the same values of F and F_S, numerical quantal calculations with sudden switching are carried out for a sufficiently higher initial value of quantum number,then a very different final state distribution such as that of Figure 4b is the result (23). The roughly exponential final state distribution is characteristic of quantal time evolution that corresponds to classical behavior in the chaotic, irregular region. The isolated-atom energy level anharmonicity is now completely overcome, and the electron's wavefunction is relatively delocalized in quantum number space after the exposure to the microwaves. Very many states must be included in numerical quantal calculations of such final state distributions (19). Indeed, it is because of the finiteness of the microwave exposure time that one can hope to include enough states. As exposure time and microwave field strength are increased, the importance of the continuum states tends to grow.

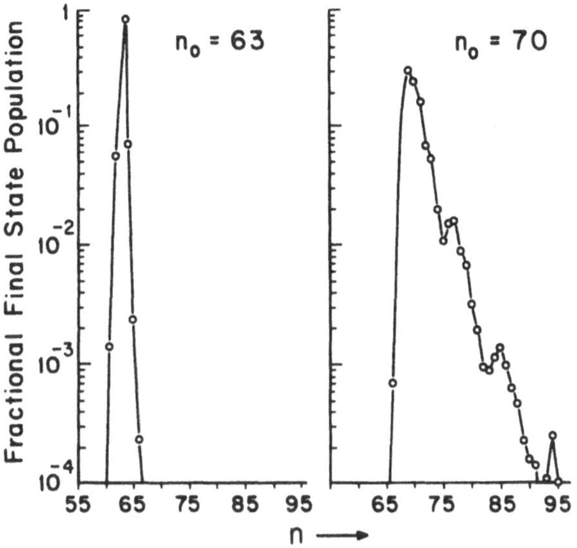

Figure 4. Quantal final bound state distribution for initial values of atom principal quantum number a) below and b) above the corresponding classical action threshold for chaos.

From Figure 4 it is clear that one empirically can find quantal threshold values of initial quantum number where classically chaotic time evolution begins to occur. Table II compares some classical and quantum threshold values obtained assuming sudden switching and the same initial value of the oscillating microwave field strength. There is good agreement.

The onset of classical behavior via strong coupling of quantum states can be estimated using a simple two-state quantal resonance- or state-overlap criterion (12). We expect semi-classical behavior when the isolated-atom quantal energy level structure is blurred out by radiative broadening. Consider the two-state time-dependent transition probability for one-photon transitions:

$$P_{1\to 2}(t) = \left[\frac{(z_{12}F)^2}{(\omega - \omega_{12})^2 + (z_{12}F)^2}\right] \sin^2\left[\frac{1}{2}\sqrt{(\omega - \omega_{12})^2 + (z_{12}F)^2}\, t\right] \qquad (7)$$

Table II. A comparison of classical and quantal thresholds for classically-chaotic behavior (17,23).

F(V/cm)	F_S(V/cm)	(GHz)	n_0	n(threshold) Classical	Quantal
9.1	0.2	7.1047	63	68	69
9.1	0.0	7.1047	63	70	70
17.3	0.0	7.1047	63	67	66
16.0	5.45	7.0061	60	64	65
11.0	0.0	57.0	66	58	58
9.8	0.0	31.58	63	55	54

For microwave frequencies ω below the one-photon resonance frequency ω_{12}, the time evolution occurs at a frequency comparable to the electron orbit and microwave frequencies when the interaction strength $z_{12}F$ is both large enough for the amplitude of P(t) to be roughly 1/2 for all frequencies ω less than ω_{12}. Empirically setting $z_{12}F$ equal to ω_{12} gives a threshold field value of $F=3/n^5$ atomic units, which results in a curve similar in shape to but some 30% lower than the ionization threshold curves to be shown in Figure 6. Although this quantal two-state overlap criterion qualitatively gives the microwave strengths necessary for semi-classical behavior, it does not correspond exactly to the Chirikov classical resonance overlap criterion for the threshold for chaos; a many-state quantal resonance overlap criterion is needed for such a correspondence.

Comparison of Numerical Calculations with Experimental Data

In the past few years experiments have been pursued in two laboratories, while numerical calculations have been carried out by five theoretical groups. During this time, a number of problems relevant to the precision comparison of experiment with theory have been initially investigated; these include the list of important parameters given early in the section on classical behavior.

Figure 5 compares the most recent numerical final state probability distributions with the available data known not to have effects from microwave harmonics. The deviations between theory and experiment might be ascribed to two causes. First, the theoretical curve (24) is for a microwave frequency of 6.24 GHz where at n_0=63 there is a resonance for transitions down to n=62; the data (15), however, are averaged results for the two frequencies 7.11 GHz and 7.49 GHz, both not resonant for the down n-changing. Thus the increased down n-changing and consequently the decreased up n-changing seen in the calculated values are not surprising. The second possible cause of residual disagreement is that there may be some enhancement of transitions involved in the experimental curve, because of residual broadband microwave noise.

Both the theoretical and experimental curves in Figure 5 exhibit a central peak superimposed upon a shoulder occuring at high values of n; this is believed due to a set of classical initial conditions containing both quasiperiodic and chaotic components, . see the horizontal line in Figure 2. As a result, one has a superposition of results of the types shown in Figures 4a and 4b. In addition, the curves of Figure 5 are averaged over sets of values associated with different initial values of the microwave field strength, each set having a somewhat different looking Figure 2.

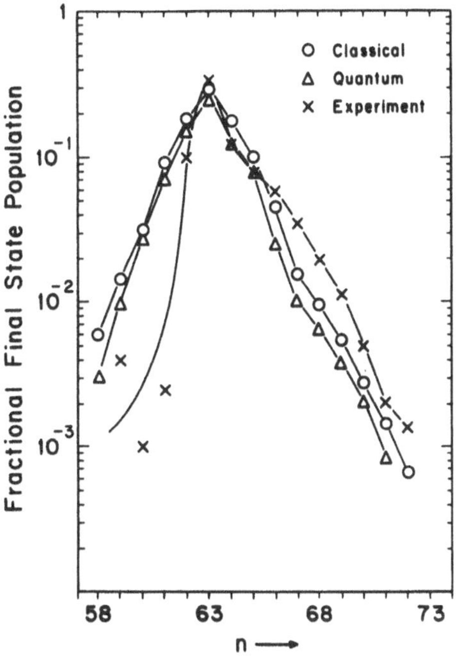

Figure 5. A comparison of classical, quantal and experimental final bound state distributions.

Figure 6 shows values of scaled microwave field strength required for 10% ionization probability in about 300 field oscillations, plotted as a function of scaled microwave frequency. In the experiments, atoms with large spreads in values of n_{10} and m_0 were used. The laboratory microwave frequency was fixed, and the scaled frequency varied by changing n_0 over the range 30 to 90 (11,25). Although the quantal numerical calculations utilized a one dimensional model (26), we again see reasonable agreement between classical calculations, quantum calculations and experiment. The structure in these ionization curves is believed due to resonance effects that arise when the classical islands (see Figure 2) must be included in the ensemble of trajectories that corresponds to the set of experimental initial conditions.

In spite of the initial qualitative agreement of experiment and quantum theory with classical predictions, some uninvestigated factors remain. Experimentally much of the work has been done without completely defined quantum numbers or with undetermined static fields; work without these limitations is only beginning. Some of the experimental data obtained with wave guide microwave structures may have involved effects induced by weak microwave system harmonics and/or weak broadband microwave noise; controlled experiments investigating this are now underway (13). The effects of enhanced microwave field strength near the edges of the beam entrance and exit holes in microwave structures are not yet known.

Quantal and classical studies of the effects of the switching-on and switching-off of the microwave field are incomplete, as is work on static field effects and on deviations from one-dimensional behavior.

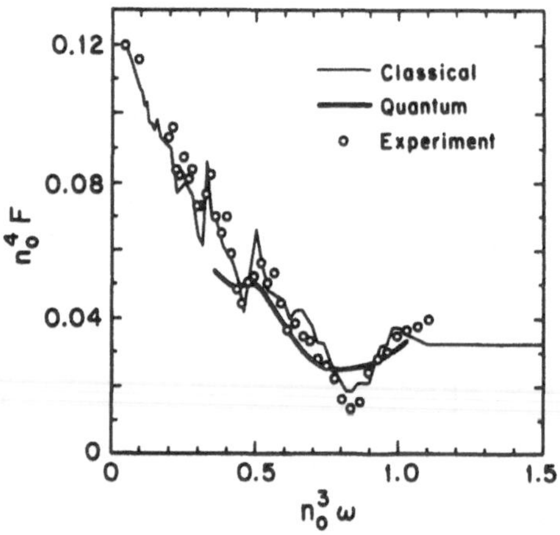

Figure 6. A comparison of classical, quantal and experimental microwave field values for 10% ionization probability.

Yet we have seen that recent work has provided increasing evidence that the real atom-plus-microwaves system behaves classically when the microwave frequency is less than the isolated-atom electron orbit frequency and when the microwave field strength is close to values necessary for observable ionization probabilities.

Future Possibilities

Although the behavior of excited hydrogen in microwaves can be quite classical, and the classical time evolution can be chaotic, one still can debate whether the quantum evolution is chaotic in some mathematical sense, or not. Heisenberg's uncertainty principle may make impossible a demonstration of "quantum chaos" having the precision of classical physics. Indeed, taking the classical limit may be the origin of mathematical deterministic chaos in our system; we do not know. At any rate, we certainly can more closely check how our quantum system at least mimics its classical counterpart.

Experiments of the types described in this paper can be improved by utilizing picosecond pulsed laser excitation of the initial state. When time-synchronized with the microwave field oscillations, such picosecond pulses should greatly reduce the spread of initial conditions in the experiments. Such excitation can produce a coherent linear combination of a number of states with adjacent principal quantum numbers to form a nonstationary bound electron wavepacket with mimimal initial uncertainty product (3,4). This would further reduce the spread in initial conditions, as indicated by the boxes in Figure 2. Thus there is hope for experimental near-separation of classical chaotic behavior from quasiperiodic behavior; this would produce data without the mixed features of Figures 5 and 6.

We also envisage new types of experiments that would probe the time evolution of the system using short-pulse probes to perturb the atoms at different times during their passage through the microwave region. This might directly reveal the diffusion-like character of the microwave energy absorption.

The hydrogen-in-microwaves system is one classically-chaotic quantum system that involves a continuum of quantum states in an essential way. Such systems are primary candidates for the discovery of some form of quantum chaos (27). In this search there are possibilities for further quantal numerical calculations that would investigate in new ways the detailed semi-classical quantal time evolution in phase space and could search for statistical measures of this evolution that might correspond to classical indicators of chaos.

The author thanks the U. S. National Science Foundation for the support of his laboratory.

References

1. R. V. Jensen, Phys. Rev. A 30, 386 (1984).
2. A. J. Lichtenberg and M. A. Liberman, "Regular and Stochastic Motion", New York: Springer-Verlag, 1983.
3. J. Parker and C. R. Stroud, Jr., Phys. Rev. Lett. 56, 716 (1986).
4. J. E. Bayfield and S.-L. Duan, Abstracts Intern. Quantum Electronics Conf., Baltimore, Md, April 27-May 1, 1987.
5. H. A. Bethe and E. E. Salpeter, "Quantum Mechanics of One- and Two-Electron Atoms", Berlin: Springer-Verlag, 1957.
6. P. J. Redmond, Phys. Rev. 133, B1352 (1964).
7. R. J. Damburg and V. V. Kolosov, in "Rydberg States of Atoms and Molecules", R. F. Stebbings and F. B. Dunning, eds., Cambridge: Cambridge Univ. Press, 1983.
8. J. E. Bayfield, in "Fundamental Aspects of Quantum Theory", V. Gorini and A. Frigerio, eds., New York: Plenum Press, 1986.
9. J. E. Bayfield and P. M. Koch, Phys. Rev. Lett. 33, 258 (1974).

10. J. E. Bayfield, L. D. Gardner and P. M. Koch, Phys. Rev. Lett. 39, 76 (1977).
11. P. M. Koch, in "Fundamental Aspects of Quantum Theory", V. Gorini and A. Frigerio, eds., New York": Plenum Press, 1986.
12. J. E. Bayfield, in "Quantum Measurement and Chaos", E. R. Pike, ed., New York: Plenum Press, 1987.
13. J. E. Bayfield and D. W. Sokol, unpublished work, 1987.
14. D. L. Shepelyansky, in "Chaotic Behavior in Quantum Systems", G. Casati, ed., New York: Plenum Press, 1985.
15. J. N. Bardsley, B. Sundaram, L. A. Pinnaduwage and J. E. Bayfield, Phys. Rev. Lett. 56, 1007 (1986).
16. R. V. Jensen, Phys. Rev. Lett. 54, 2057 (1985).
17. R. V. Jensen, Private Communications, 1986.
18. M. Wilkinson, Physica 21 D, 341 (1986).
19. G. Casati, B. V. Chirikov, D. L. Shepelyansky and I. Guarneri, Physics Reports, to be published.
20. H. G. Schuster, "Deterministic Chaos", Weinheim: Physik-Verlag, 1984.
21. T. Geisel, in "Nonequilibrium Cooperative Phenomena in Physics and Related Fields", M. G. Velarde, ed., New York: Plenum Press, 1984.
22. I. C. Percival, In "Atomic Physics 2", G. K. Woodgate and P. G. H. Sandars, eds., New York: Plenum Press, 1971.
23. B. Sundaram, Ph. D. Dissertation, University of Pittsburgh, 1986, unpublished.
24. G. Casati, Private Communications, 1987.
25. R. V. Jensen, in "Atomic Physics 10," H. Narumi and I. Shimamura, eds., Amsterdam: North-Holland, 1987.
26. J. N. Bardsley and M. J. Comella, J. Phys. B 19, L565 (1986).
27. J. E. Bayfield, Comments on Atomic and Molecular Physics, to be published.

SHIL'NIKOV CHAOS IN LASERS

F.T. Arecchi

Istituto Nazionale di Ottica
and
Dept. of Physics - University of Firenze - Italy

ABSTRACT

The onset of deterministic chaos in lasers is studied by referring to low dimensional systems, in order to isolate the characteristics of chaos from the random fluctuations due to the coupling with a thermal reservoir. For this purpose, attention is focused on single mode homogeneous line lasers, whose dynamics is ruled by a low number of coupled variables. In the examined cases, experiments and theoretical model are in close agreement. In particular I describe Shilnikov chaos, how it can be characterized, and the strong resulting coupling between nonlinear dynamics and statistical mechanics.

1. COHERENCE AND CHAOS IN LASERS

Quantum optics from its beginning was considered as the physics of coherent and intrinsically stable radiation sources. Lamb's semiclassical theory /1/ showed the role of the e.m. field in the cavity in ordering the phases of the induced atomic dipoles, thus giving rise to a macroscopic polarization and making it possible a description in terms of very few collective variables. In the case of a single mode laser and a homogeneous gain line this means just five coupled degrees of freedom, namely, a complex field amplitude E, a complex polarization P, and a population inversion N. A corresponding quantum theory, even for the simplest laser model, does not lead to a closed set of equations, however the interaction with other degrees of freedom acting as a thermal bath (atomic collisions, thermal radiation) provides truncation of high order terms in the atom-field interaction /2,3,4/. The problem may be reduced to five coupled equations (the so-called Maxwell-Bloch equations) but now they are affected by noise sources to account for the coupling with the thermal bath /5/. Being stochastic, or Langevin, equations, the corresponding solution in closed form refers to a suitable weight function or phase space density. Anyway the average motion matches the semiclassical one, and fluctuations play a negligible role if one excludes the bifurcation points where there are changes of stability in the stationary branches. Leaving out the peculiar statistical phenomena which characterize the threshold points and which suggested a formal analogy with thermodynamic phase transitions /6/ the main point of interest is that a single mode laser provides a highly stable or coherent radiation field.

From the point of view of the associated information, the standard interferometric or spectroscopic measurements of classical optics, relying on average field values or on their first order correlation functions, are insufficient. In order to characterize the statistical features of Quantum optics it was necessary to make extensive use of photon statistics /7,8/.

From a dynamical point of view, coherence is equivalent of having a stable fixed point attractor and this does not depend on details of the nonlinear coupling, but on the number of relevant degrees of freedom. Since such a number depends on the time scales on which the output field is observed, coherence becomes a question of time scales. This is the reason why for some lasers coherence is a robust quality, persistent even in presence of strong perturbations, whereas in other cases coherence is easily destroyed by the manipulations common in the laboratory use of lasers, such as modulation, feedback or injection from another laser.

Here I give a general presentation of low dimensional chaos in lasers. For a more complete approach to the problem, I refer to a recent monograph on the subject /9/.

We focus on those situations in quantum optics which permit close comparison between experiments and theory. By purpose I do not tackle the vast class of inhomogeneously braodened lasers, where it is extremely difficult to derive close correspondences between experiments and theory because of the large number of coupled degrees of freedom.

If we couple Maxwell equations with Schrödinger equations for N atoms confined in a cavity, and expand the field in cavity modes, keeping only the first mode which goes unstable, this is coupled with the collective variables P and Δ describing the atomic polarization and population inversion as follows (E being the slowly varying mode amplitude)

$$\begin{aligned} \dot{E} &= - kE + gP \\ \dot{P} &= -\gamma_{\perp} P + gE\Delta \\ \dot{\Delta} &= -\gamma_{\parallel}(\Delta-\Delta_0)-4g\ PE \end{aligned} \qquad (1)$$

For simplicity we consider the cavity frequency at resonance with the atomic resonance, so that we can take E and P as real variables and we have three coupled equations. Here, $k, \gamma_{\perp}, \gamma_{\parallel}$ are the loss rates for field, polarization and population respectively, g is a coupling constant and Δ_0 is the population inversion which would be established by the pump mechanism in the atomic medium, in the absence of coupling. While the first equation comes from Maxwell equations, the other two imply the reduction of each atom to a two-level atom resonantly coupled with the field.

The presence of loss rates means that the three relevant degrees of freedom are in contact with a "sea" of other degrees of freedom. In principle, Eqs. (1) could be deduced from microscopic equations by statistical reduction techniques /5/.

The similarity of Maxwell-Bloch equations (1) with Lorenz equations /10/ would suggest the easy appearence of chaotic instabilities in single-mode, homogeneous-line lasers. Indeed Lorenz model is a suggestive example of the general fact that a nonlinear coupling of at least three dynamical degree of fredom may induce instabilities on the motion, which in such cases becomes irregular. However time scale considerations rule out the full dynamics for most of the available lasers. Lorenz equations have damping rates within one order of magnitude. On the contrary, in most lasers the three damping rates are wildly different from one another.

The following classification has been introduced /11/

<u>Class A</u> (e.g. He-Ne, Ar , Kr , dye): $\gamma_\perp \simeq \gamma_\parallel \gg k$.
The two last equations can be solved at equilibrium (adiabatic elimination procedure) and one single nonlinear field equation describes the laser. N=1 means fixed point attractor, hence coherent emission.

<u>Class B</u> (e.g. ruby, Nd, CO): $\gamma_\perp \gg k \simeq \gamma_\parallel$.
Only polarization is adiabatically eliminated and the dynamics is ruled by two rate equations for field and population. N=2 allows also for period oscillations.

<u>Class C</u> The complete set of eqs. (1) has to be used, hence Lorenz like chaos is feasible, whenever $\gamma_\perp \simeq \gamma_\parallel \simeq k$.

We have carried a series of experiments on the birth of deterministic chaos in CO_2 lasers (Class B). In order to increase by at least 1 the number of degrees of freedom, we have tested the following configurations.

(i) Introduction of a time dependent parameter to make the system non autonomous /12/. Precisely, an eletro-optical modulator modulates the cavity losses at a frequency near the proper oscillation frequency Ω provided by a linear stability analysis, which for a CO_2 laser happens to lie in the 50-100KHz range, making it easy an accurate set of measurements.

ii) Injection of signal from an external laser detuned with respect to main one, choosing the frequency difference near the above mentioned Ω. With respect to the external reference the laser field has two quadrature components which represent two dynamical variables. Hence we reach N = 3 and observe chaos /11/.

(iii) Use a bidirectional ring, rather than a Fabry-Perot cavity /13/. In the latter case the boundary conditions constrain forward and backward wave, by phase relations on the mirror, to act as a single standing wave. In the former case forward and backward waves have just to fill the total ring length with an integer number of wavelengths but there are no mutual phase constrains, hence they act as two separate variables. Furthermore, when the field frequency is detuned with respect to the center of the gain line, a complex population grating arises from interference of the two counter-going waves, and as a result the dynamics becomes rather complex, requiring $N > 3$ dimensions.

(iv)Add an overall feedback, besides that provided by the cavity mirrors, by modulating the losses with a signal provided by the output intensity /14/. If the feedback has a time constant comparable with the population decay time, it provides a third equation sufficient to yield chaos.

Notice that while methods (i), (ii) and (iv) require an external device, (iii) provides intrinsic chaos. In any case, since feedback, injection or modulation are currently used in laser applications, the evidence of chaotic regions puts a caution on the optimistic trust in the laser coherence.

Of course, the requirement of three coupled nonlinear equations does not necessarily restrict the attention to just Lorenz equations. In fact none of the explored case i) to iv) corresponds to Lorenz chaos.

2. SHIL'NIKOV CHAOS, THE METHOD OF RETURN TIME AND THE FLUCTUATION ENHANCEMENT

Of the whole phenomenology explored in the past years, I select a single topic of particular relevance. I report experimental evidence of quasi homoclinic behavior characterized by pulses with regular shapes but chaotic in their time sequence /15/. The regularity in the shape means that the points at any Poincaré section are so closely packed that extremely precise measurements of their position would be required to yield relevant features of the motion. On the contrary, return times to a Poincaré section close to the unstable point display a large spread, due to the sensitive dependence of the motion upon the intersection coordinate. Based on such a consideration, we introduce the spread in return times as the specific indicator of homoclinic chaos. Our experimental data yield iteration maps of return times in close agreement with those arising from the theory of Shil'nikov chaos /16,17/. Thus, the test introduced in Ref. 15 appears as the most direct one to characterize chaotic dynamical situations associated with pulses almost equal in shape but having fluctuating occurrence times. Furthermore, at variance with the theory, the experimental return maps show a variable degree of thickening independent from the measurement

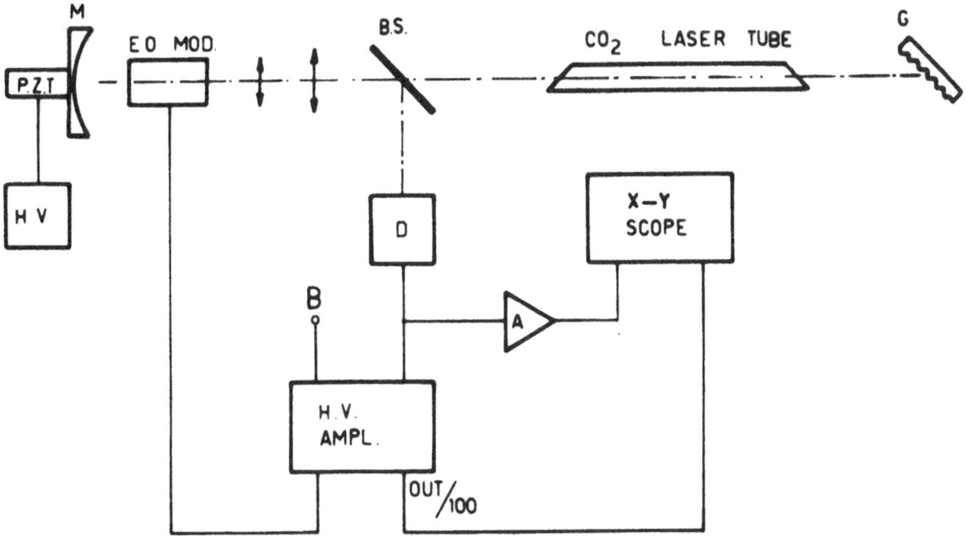

Fig. 1 Experimental set up. M - total reflecting mirror mounted on a PZT drive. E.O.MOD - electro-optic modulator. BS - ZnSe beam splitter. G - grating. D - HgCdTe detector. B - bias voltage.

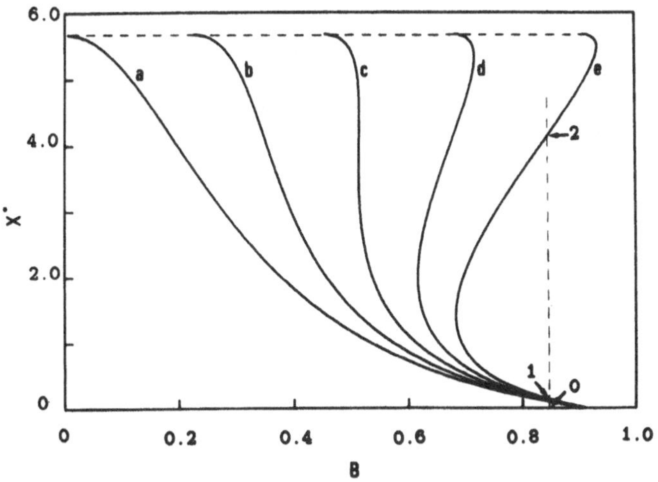

Fig. 2 Plot of the normalized stationary intensity x* versus B for a given pump rate. Call r the gain of the feedback loop. Curves a), b), c), d) and e) refer to r = 0.0, 0.04, 0.08, 0.12 and 0.16 respectively. Points 0,1 and 2 indicated by arrows, are the stationary points for B = 0.838 and r = 0.16.

accuracy. This is the phenomenon of fluctuation enhancement already described in the decay of the unstable state of a macroscopic system /18/. This phenomenon introduces unavoidable statistical features in the nonlinear dynamics of a macroscopic system. In such case the collective description in terms of a few dynamical variables breaks down, because of large fluctuations. This was first observed in the switch-on of a laser /18a/ and then in many quenching phenomena as spinodal decomposition /18b/.

Our experimental evidence of Shil'nikov type instability is based on a quantum optical system, namely a laser with an overall feedback. Precisely, we work on a single mode CO_2 laser with an intracavity electro-optic modulator yielding cavity losses proportional to the laser output intensity (Fig. 1) /14/. If the time scale of the feedback loop is of the same order of the other two relevant variables, the system becomes three dimensional. Such a system is described by three first order differential equations for the laser intensity x(t), population inversion y(t) and modulation voltage z(t) as follows /14/:

$$\begin{aligned} \dot{x} &= -K_0\, x \; (1 + \alpha \sin^2(z) - y), \\ \dot{y} &= -\gamma_{\parallel} \; (y + x\, y - A), \qquad (2) \\ \dot{z} &= -\beta\, (z - B + r\; x), \end{aligned}$$

where $K_0 = (c/L)\,T$ is the non-modulated cavity loss parameter, L is the cavity length, T is the effective transmition of the cavity, $\gamma_{\parallel}$ is the population decay rate. The intensity x(t) is normalized to the saturation intensity $I_s = \gamma_{\parallel}/2G$, with G the field-matter coupling constant. The population inversion y(t) is normalized to the threshold inversion K_0/G, z(t) is the modulation voltage normalized to the $\pi/V_{\lambda/2}$, with $V_{\lambda/2}$ the $\lambda/2$ modulator voltage, A is the normalized pump parameter, β is the damping rate of the feedback loop, r is a coupling coefficient between the detected intensity x(t) and the normalized z(t) voltage, B is the bias voltage appied to the EOM, $\alpha = (1 - T)/T$.

The stationary solution (x*, y*, z*) for the system (2) implies the condition

$$B = r\; x^* + \arcsin \left[(A/1 + x^*) - 1)/\alpha \right]^{1/2} \qquad (3)$$

In Fig. 2 we report the stationary laser intensity versus one of the control parameters (B) for different values of the second one (r). This shows the coexistence of three fixed points for a wide range of r values.

In Fig. 3 we present a schematic view of the trajectory in the three dimensional space, obtained by a linear stability analysis of the motion around the stationary points, and qualitative connections between the linear manifolds (dashed lines) /19/. Ref. 19 describes the

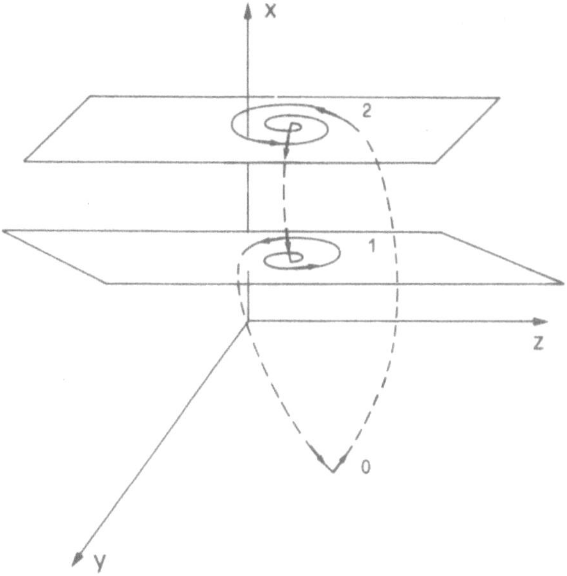

Fig. 3 Schematic view of a trajectory in the phase space (x = laser intensity, y = population inversion, z = feedback voltage) when the dynamics is affected by all the three unstable stationary points.

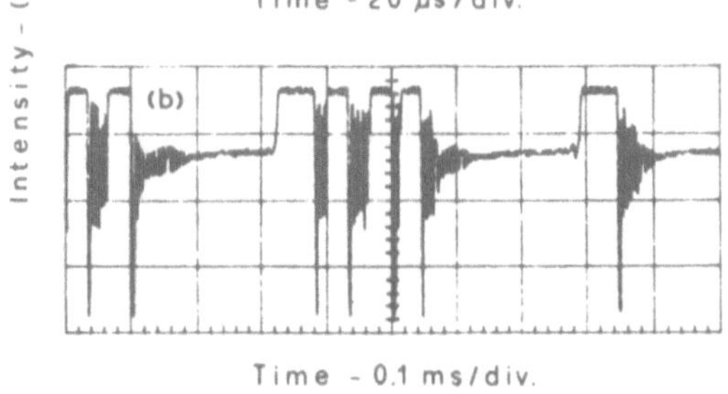

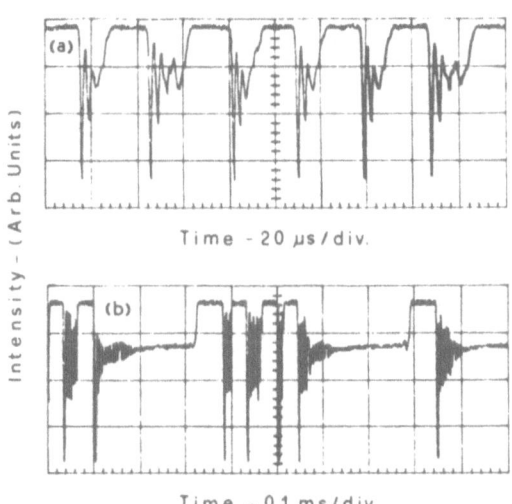

Fig. 4 Time plots of the laser intensity x(t) in the regime of Shil'nikov chaos. Intensity increases downward.a) and b) refer to the same B value, but two different gain r of the feedback loop. Fig.4b) shows two long transients corresponding to a large number of small spirals around the saddle focus (see also Fig. 5b)).

competition of the three instabilities in controlling the global features of the motion. We adjust the control parameter in order to have a dominance of the saddle focus and reduce the influence of the other two instabilities. This way, the motion consists of a homoclinic orbit asymptotic to the saddle focus. This instability provides an exponential divergence within the flow while the homoclinic orbit ensures that at least a portion of this diverging flow is reinjected into the neighbourhood of the saddle focus. This structure of the flow is one of the simplest capable of generating chaotic behavior in many automonous systems like for instance the Lorenz equations /20/ or the Belousov-Zhabotinskii reaction /21/. The figures show clear evidence of a homoclinic orbit with long transients, which provide a lengthy permanence in a phase space region of almost constant intensity. This appears more clearly in the corresponding phase space projections (Figs. 4, 5).

We measure the time spacing between successive orbits by setting a threshold circuit near the top of the largest peak of the intensity signal. A time to amplitude converter yields the sequence τ_i of successive time spacings, which is then classified as a statistical distribution by a multichannel pulse height analyser, or stored in a digitizer, so that correlation functions or return maps can be sorted out.

The statistical distribution of return times is a broad featureless curve which does not offer clues on the ordering of τ_i. On the contrary, the return map displays an extremely regular structure (Fig. 6). To check whether we are in presence of a one dimensional (1D) return map and the remaining thickness is due to the observation technique, or the map is more than 1D and its thickness hides new information, we measure also the return maps corresponding to three regular periodic situations. In the absence of fluctuations in τ_i they should be point like. In fact 1) corresponds to an electronic oscillator and it just shows the resolution of the measurement, 2) corresponds to the laser in a regular periodic regime away from the Shil'nikov instability, and 3) corresponds to the laser just on the verge of the instability but still with a regular period. In this last case, the fluctuation associated with the nerby transition shows that, even without chaos in the return time, the close approach to an instability point introduces a fluctuation enhancement, which has no theoretical counterpart in the current treatment of deterministic chaos.

To deal with this broadening, the dynamical equations should include a statistical spread in the injection coordinate to account for the macroscopic character of the experimental system.

From a theoretical point of view, a homoclinic orbit asymptotic to a saddle focus can be modelled in terms of the following 1D iteration map:

$$\zeta_{n+1} = \zeta_n^{\frac{\lambda}{\gamma}} \cos\left(\frac{\omega}{\gamma} \ln \zeta_n\right) + \epsilon \qquad (4)$$

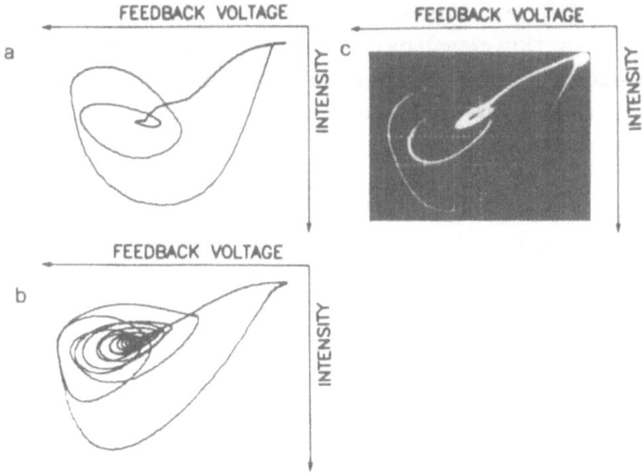

Fig. 5 Phase space projections x-z (laser intensity-feedback voltage). a) and b) are single orbits obtained by a digitizer, while c) is a photographic exposure over 1sec. a) and b) refer respectively to the same parameter situation of Fig. 4a), and b).

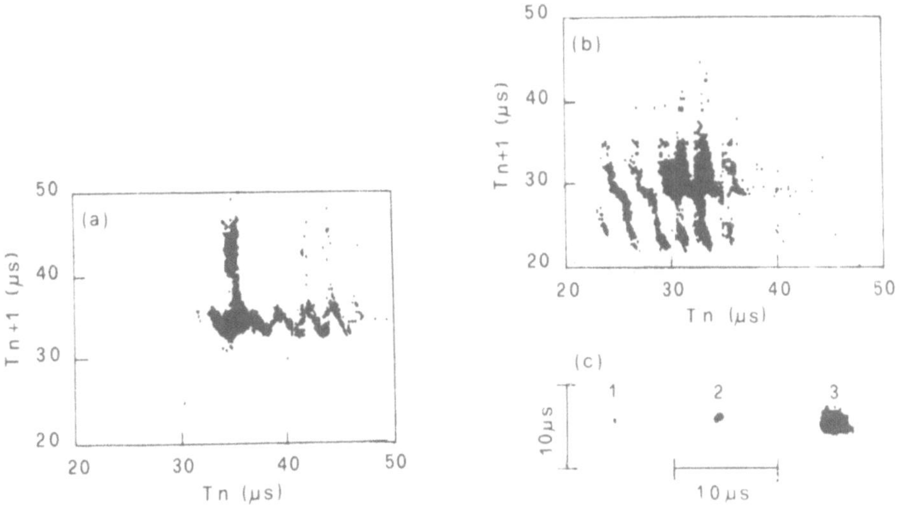

Fig. 6 Experimental return time maps. a) and b) refer to the same gain and to B values of 0.459 and 0.427 respectively. c) shows the maps corresponding to regular periodic situations, namely, 1) an electronic oscillator, 2) the laser in a regular periodic regime and 3) the laser just at the onset of the instability but still with a regular period.

where γ and $-\lambda \pm i\omega$ are the eigenvalues of the linearized flow at the saddle focus, ζ is the coordinate along the unstable manifold and ϵ is the deviation along ζ from the homoclinic orbit at the Poincaré section in the neighbourhood of the saddle point ($\epsilon = 0$ corresponds to the homoclinic condition).

If we build a small cubic box of unit side centered at the saddle focus and oriented along the eigenvectors, any tiny difference in the entrance coordinate along the expanding axis ζ will strongly influence the residence time inside the box and hence the spacing from the next re-injection.

Observing that most of the time is spent in the box around the saddle point, we relate the return time τ to the coordinate ζ of the unstable manifold by $\zeta = \zeta_0 \exp(\gamma\tau)$, thus obtaining an iteration map for the return times:

$$\tau_{n+1} = -\ln\left[e^{-\frac{\lambda}{\gamma}\tau_n} \cdot \cos\frac{\omega}{\gamma}\tau_n + \epsilon \right] \qquad (5)$$

where λ, γ, ω and ϵ are the same as above.

The main difference between the experimental maps reported in Fig. 6 and the theoretical ones (Fig. 7), is related to the finite thickening independent of the accuracy of the measurements. This spread is due to a transient fluctuation enhancement, peculiar in the decay of an unstable state of a macroscopic system as already stressed in Ref. 18. This phenomenon unavoidably introduces further statistical fluctuations into the chaotic dynamics of a macroscopic system.

As it was shown in Ref. 18, even though this spread has no relevance on the average dynamics, it contributes a large fluctuation enhancement whenever the system slides downhill. In Ref. 18 this was observed in the transient decay of an unstable state, here we report the same feature repeatea at each Poincaré cycle.

As a conclusion, low dimensional chaos is described by a small number of coupled deterministic equations (as e.g. Eq (1)), that we write in general as

$$\dot{x}_i = F_i(\{x_j\}) \quad (i,j = 1, 2, 3) \qquad (6)$$

where F_i are nonlinear functions whose power expansion implies terms as $x_i x_j$ and higher order. But whenever this low dimensional chaos is a contracted description of a large system, then x_i are collective variables corresponding to a macroscopic dynamics, and the nonlinearities of Eq (2) depend critically on whether

$$\langle x_i x_j \rangle \simeq \langle x_i \rangle \langle x_j \rangle . \qquad (7)$$

Relation (7) fails to hold in the decay situations typical of Shilnikov chaos. In such a case, we have a strong coupling between nonlinear chaotic dynamics and statistical mechanics.

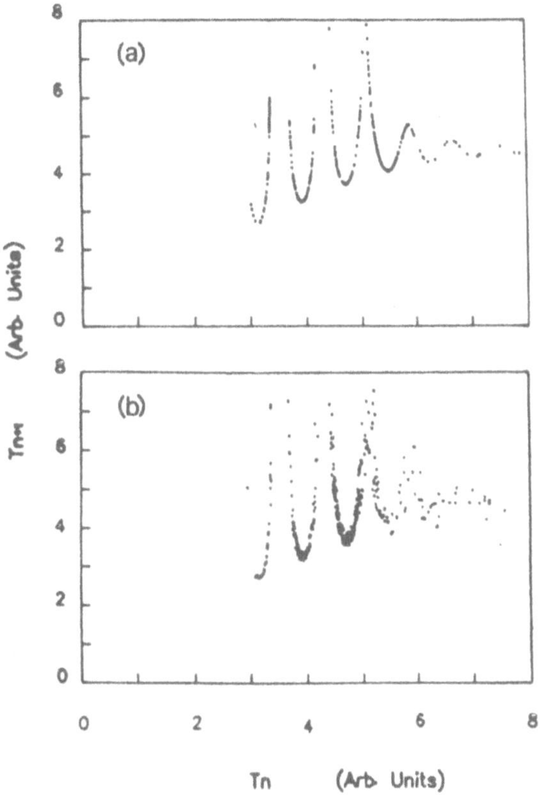

Fig. 7 Numerical return time maps for Shil'nokov chaos. a) no addition of noise or anharmonic contributions. b) with the addition of noise and 1% of second and third harmonic contributions.

/1/ W.E. Lamb Jr., Phys. Rev. 134, A1429 (1964).
/2/ H. Haken, Laser Theory, in Encyclopedia of Physics vol. XXV/2c, ed. S. Flügge, Springer 1970.
/3/ M. Scully and W.E. Lamb Jr., Phys. Rev. Letters 16, 853 (1966), Phys. Rev. 159, 208 (1967) and 166, 246 (1968).
/4/ J.P. Gordon, Phys. Rev. 161, 367 (1967).
/5/ H. Haken, Synergetics, 3d Ed., Springer 1983.
/6/ F.T. Arecchi, in Order and Fluctuations in Equilibrium and Nonequilibrium Statistical Mechanics (Proc. XVII Solvay Conf. on Physics, ed. G. Nicolis et al.), J. Wiley 1981.
/7/ R.J. Glauber, in Quantum Optics and Electronics, (ed. C. De Witt et al.) Gordon and Breach, 1965.
/8/ F.T. Arecchi, in Quantum Optics (ed. R.J. Glauber), Academic Press 1969.
/9/ F.T. Arecchi and R.G. Harrison (Eds) Instabilities and Chaos in Quantum Optics, Springer Verlag 1987.
/10/ E.M. Lorenz, J. Atmos. Sci. 20, 130 (1963).
/11/ F.T. Arecchi, G.L. Lippi, G.P. Puccioni and J.R. Tredicce, Optics Comm. 51, 308 (1984).
/12/ F.T. Arecchi, R. Meucci, G. P. Puccioni and J.R. Tredicce, Phys. Rev. Lett. 49, 1217 (1982).
/13/ G.L. Lippi, J.R. Tredicce, N.B. Abraham and F.T. Arecchi, Optics Comm. 53, 129 (1985).
/14/ F.T. Arecchi, W. Gadomski and R. Meucci Phys. Rev. A43, 1617 (1986).
/15/ F.T. Arecchi, A. Lapucci, R. Meucci, J.A. Roversi and P. Coullet (to be published).
/16/ L.P. Shilnikov, Dokl. Aka. Nauk SSSR 160, 558 (1965).
L.P. Shilnikov, Mat. Sbornik 77, (119) 461 (1968) and 81, (123) 92 (1970).
/17/ A. Arneodo, P.H. Coullet, E.A. Spiegel and C. Tresser, Physica 14D, 327 (1985).
/18/ a) F.T. Arecchi, V. Degiorgio and B. Querzola, Phys. Rev. Lett. 19, 1168 (1967); b) F. Haake, Phys. Rev. Lett. 41, 1685 (1978); c) F.T. Arecchi and A. Politi, Phys. Rev. lett. 45, 1215 (1980); F.T. Arecchi, A. Politi and L. Ulivi, Nuovo Cimento 71B, 119 (1982).
/19/ F.T. Arecchi, R. Meucci and W. Gadomski, Phys. Rev. Lett. 58, 2205 (1987).
/20/ P. Glendinning and C. Sparrow, Jour. Stat. Phys. 35, 645 (1984).
/21/ F. Argoul, A. Arneodo and P. Richetti, Phys. Lett. A120, 269 (1987).

RECENT RESULTS OF EXPERIMENTS WITH SAFFMAN-TAYLOR FLOW

Mark W. DiFrancesco*

Department of Physics and Astronomy
University of Pittsburgh
Pittsburgh, PA 15260 U.S.A.

In this paper we summarize our recent work on the viscous fingering problem[1,2]. In its initially planar form, Saffman-Taylor flow[3], viscous fingering represents the simplest of pattern formation problems. Thus the observation of its details provides a valuable opportunity for direct testing of the computer calculations which are now just becoming feasible for pattern formation[4]. Despite its extreme simplicity, Saffman-Taylor flow has much in common with the Mullins-Sekerka instability[5] which gives rise to dendritic growth in alloys. In this paper we first discuss the formal similarity between the dispersion relations for the Saffman-Taylor and Mullins-Sekerka instabilities, then we set out some of our recent results on the Saffman-Taylor problem[1].

The Saffman-Taylor instability[1] arises at the initially planar interface between two fluids flowing in a Hele-Shaw cell (a cell formed by parallel plates with a gap between them of thickness b where b is smaller than any other length scale in the problem). It is driven either by a pressure gradient advancing the less viscous fluid against the more viscous or by gravity as a result of a density difference between the fluids. For the case of gravity driven flow in a closed rectangular cell where the average velocity of the interface is zero, the dispersion relation, from linear stability analysis, takes the form:

$$i\omega\,[2\bar{\mu}/K] - (\rho_2-\rho_1)\,gk + \sigma^*\,k^3 = 0 \qquad (1)$$

where $K \equiv b^2/12$ is a mobility. The average shear viscosity, the effective interfacial surface tension, the density of fluid n, and the acceleration due to gravity are represented by $\bar{\mu}$, σ^*, ρ_n, and g respectively. This dispersion relation predicts broad band instability for all wavenumbers, k, below a critical value set by σ^*, $\Delta\rho$, $\bar{\mu}$, and b.

When a binary liquid, in its two phase region, is quenched rapidly further into that region, a concentration gradient is imposed at the interface between the phases. It has been proposed[6] that such a gradient leads to a Mullins-Sekera type instability envisioned in its simplest form, namely, the stationary symmetric model. In this picture, linear analysis reveals the dispersion relation:

$$i\omega - (2D/\ell)\,k + 2D\,\ell_c\,k^3 = 0 \qquad (2)$$

where D is the diffusion coefficient of the liquid, l_c is a capillary length, and l characterizes the thickness of the region near the interface where the concentration gradient is significant.

The two dispersion relations described above obviously share the same generic form with a simple competition between a destabilizing linear term and a stabilizing cubic term. This suggests that knowledge of one may well provide information about both, at least for the onset of the instability (where the linear stability analysis should be valid) and possibly for early stages of the nonlinear pattern formation. This hope is heightened by the formal similarity of calculations of viscous fingering and dendritic growth in the case where one of the fluids in the fingering flow has negligible viscosity in comparison to the other fluid[7]. With all these considerations in mind we present our results for relatively early stages of viscous fingering and discuss analyses of these results for features which could be directly compared with feasible but admittedly large scale computer calculations. Further work will concentrate on varying flow parameters to overlap with and then extend the study of equations 1 and 2 out of the ranges previously covered in solidification experiments.

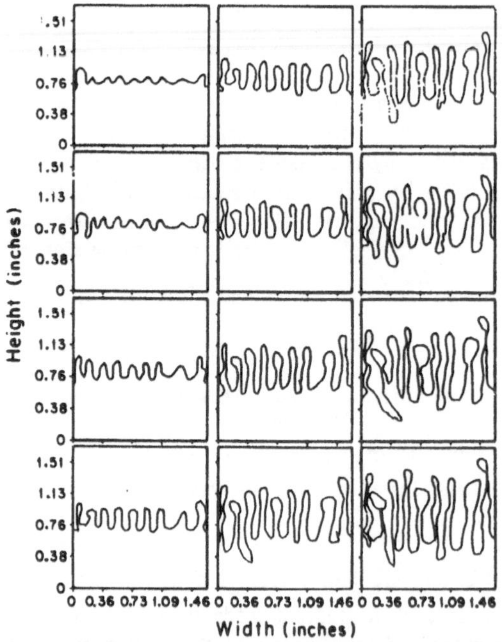

Fig. 1. Time series of fingering patterns. Time progression is down succeeding columns starting with the pattern in the upper left corner.

RESULTS OF RECENT EXPERIMENTS

As was mentioned above, a basic rectangular Hele-Shaw cell consists of two rectangular glass plates separated by a uniform gap that is very small compared with the other linear dimensions.

In such a cell, fluid flow is effectively two dimensional and described by Darcy's Law. Under the proper conditions, the interface between two immiscible liquids in a cell becomes unstable and forms patterns like the ones shown in Figure 1.

One goal of the study of patterns is to find ways to describe their shapes and development that are universal and meaningful for comparison with the results of calculation. An obvious first approach is to observe some gross feature of the flow, such as the length of the fingers as a function of time. Figure 2 shows the dimensionless length of the longest finger, θ, as a function of dimensionless time for several runs with widely varying parameters (i.e. viscosity contrast and surface tension). The lines labeled 5 and 6 correspond to flows driven by a pressure gradient while the others result from gravity driven flows. All show θ increasing with time with a power law with an exponent of ≈ 1.5. There is a formal similarity between this flow and diffusion-limited-aggregation (DLA)[7,8] so it may be no accident that the exponent observed here is the same as that seen in computer simulations of the temporal growth of two-dimensional DLA patterns.

A more detailed picture of the interface has been gained through a Fourier analysis of its shape. Figure 3 shows an example of an interface four seconds into a flow and its corresponding Fourier transform exhibiting dominant wavenumbers which agree roughly with the numbers of fingers one would judge to be in the pattern. The Fourier transforms are useful in that they enable one to follow the development of individual wavenumbers. In this way, it is possible to compare the behavior of the various length scales imbedded in the patterns. For instance, the temporal developments of the two dominant wavenumbers in Figure 3 are depicted in Figure 4.

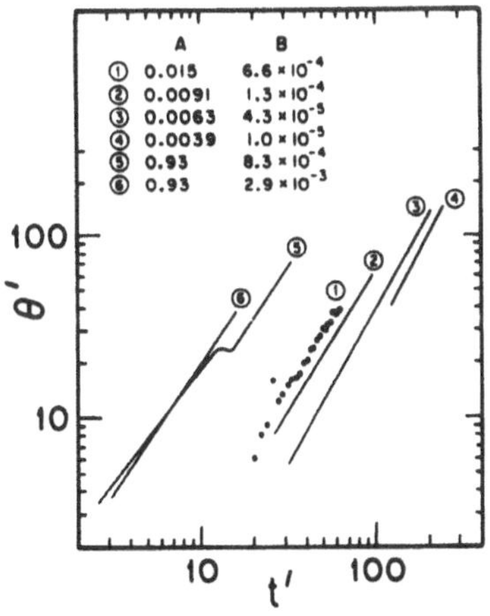

Fig. 2. Dimensionless length of the longest finger vs. dimensionless time as discussed in the text (see ref. 1).

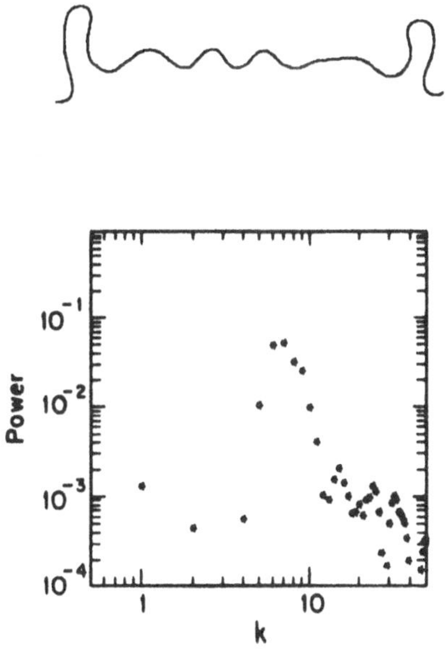

Fig. 3. Trace of a fingering pattern four seconds into a flow and its corresponding Fourier transform.

One, however, soon encounters a difficulty in forming the Fourier transform of the flow patterns as they develop beyond the first few seconds. The pattern becomes a multivalued function of position across the cell as the fingers become more complicated, forming "balloons" at their tips. We have thus decided to use, as an alternative, a plot of the local curvature vs arclength for each pattern's description. Curvature as a function of arclength has two clear advantages:

1) it remains single-valued and thus allows a modal analysis no matter how complicated the interface becomes and 2) no information about the local interface is discarded when this function replaces the spatial pattern.

A concomitant disadvantage is that we have kept all the considerable physical detail and are now dealing with a function for which we have less intuition than we had for the spatial pattern. As a first attempt at digesting the information contained in the curvatures, we have constructed "power spectra" - the square of the Fourier amplitudes resulting from a modal analysis of the curvature data.

One of our current projects is to try to scale the power spectra over a wide variety of run conditions. For the present, however, our best insight is provided by a much more crude measure, the first moments of the power spectra as they evolve in time. If we represent a power spectrum as P (k), then this moment, $\bar{k}$, is

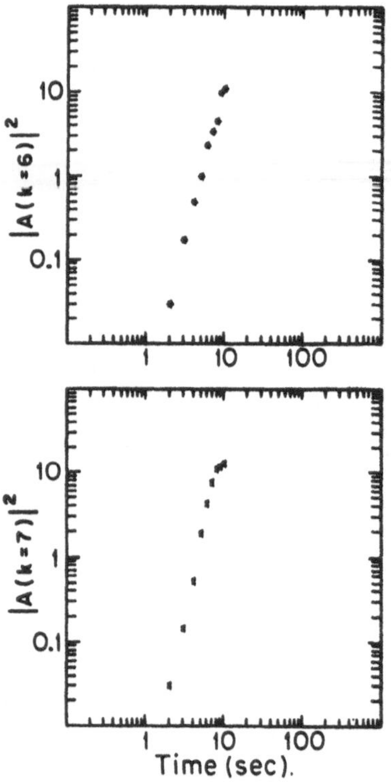

Fig. 4. Plots of the square Fourier amplitude vs time for the dominant wavenumber, k = 6 and k = 7, shown in Fig. 3.

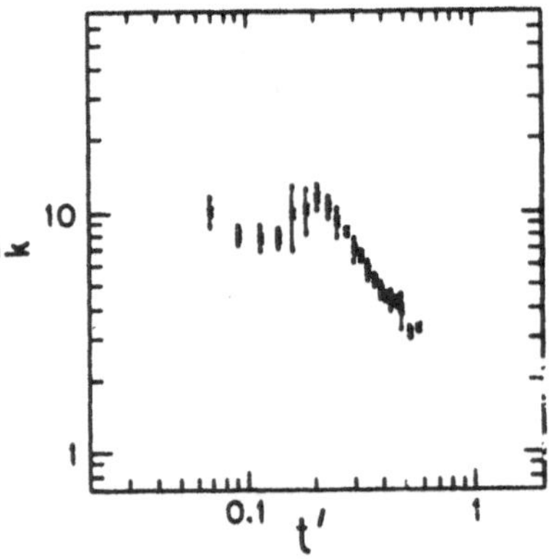

Fig. 5. The average wavenumber, $\bar{k}$, vs dimensionless time.

$$\bar{k} = \int kP(k)\, dk \;/\; \int P(k)\, dk.$$

In Figure 5 a plot of $\bar{k}$ vs dimensionless time shows, after an initial noisy period, that $\bar{k} \approx t^{-1.6}$. This sharp tendency toward longer wavelengths at late times may suggest a progressive smoothing of the patterns. Each value of $\bar{k}$ in this plot is actually itself an average from an ensemble of 14 runs under identical conditions. The approximately 5% standard deviations shown in the figure do not result from any $\sim$ 5% difficulty in controlling the initial conditions but rather represent the intrinsic noise in the flow presumably arising from very small fluctuations in the initial conditions.

In summary, the tantalizing preliminary result presented above concerns the secular development of average wavenumbers $\bar{k}$ for power spectra of curvatures from initially planar viscous fingering patterns. These $\bar{k}$, after a complicated initially ramifying flow period, decrease with time and represent a coursening of the pattern in the quiet regions behind the advancing fingers to a point where the interfacial capillary length is small in comparison to typical long smooth sections of the interface.

Acknowledgement-This work was supported by the U. S. Department of Energy under grant #DE-FG02-84ER45131.

REFERENCES

* Work done in collaboration with J. V. Maher.

1. J. V. Maher, Phys. Rev. Lett. **54**, 1498 (1985); in, The Physics of Finely Divided Matter, ed. by N. Boccara and M. Daoud (Springer-Verlag, Berlin, 1985) p. 252 - 257.
2. S. N. Rauseo, P. D. Barnes, Jr., and J. V. Maher, Phys. Rev. A **35**, 1245 (1987).
3. P. G. Saffman and G. I. Taylor, Proc. R. Soc. London Ser. A **245**, 312 (1958).
4. D. Kessler and H. Levine, Phys. Rev. A **33**, 2632 (1986); **33**, 2639 (1986); L. M. Sander, P. Ramanlal, and E. Ben-Jacob, ibid. **32**, 3160 (1985); E. Ben-Jacob, N. D. Goldenfeld, J. Koplik, H. Levine, T. Mueller, and L. M. Sander, Phys. Rev. Lett. **55**, 1315 (1985); G. Daccord, J. Nittman, and H. E. Stanley, ibid. **56**, 336 (1986); B. I. Shraiman, ibid. **56**, 2028 (1986); D. C. Hong and J. S. Langer, ibid. **56**, 2032 (1986); R. Combescot, T. Dombre, V. Hakim, Y. Pomeau, and A. Pumir, ibid. **56**, 2036 (1986); D. Bensimon, L. P. Kadanoff, S. Liang, B. I. Shraiman, and C. Tang (unpublished); A.J. DiGregoria and L. W. Schwartz, J. Fluid Mech. **164**, 383 (1986); J. Nittman, H. E. Stanley, Nature **321**, 663 (1986).
5. W. W. Mullins and R. F. Sekerka, J. Appl. Phys. **34**, 323 (1963); **35**, 444 (1964).
6. J. S. Langer and L. A. Turski, Acta Metall. **25**, 1113 (1977); hydrodynamic degrees of freedom added by D. Jasnow, D. A. Nicole, and T. Ohta, Phys. Rev. A **23**, 3192 (1981).
7. L. P. Kadanoff, J. Stat. Phys. **39**, 267 (1985).
8. T. A. Witten and L. M. Sander, Phys. Rev. Lett. **47**, 1499 (1981); Phys. Rev. B **27**, 5685 (1983).

INSTABILITES AND CHAOS IN ROTATING FLUIDS*

Harry L. Swinney

Center for Nonlinear Dynamics and Department of Physics
The University of Texas
Austin, TX 78712 USA.

0. *INTRODUCTION*

The following text summarizes the two lectures I presented at the Noto NATO Summer School. Both lectures were concerned with turbulence in rotating fluids, but from different points of view, and involving different experiments. The summaries are well enough referenced, I hope, that the interested reader may be able to use them to delve in greater depth into the subjects treated so briefly here.

I. *COUETTE FLOW*

We shall consider a system consisting of two rotating cylinders of length L. The inner cylinder, of radius a, has an angular velocity Ω_i, while the outer cylinder, of radius b, has angular velocity Ω_o. A fluid is contained in the annular region of width d=b-a between the two cylinders. Denoting the kinematic viscosity by ν, we define inner and outer Reynolds numbers by

$$R_i = \frac{ad}{\nu} \Omega_i \; , \qquad (1a)$$

$$R_o = \frac{bd}{\nu} \Omega_o \; . \qquad (1b)$$

Other important parameters include the radius ratio

$$\eta = \frac{a}{b} \qquad (1c)$$

and the aspect ratio

*Research supported by the Office of Naval Research Nonlinear Dynamics Program.

$$\Gamma = \frac{L}{d} \quad . \tag{1d}$$

Such a system is described by (non-linear) Navier-Stokes equation

$$\frac{\partial \vec{v}(\vec{r},t)}{\partial t} + (\vec{v}\cdot\vec{\nabla})\vec{v} = -\frac{1}{\rho}\vec{\nabla}p(\vec{r},t) + \nu\nabla^2\vec{v} \tag{2a}$$

where $v(\vec{r},t)$ is the local fluid velocity, p the pressure and ρ the density.

For an incompressible fluid (assumed throughout) we have as well

$$\vec{\nabla}\cdot\vec{v} = 0. \tag{2b}$$

The usual "no slip" boundary conditions then give

$$v_\theta(r=a) = a\Omega_i \ , \quad v_r(a) = v_z(a) = 0 \quad ; \tag{3a}$$

$$v_\theta(r=b) = b\Omega_i \ , \quad v_r(b) = v_z(b) = 0 \quad . \tag{3b}$$

Flow satisfying these conditions is called "basic" flow.

The simplest stability analysis of a system such as the one we have described involves purely inviscid flow ($\nu=0$) between infinitely long cylinders. Then simple centrifugal force arguments indicate that the flow will be stable if the following inequality is satisfied:

$$\Omega_i a^2 < \Omega_o b^2 \ .$$

This is the so-called Rayleigh criterion.

Any basic flow at low Reynolds numbers has the symmetry of the boundary conditions. The basic flow solution for the Couette system is

$$v_\theta = Ar + \frac{B}{r} \tag{5a}$$

with the constants A and B given by

$$A = -\Omega_i(\eta^2-\mu)/(1-\eta^2) \tag{5b}$$

and

$$B = \Omega_i a^2(1-\mu)/(1-\eta^2) \quad . \tag{5b}$$

In 1923, Taylor[1] analyzed this flow both experimentally and theoretically (using linear stability analysis). He obtained an

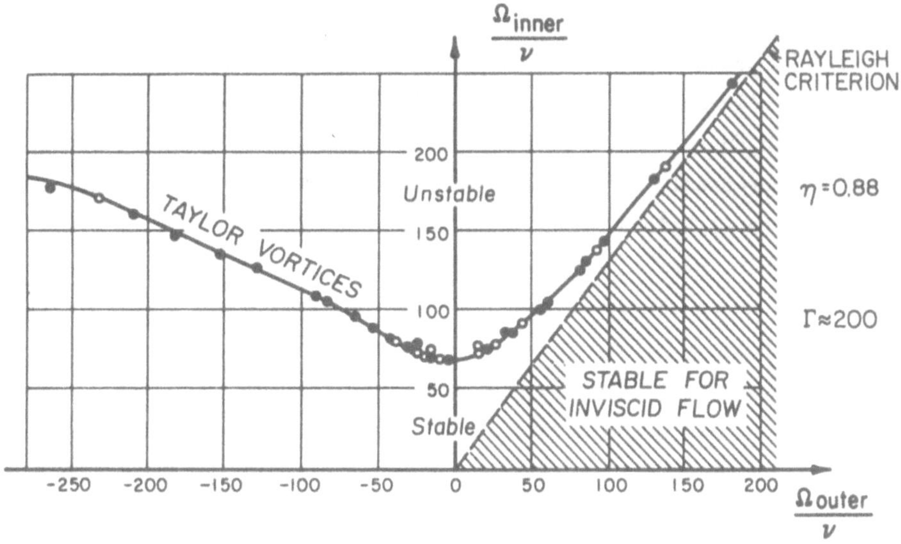

Fig. 1. Comparison between observed and calculated speeds at which instability first appears; case when R_1 = 3·55 cm., R_3 = 4·035 cm. (Taylor, 1923)

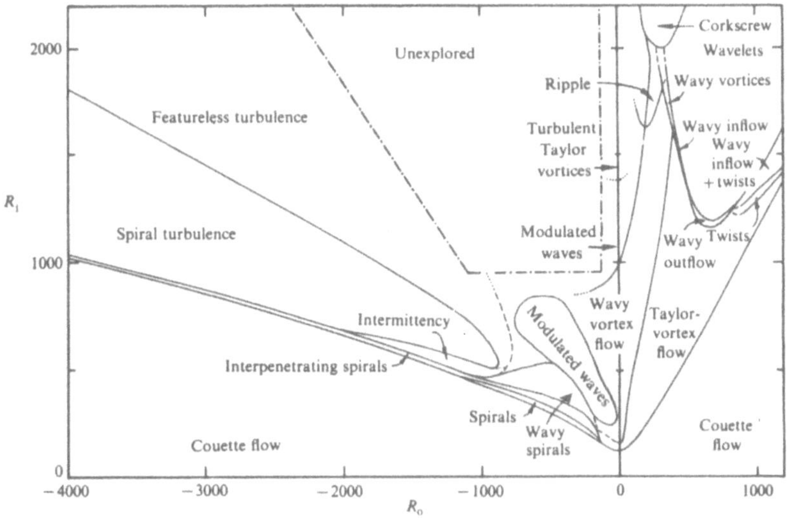

Fig. 2. Regimes observed in flow between independently rotating concentric cylinders. Dashed lines indicate the transition boundaries that are difficult to establish from visual observation alone since there is no abrupt change in the appearance. Dotted lines indicate the expected, but not yet observed, continuation of several boundaries. (From Andereck et al.[7])

improvement to the Rayleigh criterion by relaxing the inviscidity assumption. His theoretical results are compared in Fig. 1 with his measurements for $\eta = 0.88$. The theory assumed infinitely long cylinders, while in the experiment Γ=200. The shaded area indicates the region predicted to be stable according to the Rayleigh criterion, while Taylor's theoretical results are shown by the solid curve, superimposed with experimental points. This curve indicates the onset of axisymmetric vortices, now called Taylor vortices. This was the <u>first</u> direct comparison between theory (linear stability analysis) and experiment for a flow instability.

In 1966, Krueger, Gross and DiPrima[2] considered the case in which the flow could have an azimuthal dependence (i.e., $\exp(im\theta)$ for $m=1,2,\ldots$). They (analytically) reproduced the Taylor vortices, but in addition, they found that when the cylinders are counter-rotating sufficiently fast, the primary instability is a spiral flow rather than Taylor vortices. The analysis of reference 2 was perhaps motivated by experiments carried out by Coles[3] in 1965 in which the spirals were observed, and their speed was measured. (It was shown to tend to $\Omega/3$ for large R, independent of the other system parameters.[4,5]) Recently the onset of Taylor vortices and spirals has been studied for a wide range of parameters by Langford et al.[6]

These studies marked the beginning of a large number of experiments and calculations in which the system parameters were systematically varied and, in particular, in which higher rotational velocites were introduced. The result has been the discovery of an incredibly rich instability structure in the flow, which goes far beyond simple Taylor vortices and/or spirals. This is illustrated in the phase diagram of Andereck et al.[7] in Fig. 2; observations include experiments by King et al[4] and Gorman and Swinney[8] and numerical simulations by Marcus.[5]

I should now like to turn to a discussion of chaotic (weakly turbulent) flow, and the existence of strange attractors. In the experiments conducted by Fenstermacher et al.[9] and Brandstater and Swinney,[10] the parameters were η=0.875, Γ=20, and ν=0.0109 cm^2/sec. The critical Reynolds number for the onset of Taylor vortices turns out to be 118.4 (in an infinite system). The following sequence of well-defined states was observed[9] as the Reynolds number R_c increased through and beyond the critical value: basic flow; Taylor vortex flow; wavy vortex flow (periodic)[4,5]; modulated wavy vortex flow (quasi-periodic)[8]; and chaotic (weakly turbulent) flow.[9,10] Fig. 3 shows the onset of chaos which occurs at a Reynolds number of about

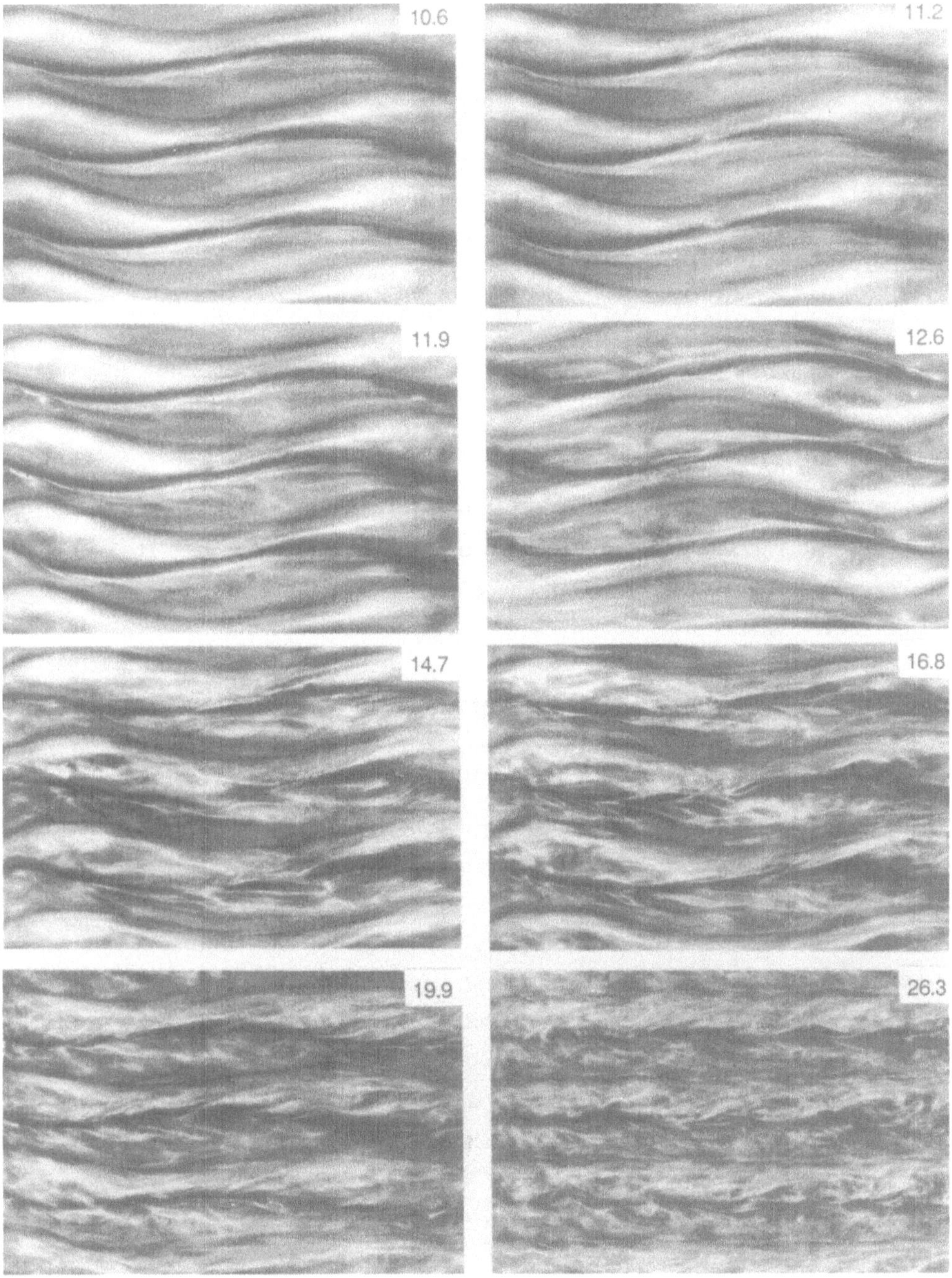

Fig. 3. Photographs of the flow as a function of Reynolds number in the region of transition from modulated wavy vortex flow to chaotic flow. The values of R/R_c are given on the photographs. The first two pictures are of modulated wavy vortex flow, while the remainder are in the chaotic regime. The flow patterns were rendered visible using a suspension of small platelets. (From Brandstater and Swinney.[10])

$12R_c$. (This agrees with an experimental determination of the Lyapunov exponents--the first one becomes positive at $11.8R_c$.)

The basic quantity measured was the velocity time series $V(t_k)$; $t_k = k\Delta t$, $1 \leq k \leq 32{,}768$. Phase space portraits were constructed by the method of time delays,[10] in which a point in an m-dimensional space is given by $\{V(t), V(t+\tau), \cdots V(t+(m-1)\tau)\}$, where τ is a time delay. The value of τ was chosen by a method introduced by Fraser and Swinney[11] (in a noiseless experiment, the results should be independent of τ, but in a practical case an optimum choice should be made.) V, the radial component of velocity, was measured by laser Doppler shifts. In Fig. 4 we show two-dimensional phase portraits of the velocity time series, i.e. $V(t+\tau)$ vs. $V(t)$ for two quasi-periodic and two chaotic situations. By reducing the data of these curves to Poincaré sections and using them to define a map on the circle (placing the origin of a polar system of co-ordinates in the interior of the Poincaré curve), we were able to show that the flow was deterministic, hence involving a (strange) attractor. The attractor dimension was calculated. It increases from 2 at the onset of chaos to about 4 for a Reynolds number 50 percent higher.[10]

II. COHERENT STRUCTURE IN TURBULENT ROTATING FLUIDS

A good example of coherent structures is found in planetary atmospheres, for example the earth's. Here we have the Gulf stream and the jet stream, both counter-rotating and both narrow and well defined; the trade winds, on the other hand, are co-rotating and diffuse. (This seems to be a general distinction between co- and counter-rotating flows.) In addition, there exist vortices, such as the various long-lived highs and lows in the earth's atmosphere and, more spectacular perhaps, the Great Red Spot of Jupiter.

The goal of the present study is to produce such phenomena in a laboratory environment, and to study their stability. As in Sec. I the flow can be studied analytically by using the Navier-Stokes equations, but in this case in a rotating rather than an inertial coordinate frame. The Coriolis force becomes an important aspect of the problem (the centrifugal force can effectively be absorbed into the pressure term). The Coriolis force is measured by the Rossby number Ro, small Ro corresponding to large Coriolis force. In addition, the Ekmann number Ek measures the ratio of viscous to Coriolis force. The experiments described below have been carried out by Sommeria at al.[12]

The experimental apparatus was a rotating annulus filled with fluid

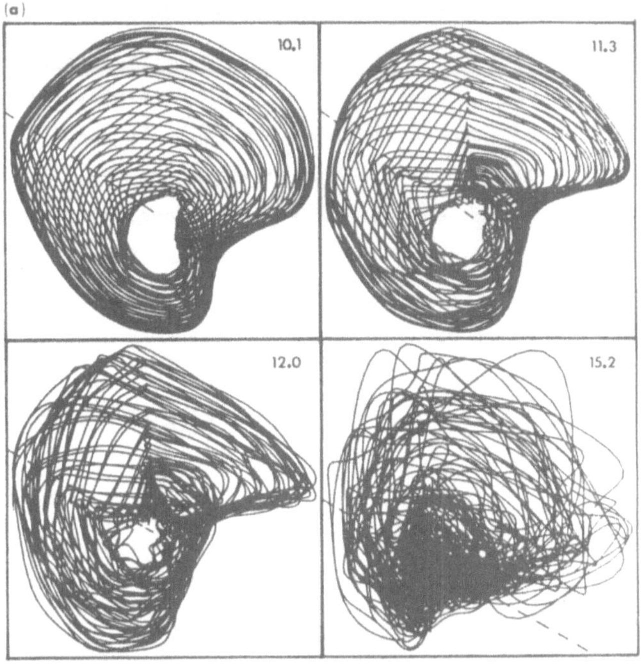

Fig. 4 (a) Two-dimensional phase portraits $V(t+\tau)$ versus $V(t)$ constructed for two Reynolds numbers (R/R_c = 10.1 and 11.3) corresponding to modulated wavy vortex flow and two Reynolds numbers (R/R_c = 12.0 and 15.2) corresponding to chaotic flow. [The delay times τ at the four Reynolds numbers are 162, 144, 144, and 108 ms, respectively; about 40 of the 300 orbits observed at each Reynolds number are shown in (a).] (From Brandstater and Swinney.[10])

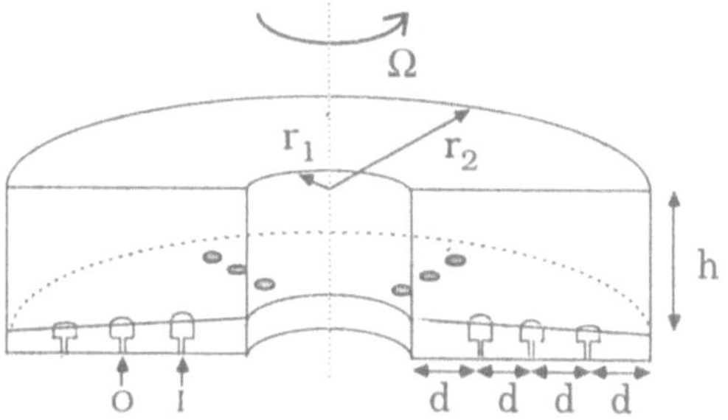

Fig. 5. Sketch of the apparatus: r_1=10.8 cm, r_2=43.2 cm, d=8.1 cm, $h(r_1)$=17.1 cm, and $h(r_2)$=20.3. The total flow rate F ranged from 15 to 370 cm^3/s, and the rotation rate Ω ranged from 0 to 25 rad/s. A counter-rotating jet was produced by the action of the Coriolis force on the fluid (water) pumped into the tank through a ring of six inlets (I) at r=18.9 cm, and from the tank through a ring of six outlets (O) at r=27.0 cm (the ring of ports at r=35.1 cm was not used). The effect of this forcing on the boundary layers was minimized by using 500 small holes (0.08 cm diameter) instead of a single hole at each inlet and outlet. (From Sommeria et al.[9])

pumped in the radial direction at Ro ≤ 0.1 and Ek $\sim 10^{-6}$. In order to mimic the latitude dependence of the Coriolis force, the tank was built with a sloping bottom, i.e. a continuously increasing depth. (A schematic of the apparatus is given in Fig. 5.) It was found that for a wide range of conditions with a counter-rotating jet, a single large vortex forms; this vortex exhibits the basic dynamical properties of the long-lived vortices of Jupiter. It seems to last indefinitely, and absorbs such smaller vortices as may form. The measurements agree with numerical simulations of P. S. Marcus.[13]

The dynamical behavior is quite different for a co-rotating jet, which is narrow and wavy. This jet has striking similarities to planetary eastward jets.

III. *REFERENCES*

1. G. I. Taylor, Phil. Trans. Roy. Soc. (London), A223, 289(1923).
2. E. R. Krueger, A. Gross and R. C. DiPrima, J. Fluid Mech. 24, 521(1966).
3. D. Coles, Ibid, 21, 385(1965).
4. G. P. King, Y. Li, W. Lee, H. L. Swinney and P. S. Marcus, Ibid 141, 365(1984).
5. P. S. Marcus, Ibid 146 45,65(1984).
6. W. Langford, R. Tagg, E. Kostelich, H. Swinney, and M. Golubitsky, Physics of Fluids, to appear.
7. C. D. Andereck, S. S. Liu, and H. L. Swinney, J. Fluid Mech. 164, 155(1986).
8. M. Gorman and H. L. Swinney, Ibid 117, 123(1982); R. S. Shaw, C. D. Andereck, L. A. Reith, and H. L. Swinney, Phys. Rev. Lett. 48, 1172(1981); M. Gorman, H. L. Swinney, and D. A. Rand, Ibid., 46, 992(1981).
9. P. R. Fenstermacher, H. L. Swinney and J. Gollub, J. Fluid Mech. 94 , 103(1979).
10. A. Brandstater and H. L. Swinney, Phys. Rev. 35A, 2207(1987).
11. A. M. Fraser and H. L. Swinney, Phys. Rev. 33A , 1134(1986).
12. J. Sommeria, S. D. Meyers and H. L. Swinney (To appear in Nature, 1987).
13. P. S. Marcus, Ibid.

FROM CHAOS TO TURBULENCE IN AN HELIUM EXPERIMENT

Albert Libchaber

Department of Physics
J. Frank Institute
University of Chicago
Chicago, Illinois 60637

Libchaber described an experiment in which a sample of Helium at 4K is made to range, by pressure changes, through ten orders of magnitude in the Reynold's number (or, better, Rayleigh number). By heat conduction measurements, the Nusselt number could be measured, and a plot of Nusselt vs. Rayleigh numbers isolated regions of heat diffusion, convection, liquid oscillation, chaos, transition to turbulence and "soft" and "hard" turbulence. He also discussed some models of turbulence developed by Kadanoff and co-workers, and expressed hope that we are beginning to obtain at least a primitive understanding of this phenomenon. He also described the measurement of wave fronts between two fluids (for example, the interface between two fluids of different viscosities, such as oil and water). He presented beautiful pictures of instabilities developing and, indeed, fully developed, and conjectured that flat fronts are always unstable, while curvature tends to lead to stability.

COMPLEX DYNAMICS IN EXPERIMENTS ON A CHEMICAL REACTION

W. D. McCormick

Center for Nonlinear Dynamics and Physics Department
The University of Texas
Austin, Texas 78712

McCormick described chemical experiments in which a number of chemicals are introduced into a stirred-flow reactor, and the concentration of one of the products emerging is measured as a function of time. Very beautiful period-doubling as well as chaotic phenomena are observed (including a specific attractor now known as the "Texas attractor"), and the phenomenon of intermittency. He also described some measurements of a chemical wave in a dish reactor, and showed pictures of the wave propagating in a gel. He went into considerable detail discussing experimental difficulties (e.g. impurities in his reactants), and also presented theoretical analyses of the experimental results.

ADDITIONAL CONTRIBUTED SEMINARS (TITLES)

SUPPRESSION OF CLASSICAL CHAOS IN A HARMONIC WAVE GUIDE

Joachim Krug

Theoretische Physik
Universität München
Theresienstr. 37
8 München
West Germany

ANDERSON LOCALIZATION

Fabio Martinelli

Universitá di Roma
Dipartimento di Matematica
P.le A. Moro
00162 Roma
Italy

REMEDIAL DYNAMICS I. A VERY, VERY, VERY ELEMENTARY TUTORIAL; REMEDIAL DYNAMICS II. A VERY, VERY ELEMENTARY TUTORIAL; AND REMEDIAL DYNAMICS III. A VERY ELEMENTARY TUTORIAL

Jack Dorning

Department of Nuclear Engineering and Engineering Physics
Nuclear Reactor Facility
School of Engineering and Applied Science
University of Virginia
Charlottesville, Virginia 22901

CONTRIBUTORS

F. T. Arecchi, Istituto Nazionale Di Ottica, Department of Physics, Largo E. Fermi 6, 50125 Firenze, Italy

James E. Bayfield, Department of Physics and Astronomy, University of Pittsburgh, Pittsburgh, Pennsylvania 15260 USA

Giancario Benettin, Dipartimento Di Fisica "G. Galil, Universita Di Padova, Via F. Marzolo 8, 35131 Padova, Italy

Michael Berry, H. H. Wills Physics Laboratory, Royal Fort, Tyndall Avenue, Bristol BS8 1TL, United Kingdom

Giulio Casati, Dipartimento Di Fisica, Via Celoia 16, 20100 Milano, Italy

Pierre Collet, Centre De Physique Theorique, Ecole Polytechnique, F91128 Palaiseau CEDEX, France

Predrag Cvitanović, Neils Bohr Institute, Bledgdamsvej 17, DK-Copenhagen 0, Denmark

B. Derrida, Service De Physique Theorique, CEN/Saclay B. P. 2, F-91191 Gif-sur-Yvetted CEDEX, France

Mark W. DiFrancesco, Department of Physics and Astronomy, University of Pittsburgh, Pittsburgh, Pennsylvania 15213 USA

Jack Dorning, Department of Nuclear Engineering and Enginnering Physics, University of Virginia, Charlottesville, Virginia 22901 USA

Detlef Dürr, Fachbereich Mathematik, Ruhr Universität Bochum, und BiBoS, Universität Bielefeld, 4800 Bielefeld, West Germany

Henri Epstein, CNRS and IHES, 35 Rue De Chartes, F-91440 Bures-sur-Yvette, France

Mitchell Feigenbaum, Rockefeller University, Department of Physics, New York, New York 10021 USA

V. Franceschini, Departimento Di Matematica, Universita Di Modena, Via Campi 213/13, 41100 Modena, Italy

Luigi Galgani, Dipartimento Di Matematica Dell'Universita, Via Saldini 50, 20133 Milano, Italy

Giovanni Gallavotti, Centro Linceo Interdisciplinare 'B. Segre', Accademia dei Lincei, Via della Lungara 10, 00100 Roma, Italia.

R. Gallimbeni, Dipartimento Di Fisica Teorica, Universita Di Torino, C.so D'Azeglio 46, 10125 Torino, Italy

Antonio Giorgilli, Dipartimento Di Fisica Dell'Universita, Via Celoria 16, 20133 Milano, Italy

Jonathan Keating, H. H. Wills Physics Laboratory, Royal Fort, Tyndall Avenue, Bristol BS8 1TL, United Kingdom

Hans Koch, Department of Mathematics, University of Texas at Austin, Austin, Texas 78712 USA

Joachim Krug, Theoretische Physik, Universität München, Theresienstr 37, 8 München, West Germany

Oscar Lanford III, IHES, 91440 Bures-sur-Yvette, France

Albert Libchaber, Department of Physics, J. Frank Institute, University of Chicago, Chicago, Illinois 60637 USA

Fabio Martinelli, Universita Di Roma, Dipartimento Di Matematica, P.le A. Moro, 00162 Roma, Italy

W. D. McCormick, Center for Nonlinear Dynamics and Physics Department, The University of Texas at Austin, Austin, Texas 78712 USA

M. Miari, Dipartimento Di Matematica Universita Di Milano, Via Celoria 16, 20133 Milano, Italy

Paul Mondragon, H. H. Wills Physics Laboratory, Royal Fort, Tyndall Avenue, Bristol BS8 1TL, United Kingdom

L. Sertorio, Dipartimento Di Fisica Teorica, Universita Di Torino, C.so M. D'Azeglio 46, 10125 Torino, Italy

Eric D. Siggia, Laboratory of Atomic and Solid State Physics, Clark Hall, Cornell University, Ithaca, New York 14853-2501 USA

Herbert Spohn, Theoretische Physik, Theresienstr 37, 8 München 2, West Germany

Dennis Sullivan, City University of New York, Graduate School and University Center, Graduate Center: 33 West 42nd Street, New York, New York 10036-8009 USA

Harry L. Swinney, Center for Nonlinear Dynamics and Physics Department, University of Texas at Austin, Austin, Texas 78712 USA

Michel Vittot, Centre De Physique Theorique, CNRS-Luminy, Case 907, F-13288 Marseille CEDEX 09, France

Jack Wisdom, Department of Earth, Atmospheric and Planetary Sciences, Massachusetts Institute of Technology, Cambridge, Massachusetts 02139 USA

Peter Wittwer, Rutgers University, Department of Mathematics, Hill Center - Busch Campus, New Brunswick, New Jersey 08903 USA

INDEX